KT-469-015

Sustainable Energy — without the hot air

NEWMAN UNIVERSITY COLLEGE BARTLEY GREEN BIRMINGHAM B32 3NT	
CLASS	333.794
BARCODE	01445138
AUTHOR	MAC

NW 10001338   BROAD   £16.36

The quest for safe, secure and sustainable energy poses one of the most critical challenges of our age. But how much energy do we need, and can we get it all from renewable sources? David MacKay sets out to find the answer through a forensic numerical analysis of what we use and what we can produce. His conclusions starkly reveal the difficult choices that must urgently be taken and readers interested in how we will power our society in the future will find this an illuminating read. For anyone with influence on energy policy, whether in government, business or a campaign group, this book should be compulsory reading. This is a technically precise and readable account of the challenges ahead. It will be a core reference on my shelf for many years to come.

Tony Juniper
Former Executive Director, Friends of the Earth

This is a really valuable contribution to the continuing discussion of energy policy. The author uses a potent mixture of arithmetic and common sense to dispel some myths and slay some sacred cows. The book is an essential reference work for anyone with an interest in energy who really wants to understand the numbers.

Lord Oxburgh KBE FRS
Former Chairman, Royal Dutch Shell

This remarkable book from an expert in the energy field sets out, with enormous clarity and objectivity, the various alternative low-carbon pathways that are open to us. Policy makers, researchers, private sector decision makers, and NGOs, all will benefit from these words of wisdom.

Sir David King FRS
Chief Scientific Adviser
to the UK Government, 2000–08

Started reading your book yesterday. Took the day off work today so that I could continue reading it. It is a fabulous, witty, no-nonsense, valuable piece of work, and I am busy sending it to everyone I know.

Matthew Sullivan
Carbon Advice Group Plc

A total delight to read. Extraordinarily clear and engaging.

Chris Goodell
Author of Ten Technologies to Sav

"Sustainable Energy – without the hot air" makes clear the science behind the headlines on energy issues. It is a fine guide for both experts and beginners.

Prof Daniel Kammen
Co-Director, Berkeley Institute of the Environment

MacKay's book shows how, when it comes to energy, you too can do the simple arithmetic and learn the simple scientific facts needed to work out what energy you need and where it might come from.

Prof David Mumford
Professor of Applied Mathematics
Brown University
Member of the US National Academy of Sciences

Common sense, technology literacy, and a little calculation go a long way in helping the reader sort sense from nonsense in the challenges of developing alternatives to fossil fuels. MacKay has provided a high priority book on a high priority problem.

Professor William W. Hogan
Raymond Plank Professor of Global Energy Policy
John F. Kennedy School of Government
Harvard University

This is a complete resource for assessing the many options for choosing between different energy options and for using energy more efficiently. Teachers, students, and any intelligent citizen will find here all the tools needed to think intelligently about sustainability. This is the most important book about applying science to public problems that I have read this year.

Prof Jerry Gollub
Professor of Physics, Haverford College
and University of Pennsylvania
Member of the US National Academy of Sciences

MacKay's book is the most practical, solidly analytical, and enjoyable book on energy that I have seen. This heroic work gets the energy story straight, assessing the constraints imposed by physical reality that we must work within.

Prof Tom Murphy
Assistant Professor of Physics, UC San Diego

WITHDRAWN

N 0144513 8

continued on next page

David MacKay's book is an intellectually satisfying, refreshing contribution to really understanding the complex issues of energy supply and use. It debunks the emotional claptrap which passes for energy policy and puts real numbers into the equations. It should be read by everyone, especially politicians.

Prof Ian Fells CBE
Founder chairman of NaREC,
the New and Renewable Energy Centre

Preventing climate chaos will require sophisticated and well informed social, economic and technological choices. Economic and social 'laws' are not immutable – politicians can and should reshape economics to deliver renewable energy and lead cultural change to save energy – but MacKay reminds us that even they "canna change the laws of physics"! MacKay's book alone doesn't have all the answers, but it provides a solid foundation to help us make well-informed choices, as individuals and more importantly as societies.

Duncan McLaren
Chief Executive, Friends of the Earth Scotland

By focusing on the metrics of energy consumption and production, in addition to the aspiration we all share for viable renewable energy, David MacKay's book provides a welcome addition to the energy literature. "Sustainable Energy – without the hot air" is a vast undertaking that provides both a practical guide and a reference manual. Perhaps ironically for a book on sustainable energy, MacKay's account of the numbers illustrates just how challenging replacing fossil fuel will be, and why both energy conservation and new energy technology are necessary.

Darran Messem
Vice President Fuel Development
Royal Dutch Shell

This is a must read for anyone who wants to help heal our world.

Carol Atkinson
Chief Executive of BRE Global

Beautifully clear and amazingly readable.

Prof Willy Brown CBE

At last a book that comprehensively reveals the true facts about sustainable energy in a form that is both highly readable and entertaining. A "must read" for all those who have a part to play in addressing our climate crisis.

Robert Sansom
Director of Strategy and Sustainable Development
EDF Energy

So much has been written about meeting future energy needs that it hardly seems possible to add anything useful, but David MacKay has managed it. His new book is a delight to read and will appeal especially to practical people who want to understand what is important in energy and what is not. Like Lord Kelvin before him, Professor MacKay realises that in many fields, and certainly in energy, unless you can quantify something you can never properly understand it. As a result, his fascinating book is also a mine of quantitative information for those of us who sometimes talk to our friends about how we supply and use energy, now and in the future.

Dr Derek Pooley CBE
Former Chief Scientist at the Department of Energy,
Chief Executive of the UK Atomic Energy Authority
and Member of the European Union Advisory
Group on Energy

The need to reduce our dependence on fossil fuels and to find sustainable sources of energy is desperate. But much of the discussion has not been based on data on how energy is consumed and how it is produced. This book fills that need in an accessible form, and a copy should be in every household.

Prof Robert Hinde CBE FRS FBA
Executive Committee, Pugwash UK
Department of Zoology, University of Cambridge

MacKay brings a welcome dose of common sense into the discussion of energy sources and use. Fresh air replacing hot air.

Prof Mike Ashby FRS
Author of Materials and the environment

I took it to the loo and almost didn't come out again.

Matthew Moss

Sustainable Energy — without the hot air

David JC MacKay

UIT
CAMBRIDGE, ENGLAND

First published in England in 2009.
UIT Cambridge Ltd.
PO Box 145
Cambridge
CB4 1GQ
England

Tel: +44 1223 302 041
Web: www.uit.co.uk

Copyright © 2009 David JC MacKay
All rights reserved.

ISBN 978-0-9544529-3-3 (paperback)
ISBN 978-1-906860-01-1 (hardback)

The right of David JC MacKay to be identified as
the author of this work has been asserted by him
in accordance with the Copyright, Designs and
Patents Act 1988.

While this publication intends to provide
accurate and authoritative information in regard
to the subject matter covered, neither the
publisher nor the author makes any
representation, express or implied, with regard
to the accuracy of information contained in this
book, nor do they accept any legal responsibility
or liability for any errors or omissions that may
be made. This work is supplied with the
understanding that UIT Cambridge Ltd and its
authors are supplying information, but are not
attempting to render engineering or other
professional services. If such services are
required, the assistance of an appropriate
professional should be sought.

Many of the designations used by manufacturers
and sellers to distinguish their products are
claimed as trade-marks. UIT Cambridge Ltd
acknowledges trademarks as the property of
their respective owners.

10 9 8 7 6 5 4

3.5.2 e1.3

to those who will not have the benefit
of two billion years' accumulated energy reserves

Preface

What's this book about?

I'm concerned about cutting UK emissions of twaddle – twaddle about sustainable energy. Everyone says getting off fossil fuels is important, and we're all encouraged to "make a difference," but many of the things that allegedly make a difference don't add up.

Twaddle emissions are high at the moment because people get emotional (for example about wind farms or nuclear power) and no-one talks about numbers. Or if they do mention numbers, they select them to sound big, to make an impression, and to score points in arguments, rather than to aid thoughtful discussion.

This is a straight-talking book about the numbers. The aim is to guide the reader around the claptrap to actions that really make a difference and to policies that add up.

This is a free book

I didn't write this book to make money. I wrote it because sustainable energy is important. If you would like to have the book for free for your own use, please help yourself: it's on the internet at www.withouthotair.com.

This is a free book in a second sense: you are free to use *all* the material in this book, *except* for the cartoons and the photos with a named photographer, under the Creative Commons Attribution-Non-Commercial-Share-Alike 2.0 UK: England & Wales Licence. (The cartoons and photos are excepted because the authors have generally given me permission only to include their work, *not* to share it under a Creative Commons license.) You are especially welcome to use my materials for educational purposes. My website includes separate high-quality files for each of the figures in the book.

How to operate this book

Some chapters begin with a quotation. Please don't assume that my quoting someone means that I agree with them; think of these quotes as provocations, as hypotheses to be critically assessed.

Many of the early chapters (numbered 1, 2, 3, . . .) have longer technical chapters (A, B, C, . . .) associated with them. These technical chapters start on page 254.

At the end of each chapter are further notes and pointers to sources and references. I find footnote marks distracting if they litter the main text of the book, so the book has no footnote marks. If you love footnote marks, you can usefully add them – almost every substantive assertion in the text will have an associated note at the end of its chapter giving sources or further information.

The text also contains pointers to web resources. When a web-pointer is monstrously long, I've used the TinyURL service, and put the tiny code in the text like this – [yh8xse] – and the full pointer at the end of the book on page 344. yh8xse is a shorthand for a tiny URL, in this case: http://tinyurl.com/yh8xse. A complete list of all the URLs in this book is provided at http://tinyurl.com/yh8xse.

I welcome feedback and corrections. I am aware that I sometimes make booboos, and in earlier drafts of this book some of my numbers were off by a factor of two. While I hope that the errors that remain are smaller than that, I expect to further update some of the numbers in this book as I continue to learn about sustainable energy.

How to cite this book:

David J.C. MacKay. *Sustainable Energy – without the hot air.*
UIT Cambridge, 2008. ISBN 978-0-9544529-3-3. Available free online from www.withouthotair.com.

Contents

Part I

Numbers, not adjectives

1 Motivations

We live at a time when emotions and feelings count more than truth, and there is a vast ignorance of science.

James Lovelock

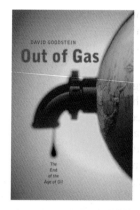

I recently read two books, one by a physicist, and one by an economist. In *Out of Gas*, Caltech physicist David Goodstein describes an impending energy crisis brought on by The End of the Age of Oil. This crisis is coming soon, he predicts: the crisis will bite, not when the last drop of oil is extracted, but when oil extraction can't meet demand – perhaps as soon as 2015 or 2025. Moreover, even if we magically switched all our energy-guzzling to nuclear power right away, Goodstein says, the oil crisis would simply be replaced by a *nuclear* crisis in just twenty years or so, as uranium reserves also became depleted.

David Goodstein's *Out of Gas* (2004).

In *The Skeptical Environmentalist*, Bjørn Lomborg paints a completely different picture. "Everything is fine." Indeed, "everything is getting better." Furthermore, "we are not headed for a major energy crisis," and "there is plenty of energy."

How could two smart people come to such different conclusions? I had to get to the bottom of this.

Energy made it into the British news in 2006. Kindled by tidings of great climate change and a tripling in the price of natural gas in just six years, the flames of debate are raging. How should Britain handle its energy needs? And how should the world?

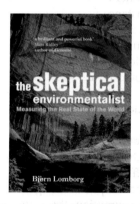

Bjørn Lomborg's *The Skeptical Environmentalist* (2001).

"Wind or nuclear?", for example. Greater polarization of views among smart people is hard to imagine. During a discussion of the proposed expansion of nuclear power, Michael Meacher, former environment minister, said "if we're going to cut greenhouse gases by 60% ... by 2050 there is no other possible way of doing that except through renewables;" Sir Bernard Ingham, former civil servant, speaking in favour of nuclear expansion, said "anybody who is relying upon renewables to fill the [energy] gap is living in an utter dream world and is, in my view, an enemy of the people."

Similar disagreement can be heard within the ecological movement. All agree that *something* must be done urgently, but *what*? Jonathan Porritt, chair of the Sustainable Development Commission, writes: "there is no justification for bringing forward plans for a new nuclear power programme at this time, and ... any such proposal would be incompatible with [the Government's] sustainable development strategy;" and "a non-nuclear strategy could and should be sufficient to deliver all the carbon savings we shall need up to 2050 and beyond, and to ensure secure access to reliable sources of energy." In contrast, environmentalist James Lovelock writes in his book, *The Revenge of Gaia*: "Now is much too late to establish sustainable development." In his view, power from nuclear fission, while

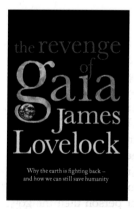

The Revenge of Gaia: Why the earth is fighting back – and how we can still save humanity. James Lovelock (2006). © Allen Lane.

2

not recommended as the long-term panacea for our ailing planet, is "the only effective medicine we have now." Onshore wind turbines are "merely ... a gesture to prove [our leaders'] environmental credentials."

This heated debate is fundamentally about numbers. How much energy could each source deliver, at what economic and social cost, and with what risks? But actual numbers are rarely mentioned. In public debates, people just say "Nuclear is a money pit" or "We have a *huge* amount of wave and wind." The trouble with this sort of language is that it's not sufficient to know that something is huge: we need to know how the one "huge" compares with another "huge," namely *our huge energy consumption*. To make this comparison, we need numbers, not adjectives.

Where numbers are used, their meaning is often obfuscated by enormousness. Numbers are chosen to impress, to score points in arguments, rather than to inform. "Los Angeles residents drive 142 million miles – the distance from Earth to Mars – every single day." "Each year, 27 million acres of tropical rainforest are destroyed." "14 billion pounds of trash are dumped into the sea every year." "British people throw away 2.6 billion slices of bread per year." "The waste paper buried each year in the UK could fill 103 448 double-decker buses."

If all the ineffective ideas for solving the energy crisis were laid end to end, they would reach to the moon and back.... I digress.

The result of this lack of meaningful numbers and facts? We are inundated with a flood of crazy innumerate codswallop. The BBC doles out advice on how we can do our bit to save the planet – for example "switch off your mobile phone charger when it's not in use;" if anyone objects that mobile phone chargers are not *actually* our number one form of energy consumption, the mantra "every little helps" is wheeled out. Every little helps? A more realistic mantra is:

if everyone does a little, we'll achieve only a little.

For the benefit of readers who speak American, rather than English, the translation of "every little helps" into American is "every little bit helps."

Companies also contribute to the daily codswallop as they tell us how wonderful they are, or how they can help us "do our bit." BP's website, for example, celebrates the reductions in carbon dioxide (CO_2) pollution they hope to achieve by changing the paint used for painting BP's ships. Does anyone fall for this? Surely everyone will guess that it's not the exterior paint job, it's the stuff *inside* the tanker that deserves attention, if society's CO_2 emissions are to be significantly cut? BP also created a web-based carbon absolution service, "`targetneutral.com`," which claims that they can "neutralize" all your carbon emissions, and that it "doesn't cost the earth" – indeed, that your CO_2 pollution can be cleaned up for just £40 per year. How can this add up? – if the true cost of fixing climate change were £40 per person then the government could fix it with the loose change in the Chancellor's pocket!

Even more reprehensible are companies that exploit the current concern for the environment by offering "water-powered batteries," "biodegradable mobile phones," "portable arm-mounted wind-turbines," and other pointless tat.

Campaigners also mislead. People who want to promote renewables over nuclear, for example, say "offshore wind power could power all UK homes;" then they say "new nuclear power stations will do little to tackle climate change" because 10 new nuclear stations would "reduce emissions only by about 4%." This argument is misleading because the playing field is switched half-way through, from the "number of homes powered" to "reduction of emissions." The truth is that the amount of electrical power generated by the wonderful windmills that "could power all UK homes" is *exactly the same* as the amount that would be generated by the 10 nuclear power stations! "Powering all UK homes" accounts for just 4% of UK emissions.

Perhaps the worst offenders in the kingdom of codswallop are the people who really should know better – the media publishers who promote the codswallop – for example, New Scientist with their article about the "water-powered car."*

In a climate where people don't understand the numbers, newspapers, campaigners, companies, and politicians can get away with murder.

We need simple numbers, and we need the numbers to be comprehensible, comparable, and memorable.

With numbers in place, we will be better placed to answer questions such as these:

1. Can a country like Britain conceivably live on its own renewable energy sources?

2. If everyone turns their thermostats one degree closer to the outside temperature, drives a smaller car, and switches off phone chargers when not in use, will an energy crisis be averted?

3. Should the tax on transportation fuels be significantly increased? Should speed-limits on roads be halved?

4. Is someone who advocates windmills over nuclear power stations "an enemy of the people"?

5. If climate change is "a greater threat than terrorism," should governments criminalize "the glorification of travel" and pass laws against "advocating acts of consumption"?

6. Will a switch to "advanced technologies" allow us to eliminate carbon dioxide pollution without changing our lifestyle?

7. Should people be encouraged to eat more vegetarian food?

8. Is the population of the earth six times too big?

*See this chapter's notes (p19) for the awful details. (Every chapter has endnotes giving references, sources, and details of arguments. To avoid distracting the reader, I won't include any more footnote marks in the text.)

Figure 1.1. This Greenpeace leaflet arrived with my junk mail in May 2006. Do beloved windmills have the capacity to displace hated cooling towers?

Why are we discussing energy policy?

Three different motivations drive today's energy discussions.

First, fossil fuels are a finite resource. It seems possible that cheap oil (on which our cars and lorries run) and cheap gas (with which we heat many of our buildings) will run out in our lifetime. So we seek alternative energy sources. Indeed given that fossil fuels are a valuable resource, useful for manufacture of plastics and all sorts of other creative stuff, perhaps we should save them for better uses than simply setting fire to them.

Second, we're interested in security of energy supply. Even if fossil fuels are still available somewhere in the world, perhaps we don't want to depend on them if that would make our economy vulnerable to the whims of untrustworthy foreigners. (I hope you can hear my tongue in my cheek.) Going by figure 1.2, it certainly looks as if "our" fossil fuels have peaked. The UK has a particular security-of-supply problem looming, known as the "energy gap." A substantial number of old coal power stations and nuclear power stations will be closing down during the next decade (figure 1.3), so there is a risk that electricity demand will sometimes exceed electricity supply, if adequate plans are not implemented.

Third, it's very probable that using fossil fuels changes the climate. Climate change is blamed on several human activities, but the biggest contributor to climate change is the increase in greenhouse effect produced by carbon dioxide (CO_2). Most of the carbon dioxide emissions come from fossil-fuel burning. And the main reason we burn fossil fuels is for energy. So to fix climate change, we need to sort out a new way of getting energy. The climate problem is mostly an energy problem.

Whichever of these three concerns motivates you, we need energy numbers, and policies that add up.

The first two concerns are straightforward selfish motivations for drastically reducing fossil fuel use. The third concern, climate change, is a more altruistic motivation – the brunt of climate change will be borne not by us but by future generations over many hundreds of years. Some people feel that climate change is not their responsibility. They say things like "What's the point in my doing anything? China's out of control!" So I'm going to discuss climate change a bit more now, because while writing this book I learned some interesting facts that shed light on these ethical questions. If you have no interest in climate change, feel free to fast-forward to the next section on page 16.

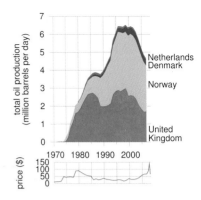

Figure 1.2. Are "our" fossil fuels running out? Total crude oil production from the North Sea, and oil price in 2006 dollars per barrel.

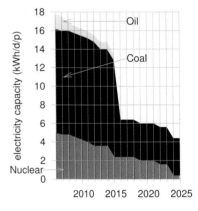

Figure 1.3. The energy gap created by UK power station closures, as projected by energy company EdF. This graph shows the predicted capacity of nuclear, coal, and oil power stations, in kilowatt-hours per day per person. The capacity is the maximum deliverable power of a source.

The climate-change motivation

The climate-change motivation is argued in three steps: one: human fossil-fuel burning causes carbon dioxide concentrations to rise; two: carbon dioxide is a greenhouse gas; three: increasing the greenhouse effect increases average global temperatures (and has many other effects).

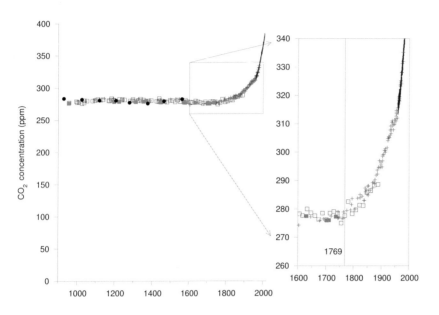

Figure 1.4. Carbon dioxide (CO_2) concentrations (in parts per million) for the last 1100 years, measured from air trapped in ice cores (up to 1977) and directly in Hawaii (from 1958 onwards).

I think something new may have happened between 1800 AD and 2000 AD. I've marked the year 1769, in which James Watt patented his steam engine. (The first practical steam engine was invented 70 years earlier in 1698, but Watt's was much more efficient.)

We start with the fact that carbon dioxide concentrations are rising. Figure 1.4 shows measurements of the CO_2 concentration in the air from the year 1000 AD to the present. Some "sceptics" have asserted that the recent increase in CO_2 concentration is a natural phenomenon. Does "sceptic" mean "a person who has not even glanced at the data"? Don't you think, just possibly, *something* may have happened between 1800 AD and 2000 AD? Something that was not part of the natural processes present in the preceding thousand years?

Something did happen, and it was called the Industrial Revolution. I've marked on the graph the year 1769, in which James Watt patented his steam engine. While the first practical steam engine was invented in 1698, Watt's more efficient steam engine really got the Industrial Revolution going. One of the steam engine's main applications was the pumping of water out of coal mines. Figure 1.5 shows what happened to British coal production from 1769 onwards. The figure displays coal production in units of billions of tons of CO_2 released when the coal was burned. In 1800, coal was used to make iron, to make ships, to heat buildings, to power locomotives and other machinery, and of course to power the pumps that enabled still more coal to be scraped up from inside the hills of England and Wales. Britain was terribly well endowed with coal: when the Revolution started, the amount of carbon sitting in coal under Britain was roughly the same as the amount sitting in oil under Saudi Arabia.

In the 30 years from 1769 to 1800, Britain's annual coal production doubled. After another 30 years (1830), it had doubled again. The next doubling of production-rate happened within *20* years (1850), and another

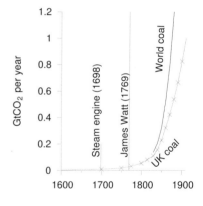

Figure 1.5. The history of UK coal production and world coal production from 1600 to 1910. Production rates are shown in billions of tons of CO_2 – an incomprehensible unit, yes, but don't worry: we'll personalize it shortly.

doubling within 20 years of that (1870). This coal allowed Britain to turn the globe pink. The prosperity that came to England and Wales was reflected in a century of unprecedented population growth:

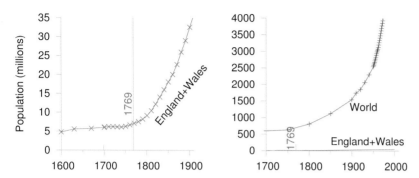

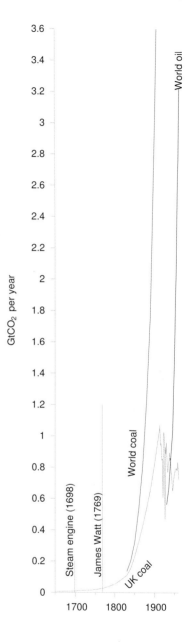

Eventually other countries got in on the act too as the Revolution spread. Figure 1.6 shows British coal production and world coal production on the same scale as figure 1.5, sliding the window of history 50 years later. British coal production peaked in 1910, but meanwhile world coal production continued to double every 20 years. It's difficult to show the history of coal production on a single graph. To show what happened in the *next* 50 years on the same scale, the book would need to be one metre tall! To cope with this difficulty, we can either scale down the vertical axis:

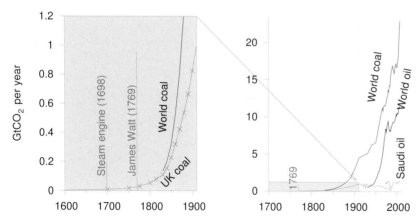

or we can squish the vertical axis in a non-uniform way, so that small quantities and large quantities can be seen at the same time on a single graph. A good way to squish the axis is called a logarithmic scale, and that's what I've used in the bottom two graphs of figure 1.7 (p9). On a logarithmic scale, all ten-fold increases (from 1 to 10, from 10 to 100, from 100 to 1000) are represented by equal distances on the page. On a logarithmic scale, a quantity that grows at a constant percentage per year (which is called "exponential growth") looks like a straight line. Logarithmic graphs are great

Figure 1.6. What happened next.

The history of UK coal production and world coal production from 1650 to 1960, on the same scale as figure 1.5.

for understanding growth. Whereas the ordinary graphs in the figures on pages 6 and 7 convey the messages that British and world coal production grew remarkably, and that British and world population grew remarkably, the relative growth rates are not evident in these ordinary graphs. The logarithmic graphs allow us to compare growth rates. Looking at the slopes of the population curves, for example, we can see that the world population's growth rate in the last 50 years was a little bigger than the growth rate of England and Wales in 1800.

From 1769 to 2006, world annual coal production increased 800-fold. Coal production is still increasing today. Other fossil fuels are being extracted too – the middle graph of figure 1.7 shows oil production for example – but in terms of CO_2 emissions, coal is still king.

The burning of fossil fuels is the principal reason why CO_2 concentrations have gone up. This is a fact, but, hang on: I hear a persistent buzzing noise coming from a bunch of climate-change inactivists. What are they saying? Here's Dominic Lawson, a columnist from the *Independent*:

> "The burning of fossil fuels sends about seven gigatons of CO_2 per year into the atmosphere, which sounds like a lot. Yet the biosphere and the oceans send about 1900 gigatons and 36 000 gigatons of CO_2 per year into the atmosphere – . . . one reason why some of us are sceptical about the emphasis put on the role of human fuel-burning in the greenhouse gas effect. Reducing man-made CO_2 emissions is megalomania, exaggerating man's significance. Politicians can't change the weather."

Now I have a lot of time for scepticism, and not everything that sceptics say is a crock of manure – but irresponsible journalism like Dominic Lawson's deserves a good flushing.

The first problem with Lawson's offering is that *all three numbers* that he mentions (seven, 1900, and 36 000) are *wrong!* The correct numbers are 26, 440, and 330. Leaving these errors to one side, let's address Lawson's main point, the relative smallness of man-made emissions.

Yes, natural flows of CO_2 *are* larger than the additional flow we switched on 200 years ago when we started burning fossil fuels in earnest. But it is terribly misleading to quantify only the large natural flows *into* the atmosphere, failing to mention the almost exactly equal flows *out* of the atmosphere back into the biosphere and the oceans. The point is that these *natural* flows in and out of the atmosphere have been almost exactly in balance for millenia. So it's not relevant at all that these natural flows are larger than human emissions. The natural flows *cancelled themselves out*. So the natural flows, large though they were, left the concentration of CO_2 in the atmosphere and ocean *constant*, over the last few thousand years. Burning fossil fuels, in contrast, creates a *new* flow of carbon that, though small, is *not cancelled*. Here's a simple analogy, set in the passport-control arrivals area of an airport. One thousand passengers arrive per hour, and

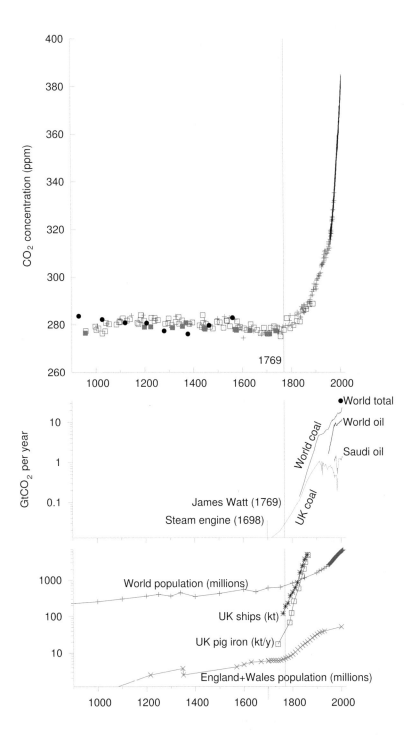

Figure 1.7. The upper graph shows carbon dioxide (CO_2) concentrations (in parts per million) for the last 1100 years – the same data that was shown in figure 1.4.

Here's a portrait of James Watt and his 1769 steam engine.

The middle graph shows (on a logarithmic scale) the history of UK coal production, Saudi oil production, world coal production, world oil production, and (by the top right point) the total of all greenhouse gas emissions in the year 2000. All production rates are expressed in units of the associated CO_2 emissions.

The bottom graph shows (on a logarithmic scale) some consequences of the Industrial Revolution: sharp increases in the population of England, and, in due course, the world; and remarkable growth in British pig-iron production (in thousand tons per year); and growth in the tonnage of British ships (in thousand tons).

In contrast to the ordinary graphs on the previous pages, the logarithmic scale allows us to show both the population of England and the population of the World on a single diagram, and to see interesting features in both.

there are exactly enough clockwork officials to process one thousand passengers per hour. There's a modest queue, but because of the match of arrival rate to service rate, the queue isn't getting any longer. Now imagine that owing to fog an extra stream of flights is diverted here from a smaller airport. This stream adds an extra 50 passengers per hour to the arrivals lobby – a small addition compared to the original arrival rate of one thousand per hour. Initially at least, the authorities don't increase the number of officials, and the officials carry on processing just one thousand passengers per hour. So what happens? Slowly but surely, *the queue grows*. Burning fossil fuels is undeniably increasing the CO_2 concentration in the atmosphere and in the surface oceans. No climate scientist disputes this fact. When it comes to CO_2 concentrations, man *is* significant.

OK. Fossil fuel burning increases CO_2 concentrations significantly. But does it matter? "Carbon is nature!", the oilspinners remind us, "Carbon is life!" If CO_2 had no harmful effects, then indeed carbon emissions would not matter. However, carbon dioxide is a greenhouse gas. Not the strongest greenhouse gas, but a significant one nonetheless. Put more of it in the atmosphere, and it does what greenhouse gases do: it absorbs infrared radiation (heat) heading out from the earth and reemits it in a random direction; the effect of this random redirection of the atmospheric heat traffic is to impede the flow of heat from the planet, just like a quilt. So carbon dioxide has a warming effect. This fact is based not on complex historical records of global temperatures but on the simple physical properties of CO_2 molecules. Greenhouse gases are a quilt, and CO_2 is one layer of the quilt.

So, if humanity succeeds in doubling or tripling CO_2 concentrations (which is where we are certainly heading, under business as usual), what happens? Here, there is a lot of uncertainty. Climate science is difficult. The climate is a complex, twitchy beast, and exactly how much warming CO_2-doubling would produce is uncertain. The consensus of the best climate models seems to be that doubling the CO_2 concentration would have roughly the same effect as increasing the intensity of the sun by 2%, and would bump up the global mean temperature by something like $3\,°C$. This would be what historians call a Bad Thing. I won't recite the whole litany of probable drastic effects, as I am sure you've heard it before. The litany begins "the Greenland icecap would gradually melt, and, over a period of a few 100 years, sea-level would rise by about 7 metres." The brunt of the litany falls on future generations. Such temperatures have not been seen on earth for at least 100 000 years, and it's conceivable that the ecosystem would be so significantly altered that the earth would stop supplying some of the goods and services that we currently take for granted.

Climate modelling is difficult and is dogged by uncertainties. But uncertainty about exactly how the climate will respond to extra greenhouse gases is no justification for inaction. If you were riding a fast-moving motorcycle in fog near a cliff-edge, and you didn't have a good map of the cliff, would the lack of a map justify *not* slowing the bike down?

So, who should slow the bike down? Who should clean up carbon emissions? Who is responsible for climate change? This is an ethical question, of course, not a scientific one, but ethical discussions must be founded on facts. Let's now explore the facts about greenhouse gas emissions. First, a word about the units in which they are measured. Greenhouse gases include carbon dioxide, methane, and nitrous oxide; each gas has different physical properties; it's conventional to express all gas emissions in "equivalent amounts of carbon dioxide," where "equivalent" means "having the same warming effect over a period of 100 years." One ton of carbon-dioxide-equivalent may be abbreviated as "$1\,t\,CO_2e$," and one billion tons (one thousand million tons) as "$1\,Gt\,CO_2e$" (one gigaton). In this book $1\,t$ means one metric ton (1000 kg). I'm not going to distinguish imperial tons, because they differ by less than 10% from the metric ton or tonne.

In the year 2000, the world's greenhouse gas emissions were about 34 billion tons of CO_2-equivalent per year. An incomprehensible number. But we can render it more comprehensible and more personal by dividing by the number of people on the planet, 6 billion, so as to obtain the greenhouse-gas pollution *per person*, which is about $5^{1}/_{2}$ tons CO_2e per year per person. We can thus represent the world emissions by a rectangle whose width is the population (6 billion) and whose height is the per-capita emissions.

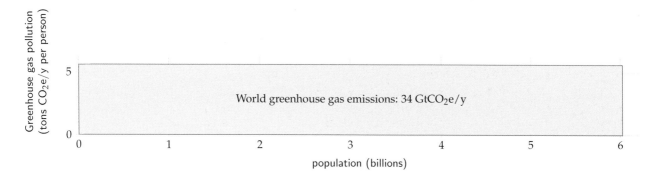

Now, all people are created equal, but we don't all emit $5^{1/2}$ tons of CO_2 per year. We can break down the emissions of the year 2000, showing how the 34-billion-ton rectangle is shared between the regions of the world:

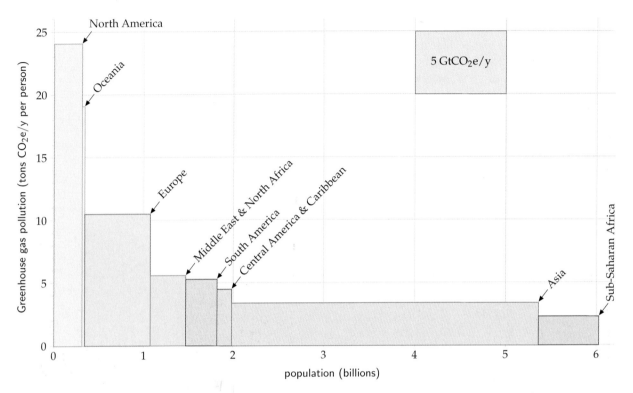

This picture, which is on the same scale as the previous one, divides the world into eight regions. Each rectangle's area represents the greenhouse gas emissions of one region. The width of the rectangle is the population of the region, and the height is the average per-capita emissions in that region.

In the year 2000, Europe's per-capita greenhouse gas emissions were twice the world average; and North America's were four times the world average.

We can continue subdividing, splitting each of the regions into countries. This is where it gets really interesting:

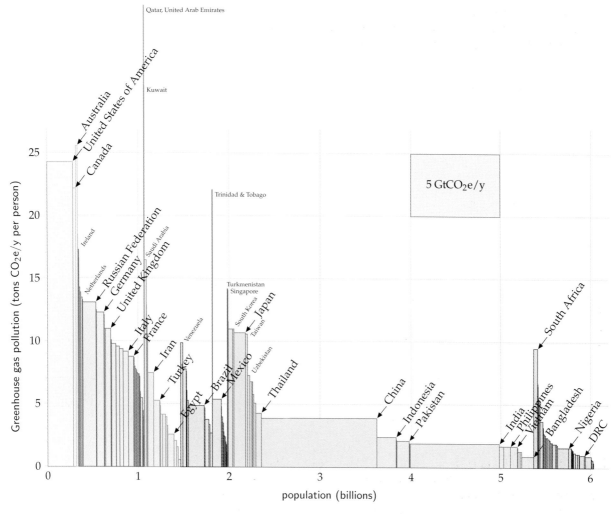

The major countries with the biggest per-capita emissions are Australia, the USA, and Canada. European countries, Japan, and South Africa are notable runners up. Among European countries, the United Kingdom is resolutely average. What about China, that naughty "out of control" country? Yes, the area of China's rectangle is about the same as the USA's, but the fact is that their per-capita emissions are *below* the world average. India's per-capita emissions are less than *half* the world average. Moreover, it's worth bearing in mind that much of the industrial emissions of China and India are associated with the manufacture of *stuff for rich countries*.

So, assuming that "something needs to be done" to reduce greenhouse gas emissions, who has a special responsibility to do something? As I said, that's an ethical question. But I find it hard to imagine any system of ethics that denies that the responsibility falls especially on the countries to the left hand side of this diagram – the countries whose emissions are two, three, or four times the world average. Countries that are most able to pay. Countries like Britain and the USA, for example.

Historical responsibility for climate impact

If we assume that the climate has been damaged by human activity, and that someone needs to fix it, who should pay? Some people say "the polluter should pay." The preceding pictures showed who's doing the polluting today. But it isn't the *rate* of CO_2 pollution that matters, it's the cumulative *total* emissions; much of the emitted carbon dioxide (about one third of it) will hang around in the atmosphere for at least 50 or 100 years. If we accept the ethical idea that "the polluter should pay" then we should ask how big is each country's historical footprint. The next picture shows each country's cumulative emissions of CO_2, expressed as an average emission rate over the period 1880–2004.

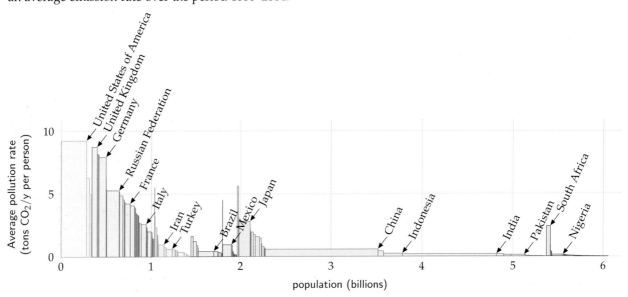

Congratulations, Britain! The UK has made it onto the winners' podium. We may be only an average European country today, but in the table of historical emitters, per capita, we are second only to the USA.

OK, that's enough ethics. What do scientists reckon needs to be done, to avoid a risk of giving the earth a 2 °C temperature rise (2 °C being the rise above which they predict lots of bad consequences)? The consensus is clear. We need to get off our fossil fuel habit, and we need to do so fast. Some countries, including Britain, have committed to at least a 60% reduction in greenhouse-gas emissions by 2050, but it must be emphasized that 60% cuts, radical though they are, are unlikely to cut the mustard. If the world's emissions were gradually reduced by 60% by 2050, climate scientists reckon it's more likely than not that global temperatures will rise by more than 2 °C. The sort of cuts we need to aim for are shown in figure 1.8. This figure shows two possibly-safe emissions scenarios presented by Baer and Mastrandrea (2006) in a report from the Institute for Public Policy Research. The lower curve assumes that a decline in emissions started in 2007, with total global emissions falling at roughly 5% per year. The upper curve assumes a brief delay in the start of the decline, and a 4% drop per year in global emissions. Both scenarios are believed to offer a modest chance of avoiding a 2 °C temperature rise above the pre-industrial level. In the lower scenario, the chance that the temperature rise will *exceed* 2 °C is estimated to be 9–26%. In the upper scenario, the chance of exceeding 2 °C is estimated to be 16–43%. These possibly-safe emissions trajectories, by the way, involve significantly sharper reductions in emissions than any of the scenarios presented by the Intergovernmental Panel on Climate Change (IPCC), or by the Stern Review (2007).

These possibly-safe trajectories require global emissions to fall by 70% or 85% by 2050. What would this mean for a country like Britain? If we subscribe to the idea of "contraction and convergence," which means that all countries aim eventually to have equal per-capita emissions, then Britain needs to aim for cuts greater than 85%: it should get down from its current 11 tons of CO_2e per year per person to roughly 1 ton per year per

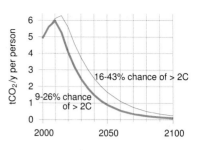

Figure 1.8. Global emissions for two scenarios considered by Baer and Mastrandrea, expressed in tons of CO_2 per year per person, using a world population of six billion. Both scenarios are believed to offer a modest chance of avoiding a 2 °C temperature rise above the pre-industrial level.

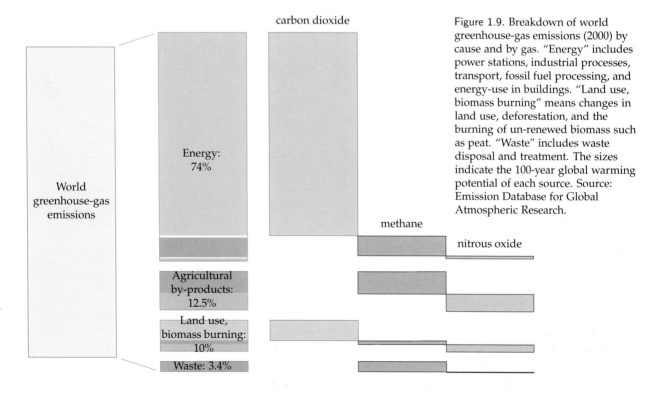

Figure 1.9. Breakdown of world greenhouse-gas emissions (2000) by cause and by gas. "Energy" includes power stations, industrial processes, transport, fossil fuel processing, and energy-use in buildings. "Land use, biomass burning" means changes in land use, deforestation, and the burning of un-renewed biomass such as peat. "Waste" includes waste disposal and treatment. The sizes indicate the 100-year global warming potential of each source. Source: Emission Database for Global Atmospheric Research.

person by 2050. This is such a deep cut, I suggest the best way to think about it is *no more fossil fuels*.

One last thing about the climate-change motivation: while a range of human activities cause greenhouse-gas emissions, the biggest cause by far is **energy use**. Some people justify not doing anything about their energy use by excuses such as "methane from burping cows causes more warming than jet travel." Yes, agricultural by-products contributed one eighth of greenhouse-gas emissions in the year 2000. But energy-use contributed three quarters (figure 1.9). The climate change problem is principally an energy problem.

Warnings to the reader

OK, enough about climate change. I'm going to assume we are motivated to get off fossil fuels. Whatever your motivation, the aim of this book is to help you figure out the numbers and do the arithmetic so that you can evaluate policies; and to lay a factual foundation so that you can see *which proposals add up*. I'm not claiming that the arithmetic and numbers in this book are new; the books I've mentioned by Goodstein, Lomborg, and Lovelock, for example, are full of interesting numbers and back-of-envelope calculations, and there are many other helpful sources on the internet too (see the notes at the end of each chapter).

What I'm aiming to do in this book is to make these numbers simple and memorable; to show you how you can figure out the numbers for yourself; and to make the situation so clear that any thinking reader will be able to draw striking conclusions. I don't want to feed you my own conclusions. Convictions are stronger if they are self-generated, rather than taught. Understanding is a creative process. When you've read this book I hope you'll have reinforced the confidence that you can figure anything out.

I'd like to emphasize that the calculations we will do are deliberately imprecise. Simplification is a key to understanding. First, by rounding the numbers, we can make them easier to remember. Second, rounded numbers allow quick calculations. For example, in this book, the population of the United Kingdom is 60 million, and the population of the world is 6 billion. I'm perfectly capable of looking up more accurate figures, but accuracy would get in the way of fluent thought. For example, if we learn that the world's greenhouse gas emissions in 2000 were 34 billion tons of CO_2-equivalent per year, then we can instantly note, without a calculator, that the average emissions per person are 5 or 6 tons of CO_2-equivalent per person per year. This rough answer is not exact, but it's accurate enough to inform interesting conversations. For instance, if you learn that a round-trip intercontinental flight emits nearly two tons of CO_2 per passenger, then knowing the average emissions yardstick (5-and-a-bit tons per year per person) helps you realize that just one such plane-trip per year corre-

"Look – it's Low Carbon Emission Man"

Figure 1.10. Reproduced by kind permission of PRIVATE EYE / Peter Dredge www.private-eye.co.uk.

sponds to over a third of the average person's carbon emissions.

I like to base my calculations on everyday knowledge rather than on trawling through impersonal national statistics. For example, if I want to estimate the typical wind speeds in Cambridge, I ask "is my cycling speed usually faster than the wind?" The answer is yes. So I can deduce that the wind speed in Cambridge is only rarely faster than my typical cycling speed of 20 km/h. I back up these everyday estimates with other peoples' calculations and with official statistics. (Please look for these in each chapter's end-notes.) This book isn't intended to be a definitive store of super-accurate numbers. Rather, it's intended to illustrate how to use approximate numbers as a part of constructive consensual conversations.

In the calculations, I'll mainly use the United Kingdom and occasionally Europe, America, or the whole world, but you should find it easy to redo the calculations for whatever country or region you are interested in.

Let me close this chapter with a few more warnings to the reader. Not only will we make a habit of approximating the numbers we calculate; we'll also neglect all sorts of details that investors, managers, and economists have to attend to, poor folks. If you're trying to launch a renewable technology, just a 5% increase in costs may make all the difference between success and failure, so in business every detail must be tracked. But 5% is too small for this book's radar. This is a book about factors of 2 and factors of 10. It's about physical limits to sustainable energy, not current economic feasibility. While economics is always changing, the fundamental limits won't ever go away. We need to understand these limits.

Debates about energy policy are often confusing and emotional because people mix together *factual* assertions and *ethical* assertions.

Examples of **factual assertions** are "global fossil-fuel burning emits 34 billion tons of carbon dioxide equivalent per year;" and "if CO_2 concentrations are doubled then average temperatures will increase by 1.5–5.8°C in the next 100 years;" and "a temperature rise of 2°C would cause the Greenland ice cap to melt within 500 years;" and "the complete melting of the Greenland ice cap would cause a 7-metre sea-level rise."

A factual assertion is either true or false; figuring out *which* may be difficult; it is a scientific question. For example, the assertions I just gave are either true or false. But we don't know whether they are all true. Some of them are currently judged "very likely." The difficulty of deciding which factual assertions are true leads to debates in the scientific community. But given sufficient scientific experiment and discussion, the truth or falsity of most factual assertions can eventually be resolved, at least "beyond reasonable doubt."

Examples of **ethical assertions** are "it's wrong to exploit global resources in a way that imposes significant costs on future generations;" and "polluting should not be free;" and "we should take steps to ensure that it's unlikely that CO_2 concentrations will double;" and "politicians should agree a cap on CO_2 emissions;" and "countries with the biggest CO_2 emissions over the last century have a duty to lead action on climate change;" and "it is fair to share CO_2 emission rights equally across the world's population." Such assertions are not "either true or false." Whether we agree with them depends on our ethical judgment, on our values. Ethical assertions may be incompatible with each other; for example, Tony Blair's government declared a radical policy on CO_2 emissions: "the United Kingdom should reduce its CO_2 emissions by 60% by 2050;" at the same time Gordon Brown, while Chancellor in that government, repeatedly urged oil-producing countries to *increase* oil production.

This book is emphatically intended to be about facts, not ethics. I want the facts to be clear, so that people can have a meaningful debate about ethical decisions. I want everyone to understand how the facts constrain the options that are open to us. Like a good scientist, I'll try to keep my views on ethical questions out of the way, though occasionally I'll blurt something out – please forgive me.

Whether it's *fair* for Europe and North America to hog the energy cake is an ethical question; I'm here to remind you of the *fact* that we can't have our cake and eat it too; to help you weed out the pointless and ineffective policy proposals; and to help you identify energy policies that are compatible with your personal values.

We need a plan that adds up!

"Okay – it's agreed; we announce – 'to do nothing is not an option!' then we wait and see how things pan out…"

Figure 1.11. Reproduced by kind permission of PRIVATE EYE / Paul Lowe www.private-eye.co.uk.

Notes and further reading

At the end of each chapter I note details of ideas in that chapter, sources of data and quotes, and pointers to further information.

page no.

2   *"…no other possible way of doing that except through renewables"; "anybody who is relying upon renewables to fill the [energy] gap is living in an utter dream world and is, in my view, an enemy of the people."* The quotes are from *Any Questions?*, 27 January 2006, BBC Radio 4 [ydoobr] . *Michael Meacher* was UK environment minister from 1997 till 2003. *Sir Bernard Ingham* was an aide to Margaret Thatcher when she was prime minister, and was Head of the Government Information Service. He is secretary of Supporters of Nuclear Energy.

–   *Jonathan Porritt* (March 2006). *Is nuclear the answer?* Section 3. Advice to Ministers. www.sd-commission.org.uk

3   *"Nuclear is a money pit", "We have a huge amount of wave and wind."* Ann Leslie, journalist. Speaking on *Any Questions?*, Radio 4, 10 February 2006.

–   *Los Angeles residents drive … from Earth to Mars* – (The Earthworks Group, 1989, page 34).

–   targetneutral.com charges just £4 per ton of CO_2 for their "neutralization." (A significantly lower price than any other "offsetting" company I have come across.) At this price, a typical Brit could have his 11 tons per year "neutralized" for just £44 per year! Evidence that BP's "neutralization" schemes don't really add up comes from the fact that its projects have not achieved the Gold Standard www.cdmgoldstandard.org (Michael Schlup, personal communication). Many "carbon offset" projects have been exposed as worthless by Fiona Harvey of the Financial Times [2jhve6].

4   *People who want to promote renewables over nuclear, for example, say "offshore wind power could power all UK homes."* At the end of 2007, the UK government announced that they would allow the building of offshore wind

turbines "enough to power all UK homes." Friends of the Earth's renewable energy campaigner, Nick Rau, said the group welcomed the government's announcement. "The potential power that could be generated by this industry is enormous," he said. [25e59w]. From the Guardian [5o7mxk]: John Sauven, the executive director of Greenpeace, said that the plans amounted to a "wind energy revolution." "And Labour needs to drop its obsession with nuclear power, which could only ever reduce emissions by about 4% at some time in the distant future." Nick Rau said: "We are delighted the government is getting serious about the potential for offshore wind, which could generate 25% of the UK's electricity by 2020." A few weeks later, the government announced that it would permit new nuclear stations to be built. "Today's decision to give the go-ahead to a new generation of nuclear power stations ... will do little to tackle climate change," Friends of the Earth warned [5c4olc].

In fact, the two proposed expansions – of offshore wind and of nuclear – would both deliver just the same amount of electricity per year. The total permitted offshore wind power of 33 GW would on average deliver 10 GW, which is 4 kWh per day per person; and the replacement of all the retiring nuclear power stations would deliver 10 GW, which is 4 kWh per day per person. Yet in the same breath, anti-nuclear campaigners say that the nuclear option would "do little," while the wind option would "power all UK homes." The fact is, "powering all UK homes" and "only reducing emissions by about 4%" are the same thing.

4 *"water-powered car"* New Scientist, 29th July 2006, p. 35. This article, headlined "Water-powered car might be available by 2009," opened thus:

"Forget cars fuelled by alcohol and vegetable oil. Before long, you might be able to run your car with nothing more than water in its fuel tank. It would be the ultimate zero-emissions vehicle.

"While water is not at first sight an obvious power source, it has a key virtue: it is an abundant source of hydrogen, the element widely touted as the green fuel of the future."

The work New Scientist was describing was not ridiculous – it was actually about a car using *boron* as a fuel, with a boron/water reaction as one of the first chemical steps. Why did New Scientist feel the urge to turn this into a story suggesting that water was the fuel? Water is not a fuel. It never has been, and it never will be. It is already burned! The first law of thermodynamics says you can't get energy for nothing; you can only convert energy from one form to another. The energy in any engine must come from somewhere. Fox News peddled an even more absurd story [2fztd3].

– *Climate change is a far greater threat to the world than international terrorism.* Sir David King, Chief Scientific Advisor to the UK government, January, 2004. [26e8z]

– *the glorification of travel* – an allusion to the offence of "glorification" defined in the UK's Terrorism Act which came into force on 13 April, 2006. [ykhayj]

5 *Figure 1.2.* This figure shows production of crude oil including lease condensate, natural gas plant liquids, and other liquids, and refinery processing gain. Sources: EIA, and BP statistical review of world energy.

6 *The first practical steam engine was invented in 1698.* In fact, Hero of Alexandria described a steam engine, but given that Hero's engine didn't catch on in the following 1600 years, I deem Savery's 1698 invention the first *practical* steam engine.

– *Figures 1.4 and 1.7: Graph of carbon dioxide concentration.* The data are collated from Keeling and Whorf (2005) (measurements spanning 1958–2004); Neftel et al. (1994) (1734–1983); Etheridge et al. (1998) (1000–1978); Siegenthaler et al. (2005) (950–1888 AD); and Indermuhle et al. (1999) (from 11 000 to 450 years before present). This graph, by the way, should not be confused with the "hockey stick graph", which shows the history of global *temperatures*. Attentive readers will have noticed that the climate-change argument I presented makes no mention of *historical* temperatures. *Figures 1.5–1.7: Coal production* numbers are from Jevons (1866), Malanima (2006), Netherlands Environmental Assessment Agency (2006), National Bureau of Economic Research (2001), Hatcher (1993), Flinn and Stoker (1984), Church et al. (1986), Supple (1987), Ashworth and Pegg (1986). Jevons was the first "Peak Oil" author. In 1865, he estimated Britain's easily-accessible coal reserves, looked at the history of exponential growth in consumption, and predicted the end of the exponential growth and the end of the British dominance of world industry. "We cannot long maintain our present rate of increase of consumption. . . . the check to our progress must become perceptible within a century from the present time. . . . the conclusion is inevitable, that our present happy progressive condition is a thing of limited duration." Jevons was right. Within a century British coal production indeed peaked, and there were two world wars.

8 *Dominic Lawson, a columnist from the Independent.* My quote is adapted from Dominic Lawson's column in the *Independent*, 8 June, 2007.
It is not a verbatim quote: I edited his words to make them briefer but took care not to correct any of his errors. *All three numbers he mentions are incorrect.* Here's how he screwed up. First, he says "carbon dioxide" but gives numbers for carbon: the burning of fossil fuels sends *26* gigatonnes of CO_2 per year into the atmosphere (not 7 gigatonnes). A common mistake. Second, he claims that the oceans send 36 000 gigatonnes of carbon per year into the atmosphere. This is a far worse error: 36 000 gigatonnes is the *total amount* of carbon in the ocean! The annual *flow* is much smaller – about 90 gigatonnes of carbon per year ($330\,Gt\,CO_2/y$), according to standard diagrams of the carbon cycle [16y5g] (I believe this $90\,GtC/y$ is the estimated flow rate, were the atmosphere suddenly to have its CO_2 concentration reduced to zero.) Similarly his "1900 gigatonne" flow from biosphere to atmosphere is wrong. The correct figure according to the standard diagrams is about 120 gigatonnes of carbon per year ($440\,Gt\,CO_2/y$).

The weights of an atom of carbon and a molecule of CO_2 are in the ratio 12 to 44, because the carbon atom weighs 12 units and the two oxygen atoms weigh 16 each. $12 + 16 + 16 = 44$.

Incidentally, the observed rise in CO_2 concentration is nicely in line with what you'd expect, assuming most of the human emissions of carbon remained in the atmosphere. From 1715 to 2004, roughly $1160\,Gt\,CO_2$ have been released to the atmosphere from the consumption of fossil fuels and cement production (Marland et al., 2007). If *all* of this CO_2 had stayed in the atmosphere, the concentration would have risen by 160 ppm (from 280 to 440 ppm). The actual rise has been about 100 ppm (from 275 to 377 ppm). So roughly 60% of what was emitted is now in the atmosphere.

10 *Carbon dioxide has a warming effect.* The over-emotional debate about this topic is getting quite tiresome, isn't it? "The science is now settled." "No it isn't!" "Yes it is!" I think the most helpful thing I can do here is direct anyone who wants a break from the shouting to a brief report written by Charney et al. (1979). This report's conclusions carry weight because the National Academy of Sciences (the US equivalent of the Royal Society) commissioned the report and selected its authors on the basis of their expertise, "and with regard for appropriate balance." The study group was convened "under the auspices of the Climate Research Board of the National Research Council to assess the scientific basis for projection of possible future climatic changes resulting from man-made releases of carbon dioxide into the atmosphere." Specifically, they were asked: "to identify the principal premises on which our current understanding of the question is based, to assess quantitatively the adequacy and uncertainty of our knowledge of these factors and processes, and to summarize in concise and objective terms our best present understanding of the carbon dioxide/climate issue for the benefit of policy-makers."
The report is just 33 pages long, it is free to download [5qfkaw], and I recommend it. It makes clear which bits of the science were already settled in 1979, and which bits still had uncertainty.
Here are the main points I picked up from this report. First, doubling the atmospheric CO_2 concentration would change the net heating of the troposphere, oceans, and land by an average power per unit area of roughly $4\,W/m^2$, if all other properties of the atmosphere remained unchanged. This heating effect can be compared with the average power absorbed by the atmosphere, land, and oceans, which is $238\,W/m^2$. So doubling CO_2 concentrations would have a warming effect equivalent to increasing the intensity of the sun by $4/238 = 1.7\%$. Second, the consequences of this CO_2-induced heating are hard to predict, on account of the complexity of the atmosphere/ocean system, but the authors predicted a global surface warming of between $2\,°C$ and $3.5\,°C$, with greater increases at high latitudes. Finally, the authors summarize: "we have tried but have been unable to find any overlooked or underestimated physical effects that could reduce the currently estimated global warmings due to a doubling of atmospheric CO_2 to negligible proportions or reverse them altogether." They warn that, thanks to the ocean, "the great and ponderous flywheel of the global climate system," it is quite possible that the warming would occur sufficiently sluggishly that it

would be difficult to detect in the coming decades. Nevertheless "warming will eventually occur, and the associated regional climatic changes ... may well be significant."

The foreword by the chairman of the Climate Research Board, Verner E. Suomi, summarizes the conclusions with a famous cascade of double negatives. "If carbon dioxide continues to increase, the study group finds no reason to doubt that climate changes will result and no reason to believe that these changes will be negligible."

10 *The litany of probable drastic effects of climate change – I'm sure you've heard it before.* See [2z2xg7] if not.

12 *Breakdown of world greenhouse gas emissions by region and by country.* Data source: Climate Analysis Indicators Tool (CAIT) Version 4.0. (Washington, DC: World Resources Institute, 2007). The first three figures show national totals of all six major greenhouse gases (CO_2, CH_4, N_2O, PFC, HFC, SF_6), excluding contributions from land-use change and forestry. The figure on p14 shows cumulative emissions of CO_2 only.

14 *Congratulations, Britain! ...in the table of historical emissions, per capita, we are second only to the USA.* Sincere apologies here to Luxembourg, whose historical per-capita emissions actually exceed those of America and Britain; but I felt the winners' podium should really be reserved for countries having both large per-capita and large total emissions. In total terms the biggest historical emitters are, in order, USA ($322\,GtCO_2$), Russian Federation ($90\,GtCO_2$), China ($89\,GtCO_2$), Germany ($78\,GtCO_2$), UK ($62\,GtCO_2$), Japan ($43\,GtCO_2$), France ($30\,GtCO_2$), India ($25\,GtCO_2$), and Canada ($24\,GtCO_2$). The per-capita order is: Luxembourg, USA, United Kingdom, Czech Republic, Belgium, Germany, Estonia, Qatar, and Canada.

–   *Some countries, including Britain, have committed to at least a 60% reduction in greenhouse-gas emissions by 2050.* Indeed, as I write, Britain's commitment is being increased to an 80% reduction relative to 1990 levels.

15 *Figure 1.8.* In the lower scenario, the chance that the temperature rise will exceed $2\,°C$ is estimated to be 9–26%; the cumulative carbon emissions from 2007 onwards are $309\,GtC$; CO_2 concentrations reach a peak of $410\,ppm$, CO_2e concentrations peak at $421\,ppm$, and in 2100 CO_2 concentrations fall back to $355\,ppm$. In the upper scenario, the chance of exceeding $2\,°C$ is estimated to be 16–43%; the cumulative carbon emissions from 2007 onwards are $415\,GtC$; CO_2 concentrations reach a peak of $425\,ppm$, CO_2e concentrations peak at $435\,ppm$, and in 2100 CO_2 concentrations fall back to $380\,ppm$. See also hdr.undp.org/en/reports/global/hdr2007-2008/.

16 *there are many other helpful sources on the internet.* I recommend, for example: BP's *Statistical Review of World Energy* [yyxq2m], the Sustainable Development Commission www.sd-commission.org.uk, the Danish Wind Industry Association www.windpower.org, Environmentalists For Nuclear Energy www.ecolo.org, Wind Energy Department, Risø University www.risoe.dk/vea, DEFRA www.defra.gov.uk/environment/statistics, especially the book *Avoiding Dangerous Climate Change* [dzcqq], the Pembina Institute www.pembina.org/publications.asp, and the DTI (now known as BERR) www.dti.gov.uk/publications/.

17 *factual assertions and ethical assertions...* Ethical assertions are also known as "normative claims" or "value judgments," and factual assertions are known as "positive claims." Ethical assertions usually contain verbs like "should" and "must," or adjectives like "fair," "right," and "wrong." For helpful further reading see Dessler and Parson (2006).

18 *Gordon Brown.* On 10th September, 2005, Gordon Brown said the high price of fuel posed a significant risk to the European economy and to global growth, and urged OPEC to raise oil production. Again, six months later, he said "we need ...more production, more drilling, more investment, more petrochemical investment" (22nd April, 2006) [y98ys5]. Let me temper this criticism of Gordon Brown by praising one of his more recent initiatives, namely the promotion of electric vehicles and plug-in hybrids. As you'll see later, one of this book's conclusions is that electrification of most transport is a good part of a plan for getting off fossil fuels.

2 The balance sheet

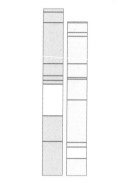

CONSUMPTION PRODUCTION

Nature cannot be fooled.

Richard Feynman

Let's talk about energy consumption and energy production. At the moment, most of the energy the developed world consumes is produced from fossil fuels; that's not sustainable. Exactly how long we could keep living on fossil fuels is an interesting question, but it's not the question we'll address in this book. I want to think about *living without fossil fuels*.

We're going to make two stacks. In the left-hand, red stack we will add up our energy consumption, and in the right-hand, green stack, we'll add up sustainable energy production. We'll assemble the two stacks gradually, adding items one at a time as we discuss them.

The question addressed in this book is "can we *conceivably* live sustainably?" So, we will add up all *conceivable* sustainable energy sources and put them in the right-hand, green stack.

In the left-hand, red stack, we'll estimate the consumption of a "typical moderately-affluent person;" I encourage you to tot up an estimate of your *own* consumption, creating your own personalized left-hand stack too. Later on we'll also find out the current *average* energy consumption of Europeans and Americans.

Some key forms of consumption for the left-hand stack will be:

- transport
    - cars, planes, freight
- heating and cooling
- lighting
- information systems and other gadgets
- food
- manufacturing

In the right-hand sustainable-production stack, our main categories will be:

- wind
- solar
    - photovoltaics, thermal, biomass
- hydroelectric
- wave
- tide
- geothermal
- nuclear? (with a question-mark, because it's not clear whether nuclear power counts as "sustainable")

As we estimate our consumption of energy for heating, transportation, manufacturing, and so forth, the aim is not only to compute a number for the left-hand stack of our balance sheet, but also to understand what each number depends on, and how susceptible to modification it is.

In the right-hand, green stack, we'll add up the sustainable production estimates for the United Kingdom. This will allow us to answer the question "can the UK conceivably live on its own renewables?"

Whether the sustainable energy sources that we put in the right-hand stack are *economically* feasible is an important question, but let's leave that question to one side, and just add up the two stacks first. Sometimes people focus too much on economic feasibility and they miss the big picture. For example, people discuss "is wind cheaper than nuclear?" and forget to ask "how *much* wind is available?" or "how much uranium is left?"

The outcome when we add everything up might look like this:

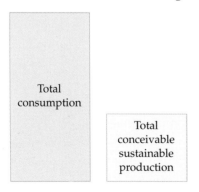

If we find consumption is much less than conceivable sustainable production, then we can say "good, *maybe* we can live sustainably; let's look into the economic, social, and environmental costs of the sustainable alternatives, and figure out which of them deserve the most research and development; if we do a good job, there *might* not be an energy crisis."

On the other hand, the outcome of our sums might look like this:

– a much bleaker picture. This picture says "it doesn't matter what the economics of sustainable power are: there's simply *not enough* sustainable power to support our current lifestyle; massive change is coming."

Energy and power

Most discussions of energy consumption and production are confusing because of the proliferation of *units* in which energy and power are measured, from "tons of oil equivalent" to "terawatt-hours" (TWh) and "exajoules" (EJ). Nobody but a specialist has a feeling for what "a barrel of oil" or "a million BTUs" means in human terms. In this book, we'll express everything in a single set of personal units that everyone can relate to.

The unit of **energy** I have chosen is the kilowatt-hour (kWh). This quantity is called "one unit" on electricity bills, and it costs a domestic user about 10p in the UK in 2008. As we'll see, most individual daily choices involve amounts of energy equal to small numbers of kilowatt-hours.

When we discuss **powers** (rates at which we use or produce energy), the main unit will be the kilowatt-hour per day (kWh/d). We'll also occasionally use the watt ($40\,W \simeq 1\,kWh/d$) and the kilowatt ($1\,kW = 1000\,W = 24\,kWh/d$), as I'll explain below. The kilowatt-hour per day is a nice human-sized unit: most personal energy-guzzling activities guzzle at a rate of a small number of kilowatt-hours per day. For example, one $40\,W$ lightbulb, kept switched on all the time, uses **one** kilowatt-hour per day. Some electricity companies include graphs in their electricity bills, showing energy consumption in kilowatt-hours per day. I'll use the same unit for all forms of power, not just electricity. Petrol consumption, gas consumption, coal consumption: I'll measure all these powers in kilowatt-hours per day. Let me make this clear: for some people, the word "power" means only *electrical* energy consumption. But this book concerns *all* forms of energy consumption and production, and I will use the word "power" for all of them.

One kilowatt-hour per day is roughly the power you could get from one human servant. The number of kilowatt-hours per day you use is thus the effective number of servants you have working for you.

People use the two terms energy and power interchangeably in ordinary speech, but in this book we must stick rigorously to their scientific definitions. *Power is the rate at which something uses energy.*

Maybe a good way to explain energy and power is by an analogy with water and water-flow from taps. If you want a drink of water, you want a *volume* of water – one litre, perhaps (if you're thirsty). When you turn on a tap, you create a *flow* of water – one litre per minute, say, if the tap yields only a trickle; or 10 litres per minute, from a more generous tap. You can get the same volume (one litre) either by running the trickling tap for one minute, or by running the generous tap for one tenth of a minute. The *volume* delivered in a particular time is equal to the *flow* multiplied by the

Figure 2.1. Distinguishing energy and power. Each of these $60\,W$ light bulbs has a *power* of $60\,W$ when switched on; it doesn't have an "energy" of $60\,W$. The bulb uses $60\,W$ of electrical *power* when it's on; it emits $60\,W$ of *power* in the form of light and heat (mainly the latter).

volume	flow
is measured in	is measured in
litres	litres per minute

energy	power
is measured in	is measured in
kWh	kWh per day

time:

$$\text{volume} = \text{flow} \times \text{time}.$$

We say that a *flow* is a *rate* at which *volume* is delivered. If you know the volume delivered in a particular time, you get the flow by dividing the volume by the time:

$$\text{flow} = \frac{\text{volume}}{\text{time}}.$$

Here's the connection to energy and power. *Energy* is like water *volume*: *power* is like water *flow*. For example, whenever a toaster is switched on, it starts to consume *power* at a rate of one kilowatt. It continues to consume one kilowatt until it is switched off. To put it another way, the toaster (if it's left on permanently) consumes one kilowatt-hour (kWh) of energy per hour; it also consumes 24 kilowatt-hours per day.

The longer the toaster is on, the more energy it uses. You can work out the energy used by a particular activity by multiplying the power by the duration:

$$\text{energy} = \text{power} \times \text{time}.$$

The joule is the standard international unit of energy, but sadly it's far too small to work with. The kilowatt-hour is equal to 3.6 million joules (3.6 megajoules).

Powers are so useful and important, they have something that water flows don't have: they have their own special units. When we talk of a flow, we might measure it in "litres per minute," "gallons per hour," or "cubic-metres per second;" these units' names make clear that the flow is "a volume per unit time." A power of *one joule per second* is called *one watt*. 1000 joules per second is called one kilowatt. Let's get the terminology straight: the toaster uses one kilowatt. It doesn't use "one kilowatt per second." The "per second" is already built in to the definition of the kilowatt: one kilowatt means "one kilojoule per second." Similarly we say "a nuclear power station generates one gigawatt." One gigawatt, by the way, is one billion watts, one million kilowatts, or 1000 megawatts. So one gigawatt is a million toasters. And the "g"s in gigawatt are pronounced hard, the same as in "giggle." And, while I'm tapping the blackboard, we capitalize the "g" and "w" in "gigawatt" only when we write the abbreviation "GW."

Please, never, ever say "one kilowatt per second," "one kilowatt per hour," or "one kilowatt per day;" none of these is a valid measure of power. The urge that people have to say "per something" when talking about their toasters is one of the reasons I decided to use the "kilowatt-hour per day" as my unit of power. I'm sorry that it's a bit cumbersome to say and to write.

Here's one last thing to make clear: if I say "someone used a gigawatt-hour of energy," I am simply telling you *how much* energy they used, not *how fast* they used it. Talking about a gigawatt-hour *doesn't* imply the

energy	power
is measured in	is measured in
kWh	kWh per day
or	or
MJ	kW
	or
	W (watts)
	or
	MW (megawatts)
	or
	GW (gigawatts)
	or
	TW (terawatts)

energy was used *in one hour*. You could use a gigawatt-hour of energy by switching on one million toasters for one hour, or by switching on 1000 toasters for 1000 hours.

As I said, I'll usually quote powers in kWh/d *per person*. One reason for liking these personal units is that it makes it much easier to move from talking about the UK to talking about other countries or regions. For example, imagine we are discussing waste incineration and we learn that UK waste incineration delivers a power of 7 TWh per year and that Denmark's waste incineration delivers 10 TWh per year. Does this help us say whether Denmark incinerates "more" waste than the UK? While the total power produced from waste in each country may be interesting, I think that what we usually want to know is the waste incineration *per person*. (For the record, that is: Denmark, 5 kWh/d per person; UK, 0.3 kWh/d per person. So Danes incinerate about 13 times as much waste as Brits.) To save ink, I'll sometimes abbreviate "per person" to "/p". By discussing everything per-person from the outset, we end up with a more transportable book, one that will hopefully be useful for sustainable energy discussions worldwide.

1 TWh (one terawatt-hour) is equal to one billion kWh.

Picky details

Isn't energy conserved? We talk about "using" energy, but doesn't one of the laws of nature say that energy can't be created or destroyed?

Yes, I'm being imprecise. This is really a book about *entropy* – a trickier thing to explain. When we "use up" one kilojoule of energy, what we're really doing is taking one kilojoule of energy in a form that has *low entropy* (for example, electricity), and *converting* it into an exactly equal amount of energy in another form, usually one that has much higher entropy (for example, hot air or hot water). When we've "used" the energy, it's still there; but we normally can't "use" the energy over and over again, because only *low entropy* energy is "useful" to us. Sometimes these different grades of energy are distinguished by adding a label to the units: one kWh(e) is one kilowatt-hour of electrical energy – the highest grade of energy. One kWh(th) is one kilowatt-hour of thermal energy – for example the energy in ten litres of boiling-hot water. Energy lurking in higher-temperature things is more useful (lower entropy) than energy in tepid things. A third grade of energy is chemical energy. Chemical energy is high-grade energy like electricity.

It's a convenient but sloppy shorthand to talk about the energy rather than the entropy, and that is what we'll do most of the time in this book. Occasionally, we'll have to smarten up this sloppiness; for example, when we discuss refrigeration, power stations, heat pumps, or geothermal power.

Are you comparing apples and oranges? Is it valid to compare different forms of energy such as the chemical energy that is fed into a petrol-powered car and the electricity from a wind turbine?

By comparing consumed energy with conceivable produced energy, I do not wish to imply that all forms of energy are equivalent and interchangeable. The electrical energy produced by a wind turbine is of no use to a petrol engine; and petrol is no use if you want to power a television. In principle, energy can be converted from one form to another, though conversion entails losses. Fossil-fuel power stations, for example, guzzle *chemical energy* and produce *electricity* (with an efficiency of 40% or so). And aluminium plants guzzle *electrical energy* to create a product with high *chemical energy* – aluminium (with an efficiency of 30% or so).

In some summaries of energy production and consumption, all the different forms of energy are put into the same units, but multipliers are introduced, rating electrical energy from hydroelectricity for example as being worth 2.5 times more than the chemical energy in oil. This bumping up of electricity's effective energy value can be justified by saying, "well, 1 kWh of electricity is equivalent to 2.5 kWh of oil, because if we put that much oil into a standard power station it would deliver 40% of 2.5 kWh, which is 1 kWh of electricity." In this book, however, I will usually use a one-to-one conversion rate when comparing different forms of energy. It is *not* the case that 2.5 kWh of oil is inescapably equivalent to 1 kWh of electricity; that just happens to be the perceived exchange rate in a worldview where oil is used to make electricity. Yes, conversion of chemical energy to electrical energy is done with this particular inefficient exchange rate. But electrical energy can also be converted to chemical energy. In an alternative world (perhaps not far-off) with relatively plentiful electricity and little oil, we might use electricity to make liquid fuels; in that world we would surely not use the same exchange rate – each kWh of gasoline would then cost us something like 3 kWh of electricity! I think the timeless and scientific way to summarize and compare energies is to hold 1 kWh of chemical energy equivalent to 1 kWh of electricity. My choice to use this one-to-one conversion rate means that some of my sums will look a bit different from other people's. (For example, BP's *Statistical Review of World Energy* rates 1 kWh of electricity as equivalent to $100/38 \simeq 2.6$ kWh of oil; on the other hand, the government's *Digest of UK Energy Statistics* uses the same one-to-one conversion rate as me.) And I emphasize again, this choice does not imply that I'm suggesting you could convert either form of energy directly into the other. Converting chemical energy into electrical energy always wastes energy, and so does converting electrical into chemical energy.

Physics and equations

Throughout the book, my aim is not only to work out numbers indicating our current energy consumption and conceivable sustainable production,

but also to make clear *what these numbers depend on.* Understanding what
the numbers depend on is essential if we are to choose sensible policies
to change any of the numbers. Only if we understand the physics behind
energy consumption and energy production can we assess assertions such
as "cars waste 99% of the energy they consume; we could redesign cars so
that they use 100 times less energy." Is this assertion true? To explain the
answer, I will need to use equations like

$$\text{kinetic energy} = \frac{1}{2}mv^2.$$

However, I recognize that to many readers, such formulae are a foreign language. So, here's my promise: *I'll keep all this foreign-language stuff in technical chapters at the end of the book.* Any reader with a high-school/secondary-
school qualification in maths, physics, or chemistry should enjoy these
technical chapters. The main thread of the book (from page 2 to page 250)
is intended to be accessible to everyone who can add, multiply, and divide.
It is especially aimed at our dear elected and unelected representatives, the
Members of Parliament.

One last point, before we get rolling: I don't know everything about
energy. I don't have all the answers, and the numbers I offer are open to
revision and correction. (Indeed I expect corrections and will publish them
on the book's website.) The one thing I *am* sure of is that the answers to
our sustainable energy questions will involve *numbers*; any sane discussion
of sustainable energy requires numbers. This book's got 'em, and it shows
how to handle them. I hope you enjoy it!

Notes and further reading

page no.

25   *The "per second" is already built in to the definition of the kilowatt.* Other examples of units that, like the watt, already
     have a "per time" built in are the knot – "our yacht's speed was ten knots!" (a knot is one nautical mile *per* hour); the
     hertz – "I could hear a buzzing at 50 hertz" (one hertz is a frequency of one cycle *per* second); the ampere – "the fuse
     blows when the current is higher than 13 amps" (*not* 13 amps per second); and the horsepower – "that stinking engine
     delivers 50 horsepower" (*not* 50 horsepower per second, nor 50 horsepower per hour, nor 50 horsepower per day, just
     50 horsepower).

 –   *Please, never, ever say "one kilowatt per second."* There are specific, rare exceptions to this rule. If talking about a
     growth in demand for power, we might say "British demand is growing at one gigawatt per year." In Chapter 26 when
     I discuss fluctuations in wind power, I will say "one morning, the power delivered by Irish windmills fell at a rate of
     84 MW per hour." Please take care! Just one accidental syllable can lead to confusion: for example, your electricity
     meter's reading is in kilowatt-hours (kWh), *not* 'kilowatts-per-hour'.

I've provided a chart on p368 to help you translate between kWh per day per person and the other major units in which
powers are discussed.

3 Cars

For our first chapter on consumption, let's study that icon of modern civilization: the car with a lone person in it.

How much power does a regular car-user consume? Once we know the conversion rates, it's simple arithmetic:

$$\frac{\text{energy used}}{\text{per day}} = \frac{\text{distance travelled per day}}{\text{distance per unit of fuel}} \times \text{energy per unit of fuel}.$$

For the **distance travelled per day**, let's use 50 km (30 miles).

For the **distance per unit of fuel**, also known as the **economy** of the car, let's use 33 miles per UK gallon (taken from an advertisement for a family car):

$$33 \text{ miles per imperial gallon} \simeq 12 \text{ km per litre}.$$

(The symbol "$\simeq$" means "is approximately equal to.")

What about the **energy per unit of fuel** (also called the **calorific value** or **energy density**)? Instead of looking it up, it's fun to estimate this sort of quantity by a bit of lateral thinking. Automobile fuels (whether diesel or petrol) are all hydrocarbons; and hydrocarbons can also be found on our breakfast table, with the calorific value conveniently written on the side: roughly 8 kWh per kg (figure 3.2). Since we've estimated the economy of the car in miles per unit *volume* of fuel, we need to express the calorific value as an energy per unit *volume*. To turn our fuel's "8 kWh per kg" (an energy per unit *mass*) into an energy per unit volume, we need to know the density of the fuel. What's the density of butter? Well, butter just floats on water, as do fuel-spills, so its density must be a little less than water's, which is 1 kg per litre. If we guess a density of 0.8 kg per litre, we obtain a calorific value of:

$$8 \text{ kWh per kg} \times 0.8 \text{ kg per litre} \simeq 7 \text{ kWh per litre}.$$

Rather than willfully perpetuate an inaccurate estimate, let's switch to the actual value, for petrol, of 10 kWh per litre.

$$\begin{aligned} \text{energy per day} &= \frac{\text{distance travelled per day}}{\text{distance per unit of fuel}} \times \text{energy per unit of fuel} \\ &= \frac{50 \text{ km/day}}{12 \text{ km/litre}} \times 10 \text{ kWh/litre} \\ &\simeq 40 \text{ kWh/day}. \end{aligned}$$

Congratulations! We've made our first estimate of consumption. I've displayed this estimate in the left-hand stack in figure 3.3. The red box's height represents 40 kWh per day per person.

Figure 3.1. Cars. A red BMW dwarfed by a spaceship from the planet Dorkon.

Figure 3.2. Want to know the energy in car fuel? Look at the label on a pack of butter or margarine. The calorific value is 3000 kJ per 100 g, or about 8 kWh per kg.

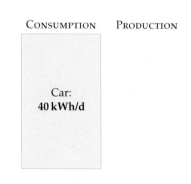

Figure 3.3. Chapter 3's conclusion: a typical car-driver uses about 40 kWh per day.

This is the estimate for a typical car-driver driving a typical car today. Later chapters will discuss the *average* consumption of all the people in Britain, taking into account the fact that not everyone drives. We'll also discuss in Part II what the consumption *could* be, with the help of other technologies such as electric cars.

Why does the car deliver 33 miles per gallon? Where's that energy going? Could we manufacture cars that do 3300 miles per gallon? If we are interested in trying to reduce cars' consumption, we need to understand the physics behind cars' consumption. These questions are answered in the accompanying technical chapter A (p254), which provides a cartoon theory of cars' consumption. I encourage you to read the technical chapters if formulae like $\frac{1}{2}mv^2$ don't give you medical problems.

Chapter 3's conclusion: a typical car-driver uses about 40 kWh per day. Next we need to get the sustainable-production stack going, so we have something to compare this estimate with.

Queries

*What about the energy-cost of **producing** the car's fuel?*

Good point. When I estimate the energy consumed by a particular activity, I tend to choose a fairly tight "boundary" around the activity. This choice makes the estimation easier, but I agree that it's a good idea to try to estimate the full energy impact of an activity. It's been estimated that making each unit of petrol requires an input of 1.4 units of oil and other primary fuels (Treloar et al., 2004).

*What about the energy-cost of manufacturing the **car**?*

Yes, that cost fell outside the boundary of this calculation too. We'll talk about car-making in Chapter 15.

Notes and further reading

page no.

29 *For the distance travelled per day, let's use 50 km.* This corresponds to 18 000 km (11 000 miles) per year. Roughly half of the British population drive to work. The total amount of car travel in the UK is 686 billion passenger-km per year, which corresponds to an "average distance travelled by car per British person" of 30 km per day. Source: Department for Transport [5647rh]. As I said on p22, I aim to estimate the consumption of a "typical moderately-affluent person" – the consumption that many people aspire to. Some people don't drive much. In this chapter, I want to estimate the energy consumed by someone who chooses to drive, rather than depersonalize the answer by reporting the UK average, which mixes together the drivers and non-drivers. If I said "the average use of energy for car driving

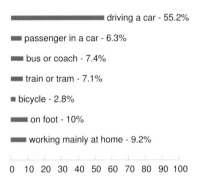

driving a car - 55.2%

passenger in a car - 6.3%

bus or coach - 7.4%

train or tram - 7.1%

bicycle - 2.8%

on foot - 10%

working mainly at home - 9.2%

0	10	20	30	40	50	60	70	80	90	100

Figure 3.4. How British people travel to work, according to the 2001 census.

in the UK is 24 kWh/d per person," I bet some people would misunderstand and say: "I'm a car driver so I guess I use 24 kWh/d."

29   *... let's use 33 miles per UK gallon.* In the European language, this is 8.6 litres per 100 km. 33 miles per gallon was the average for UK cars in 2005 [27jdc5]. Petrol cars have an average fuel consumption of 31 mpg; diesel cars, 39 mpg; new petrol cars (less than two years old), 32 mpg (Dept. for Transport, 2007). Honda, "the most fuel-efficient auto company in America," records that its fleet of new cars sold in 2005 has an average top-level fuel economy of 35 miles per UK gallon [28abpm].

29   *Let's guess a density of 0.8 kg per litre.* Petrol's density is 0.737. Diesel's is 0.820–0.950 [nmn41].

–   *... the actual value of 10 kWh per litre.* ORNL [2hcgdh] provide the following calorific values: diesel: 10.7 kWh/l; jet fuel: 10.4 kWh/l; petrol: 9.7 kWh/l. When looking up calorific values, you'll find "gross calorific value" and "net calorific value" listed (also known as "high heat value" and "low heat value"). These differ by only 6% for motor fuels, so it's not crucial to distinguish them here, but let me explain anyway. The gross calorific value is the actual chemical energy released when the fuel is burned. One of the products of combustion is water, and in most engines and power stations, part of the energy goes into vaporizing this water. The net calorific value measures how much energy is left over assuming this energy of vaporization is discarded and wasted.

When we ask "how much energy does my lifestyle consume?" the gross calorific value is the right quantity to use. The net calorific value, on the other hand, is of interest to a power station engineer, who needs to decide which fuel to burn in his power station. Throughout this book I've tried to use gross calorific values.

A final note for party-pooping pedants who say "butter is not a hydrocarbon": OK, butter is not a *pure* hydrocarbon; but it's a good approximation to say that the main component of butter is long hydrocarbon chains, just like petrol. The proof of the pudding is, this approximation got us within 30% of the correct answer. Welcome to guerrilla physics.

calorific values	
petrol	10 kWh per litre
diesel	11 kWh per litre

4 Wind

The UK has the best wind resources in Europe.

Sustainable Development Commission

Wind farms will devastate the countryside pointlessly.

James Lovelock

How much wind power could we plausibly generate?

We can make an estimate of the potential of *on-shore* (land-based) wind in the United Kingdom by multiplying the average power per unit land-area of a wind farm by the area per person in the UK:

power per person = wind power per unit area × area per person.

Chapter B (p263) explains how to estimate the power per unit area of a wind farm in the UK. If the typical windspeed is 6 m/s (13 miles per hour, or 22 km/h), the power per unit area of wind farm is about $2\,W/m^2$.

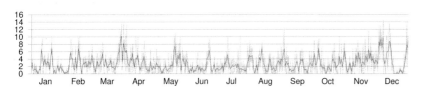

Figure 4.1. Cambridge mean wind speed in metres per second, daily (red line), and half-hourly (blue line) during 2006. See also figure 4.6.

This figure of 6 m/s is probably an over-estimate for many locations in Britain. For example, figure 4.1 shows daily average windspeeds in Cambridge during 2006. The daily average speed reached 6 m/s on only about 30 days of the year – see figure 4.6 for a histogram. But some spots do have windspeeds above 6 m/s – for example, the summit of Cairngorm in Scotland (figure 4.2).

Plugging in the British population density: 250 people per square kilometre, or 4000 square metres per person, we find that wind power could

100 km

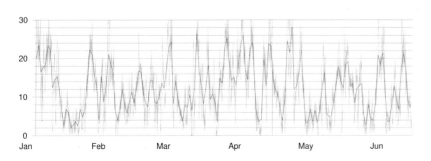

Figure 4.2. Cairngorm mean wind speed in metres per second, during six months of 2006.

generate

$$2\,\text{W/m}^2 \times 4000\,\text{m}^2/\text{person} = 8000\,\text{W per person},$$

if wind turbines were packed across the *whole* country, and assuming $2\,\text{W/m}^2$ is the correct power per unit area. Converting to our favourite power units, that's 200 kWh/d per person.

Let's be realistic. What fraction of the country can we really imagine covering with windmills? Maybe 10%? Then we conclude: if we covered the windiest 10% of the country with windmills (delivering $2\,\text{W/m}^2$), we would be able to generate 20 kWh/d per person, which is **half** of the power used by driving an average fossil-fuel car 50 km per day.

Britain's onshore wind energy resource may be "huge," but it's evidently not as huge as our huge consumption. We'll come to offshore wind later.

I should emphasize how generous an assumption I'm making. Let's compare this estimate of British wind potential with current installed wind power worldwide. The windmills that would be required to provide the UK with 20 kWh/d per person amount to 50 times the entire wind hardware of Denmark; 7 times all the wind farms of Germany; and double the entire fleet of all wind turbines in the world.

Please don't misunderstand me. Am I saying that we shouldn't bother building wind farms? Not at all. I'm simply trying to convey a helpful fact, namely that if we want wind power to truly make a difference, the wind farms must cover a very large area.

This conclusion – that the maximum contribution of onshore wind, albeit "huge," is much less than our consumption – is important, so let's check the key figure, the assumed power per unit area of wind farm ($2\,\text{W/m}^2$), against a real UK wind farm.

The Whitelee wind farm being built near Glasgow in Scotland has 140 turbines with a combined *peak* capacity of 322 MW in an area of 55 km^2. That's $6\,\text{W/m}^2$, *peak*. The average power produced is smaller because the turbines don't run at peak output all the time. The ratio of the average power to the peak power is called the "load factor" or "capacity factor," and it varies from site to site, and with the choice of hardware plopped on the site; a typical factor for a good site with modern turbines is 30%. If we assume Whitelee has a load factor of 33% then the average power production per unit land area is $2\,\text{W/m}^2$ – exactly the same as the power density we assumed above.

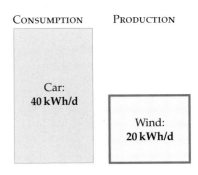

Figure 4.3. Chapter 4's conclusion: the maximum plausible production from on-shore windmills in the United Kingdom is 20 kWh per day per person.

POWER PER UNIT AREA	
wind farm (speed 6 m/s)	$2\,\text{W/m}^2$

Table 4.4. Facts worth remembering: wind farms.

POPULATION DENSITY OF BRITAIN
250 per km^2 $\leftrightarrow$ 4000 m^2 per person

Table 4.5. Facts worth remembering: population density. See page 338 for more population densities.

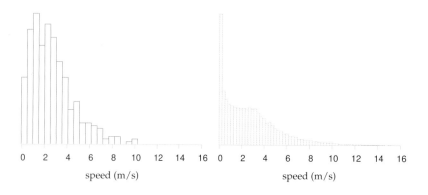

Figure 4.6. Histogram of Cambridge average wind speed in metres per second: daily averages (left), and half-hourly averages (right).

Queries

Wind turbines are getting bigger all the time. Do bigger wind turbines change this chapter's answer?

Chapter B explains. Bigger wind turbines deliver financial economies of scale, but they don't greatly increase the total power per unit land area, because bigger windmills have to be spaced further apart. A wind farm that's twice as tall will deliver roughly 30% more power.

Wind power fluctuates all the time. Surely that makes wind less useful?

Maybe. We'll come back to this issue in Chapter 26, where we'll look at wind's intermittency and discuss several possible solutions to this problem, including energy storage and demand management.

Notes and further reading

page no.

32 *Figure 4.1 and figure 4.6.* Cambridge wind data are from the Digital Technology Group, Computer Laboratory, Cambridge [vxhhj]. The weather station is on the roof of the Gates building, roughly 10 m high. Wind speeds at a height of 50 m are usually about 25% bigger. Cairngorm data (*figure 4.2*) are from Heriot–Watt University Physics Department [tdvml].

33 *The windmills required to provide the UK with 20 kWh/d per person are 50 times the entire wind power of Denmark.* Assuming a load factor of 33%, an average power of 20 kWh/d per person requires an installed capacity of 150 GW. At the end of 2006, Denmark had an installed capacity of 3.1 GW; Germany had 20.6 GW. The world total was 74 GW (wwindea.org). Incidentally, the load factor of the Danish wind fleet was 22% in 2006, and the average power it delivered was 3 kWh/d per person.

5 Planes

Imagine that you make one intercontinental trip per year by plane. How much energy does that cost?

A Boeing 747-400 with 240 000 litres of fuel carries 416 passengers about 8 800 miles (14 200 km). And fuel's calorific value is 10 kWh per litre. (We learned that in Chapter 3.) So the energy cost of one full-distance round-trip on such a plane, if divided equally among the passengers, is

$$\frac{2 \times 240\,000\,\text{litre}}{416\,\text{passengers}} \times 10\,\text{kWh/litre} \simeq 12\,000\,\text{kWh per passenger}.$$

If you make one such trip per year, then your average energy consumption per day is

$$\frac{12\,000\,\text{kWh}}{365\,\text{days}} \simeq 33\,\text{kWh/day}.$$

14 200 km is a little further than London to Cape Town (10 000 km) and London to Los Angeles (9 000 km), so I think we've slightly overestimated the distance of a typical long-range intercontinental trip; but we've also overestimated the fullness of the plane, and the energy cost per person is more if the plane's not full. Scaling down by 10 000 km/14 200 km to get an estimate for Cape Town, then up again by 100/80 to allow for the plane's being 80% full, we arrive at 29 kWh per day. For ease of memorization, I'll round this up to 30 kWh per day.

Let's make clear what this means. Flying once per year has an energy cost slightly bigger than leaving a 1 kW electric fire on, non-stop, 24 hours a day, all year.

Just as Chapter 3, in which we estimated consumption by cars, was accompanied by Chapter A, offering a model of where the energy goes in cars, this chapter's technical partner (Chapter C, p269), discusses where the energy goes in planes. Chapter C allows us to answer questions such as "would air travel consume significantly less energy if we travelled in slower planes?" The answer is **no**: in contrast to wheeled vehicles, which *can* get more efficient the slower they go, planes are already almost as energy-efficient as they could possibly be. Planes unavoidably have to use energy for two reasons: they have to throw air down in order to stay up, and they need energy to overcome air resistance. No redesign of a plane is going to radically improve its efficiency. A 10% improvement? Yes, possible. A doubling of efficiency? I'd eat my complimentary socks.

Queries

*Aren't turboprop aircraft **far** more energy-efficient?*

No. The "comfortably greener" Bombardier Q400 NextGen, "the most technologically advanced turboprop in the world," according to its manu-

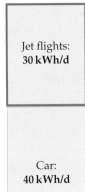

Figure 5.1. Taking one intercontinental trip per year uses about 30 kWh per day.

Figure 5.2. Bombardier Q400 NextGen. www.q400.com.

facturers [www.q400.com], uses 3.81 litres per 100 passenger-km (at a cruise speed of 667 km/h), which is an energy cost of 38 kWh per 100 p-km. The full 747 has an energy cost of 42 kWh per 100 p-km. So both planes are twice as fuel-efficient as a single-occupancy car. (The car I'm assuming here is the average European car that we discussed in Chapter 3.)

Is flying extra-bad for climate change in some way?

Yes, that's the experts' view, though uncertainty remains about this topic [3fbufz]. Flying creates other greenhouse gases in addition to CO_2, such as water and ozone, and indirect greenhouse gases, such as nitrous oxides. If you want to estimate your carbon footprint in tons of CO_2-equivalent, then you should take the actual CO_2 emissions of your flights and bump them up two- or three-fold. This book's diagrams don't include that multiplier because here we are focusing on our *energy* balance sheet.

> *The best thing we can do with environmentalists is shoot them.*
>
> Michael O'Leary, CEO of Ryanair [3asmgy]

	energy per distance (kWh per 100 p-km)
Car (4 occupants)	20
Ryanair's planes, year 2007	37
Bombardier Q400, full	38
747, full	42
747, 80% full	53
Ryanair's planes, year 2000	73
Car (1 occupant)	80

Table 5.3. Passenger transport efficiencies, expressed as energy required per 100 passenger-km.

Notes and further reading

page no.

35 *Boeing 747-400* – data are from [9ehws].
    Planes today are not completely full. Airlines are proud if their average full-ness is 80%. Easyjet planes are 85% full on average. (Source: thelondonpaper Tuesday 16th January, 2007.) An 80%-full 747 uses about 53 kWh per 100 passenger-km.
    What about short-haul flights? In 2007, Ryanair, "Europe's greenest airline," delivered transportation at a cost of 37 kWh per 100 p-km [3exmgv]. This means that flying across Europe with Ryanair has much the same energy cost as having all the passengers drive to their destination in cars, two to a car. (For an indication of what other airlines might be delivering, Ryanair's fuel burn rate in 2000, before their environment-friendly investments, was above 73 kWh per 100 p-km.) London to Rome is 1430 km; London to Malaga is 1735 km. So a round-trip to Rome with the greenest airline has an energy cost of 1050 kWh, and a round-trip to Malaga costs 1270 kWh. If you pop over to Rome and to Malaga once per year, your average power consumption is 6.3 kWh/d with the greenest airline, and perhaps 12 kWh/d with a less green one.
    What about frequent flyers? To get a silver frequent flyer card from an in-tercontinental airline, it seems one must fly around 25 000 miles per year in economy class. That's about 60 kWh per day, if we scale up the opening numbers from this chapter and assume planes are 80% full.
    Here are some additional figures from the Intergovernmental Panel on Climate Change [yrnmum]: a full 747-400 travelling 10 000 km with low-density seating (262 seats) has an energy consumption of 50 kWh per 100 p-km. In a high-density seating configuration (568 seats) and travelling 4000 km, the

Figure 5.4. Ryanair Boeing 737-800. Photograph by Adrian Pingstone.

same plane has an energy consumption of 22 kWh per 100 p-km. A short-haul Tupolev-154 travelling 2235 km with 70% of its 164 seats occupied consumes 80 kWh per 100 p-km.

35   *No redesign of a plane is going to radically improve its efficiency*. Actually, the Advisory Council for Aerospace Research in Europe (ACARE) target is for an overall 50% reduction in fuel burned per passenger-km by 2020 (relative to a 2000 baseline), with 15–20% improvement expected in engine efficiency. As of 2006, Rolls Royce is half way to this engine target [36w5gz]. Dennis Bushnell, chief scientist at NASA's Langley Research Center, seems to agree with my overall assessment of prospects for efficiency improvements in aviation. The aviation industry is mature. "There is not much left to gain except by the glacial accretion of a per cent here and there over long time periods." (New Scientist, 24 February 2007, page 33.)

The radically reshaped "Silent Aircraft" [silentaircraft.org/sax40], if it were built, is predicted to be 16% more efficient than a conventional-shaped plane (Nickol, 2008).

If the ACARE target is reached, it's presumably going to be thanks mostly to having fuller planes and better air-traffic management.

Short hauls: **6 kWh/d**

Frequent flyer: **60 kWh/d**

Figure 5.5. Two short-haul trips on the greenest short-haul airline: 6.3 kWh/d. Flying enough to qualify for silver frequent flyer status: 60 kWh/d.

6 Solar

We are estimating how our consumption stacks up against conceivable sustainable production. In the last three chapters we found car-driving and plane-flying to be bigger than the plausible on-shore wind-power potential of the United Kingdom. Could solar power put production back in the lead?

The power of raw sunshine at midday on a cloudless day is 1000 W per square metre. That's 1000 W per m² of area oriented towards the sun, not per m² of land area. To get the power per m² of *land area* in Britain, we must make several corrections. We need to compensate for the tilt between the sun and the land, which reduces the intensity of midday sun to about 60% of its value at the equator (figure 6.1). We also lose out because it is not midday all the time. On a cloud-free day in March or September, the ratio of the *average* intensity to the midday intensity is about 32%. Finally, we lose power because of cloud cover. In a typical UK location the sun shines during just 34% of daylight hours.

The combined effect of these three factors and the additional complication of the wobble of the seasons is that the average raw power of sunshine per square metre of south-facing roof in Britain is roughly 110 W/m², and the average raw power of sunshine per square metre of flat ground is roughly 100 W/m².

We can turn this raw power into useful power in four ways:

1. Solar thermal: using the sunshine for direct heating of buildings or water.

2. Solar photovoltaic: generating electricity.

3. Solar biomass: using trees, bacteria, algae, corn, soy beans, or oilseed to make energy fuels, chemicals, or building materials.

4. Food: the same as solar biomass, except we shovel the plants into humans or other animals.

(In a later chapter we'll also visit a couple of other solar power techniques appropriate for use in deserts.)

Let's make quick rough estimates of the maximum plausible powers that each of these routes could deliver. We'll neglect their economic costs, and the energy costs of manufacturing and maintaining the power facilities.

Solar thermal

The simplest solar power technology is a panel making hot water. Let's imagine we cover *all* south-facing roofs with solar thermal panels – that

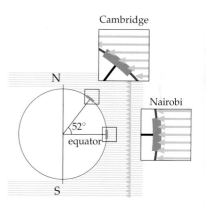

Figure 6.1. Sunlight hitting the earth at midday on a spring or autumn day. The density of sunlight per unit land area in Cambridge (latitude 52°) is about 60% of that at the equator.

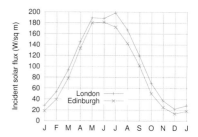

Figure 6.2. Average solar intensity in London and Edinburgh as a function of time of year. The average intensity, per unit land area, is 100 W/m².

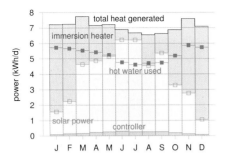

would be about $10\,\text{m}^2$ of panels per person – and let's assume these are 50%-efficient at turning the sunlight's $110\,\text{W/m}^2$ into hot water (figure 6.3). Multiplying

$$50\% \times 10\,\text{m}^2 \times 110\,\text{W/m}^2$$

we find solar heating could deliver

13 kWh per day per person.

I colour this production box white in figure 6.4 to indicate that it describes production of low-grade energy – hot water is not as valuable as the high-grade electrical energy that wind turbines produce. Heat can't be exported to the electricity grid. If you don't need it, then it's wasted. We should bear in mind that much of this captured heat would not be in the right place. In cities, where many people live, residential accommodation has less roof area per person than the national average. Furthermore, this power would be delivered non-uniformly through the year.

Solar photovoltaic

Photovoltaic (PV) panels convert sunlight into electricity. Typical solar panels have an efficiency of about 10%; expensive ones perform at 20%. (Fundamental physical laws limit the efficiency of photovoltaic systems to at best 60% with perfect concentrating mirrors or lenses, and 45% without concentration. A mass-produced device with efficiency greater than 30% would be quite remarkable.) The average power delivered by south-facing 20%-efficient photovoltaic panels in Britain would be

$$20\% \times 110\,\text{W/m}^2 = 22\,\text{W/m}^2.$$

Figure 6.5 shows data to back up this number. Let's give every person $10\,\text{m}^2$ of expensive (20%-efficient) solar panels and cover all south-facing roofs. These will deliver

5 kWh per day per person.

Figure 6.3. Solar power generated by a $3\,\text{m}^2$ hot-water panel (green), and supplementary heat required (blue) to make hot water in the test house of Viridian Solar. (The photograph shows a house with the same model of panel on its roof.) The average solar power from $3\,\text{m}^2$ was 3.8 kWh/d. The experiment simulated the hot-water consumption of an average European household – 100 litres of hot (60 °C) water per day. The 1.5–2 kWh/d gap between the total heat generated (black line, top) and the hot water used (red line) is caused by heat-loss. The magenta line shows the electrical power required to run the solar system. The average power per unit area of these solar panels is 53 W/m^2.

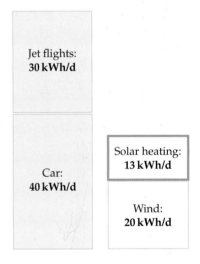

Figure 6.4. Solar thermal: a $10\,\text{m}^2$ array of thermal panels can deliver (on average) about 13 kWh per day of thermal energy.

Since the area of all south-facing roofs is $10\,\text{m}^2$ per person, there certainly isn't space on our roofs for these photovoltaic panels as well as the solar thermal panels of the last section. So we have to choose whether to have the photovoltaic contribution or the solar hot water contribution. But I'll just plop both these on the production stack anyway. Incidentally, the present cost of installing such photovoltaic panels is about four times the cost of installing solar thermal panels, but they deliver only half as much energy, albeit high-grade energy (electricity). So I'd advise a family thinking of going solar to investigate the solar thermal option first. The smartest solution, at least in sunny countries, is to make combined systems that deliver both electricity and hot water from a single installation. This is the approach pioneered by Heliodynamics, who reduce the overall cost of their systems by surrounding small high-grade gallium arsenide photovoltaic units with arrays of slowly-moving flat mirrors; the mirrors focus the sunlight onto the photovoltaic units, which deliver both electricity and hot water; the hot water is generated by pumping water past the back of the photovoltaic units.

The conclusion so far: covering your south-facing roof at home with photovoltaics may provide enough juice to cover quite a big chunk of your personal average electricity consumption; but roofs are not big enough to make a huge dent in our total *energy* consumption. To do more with PV, we need to step down to terra firma. The solar warriors in figure 6.6 show the way.

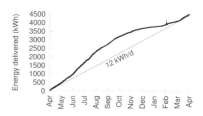

Figure 6.5. Solar photovoltaics: data from a 25-m^2 array in Cambridgeshire in 2006. The peak power delivered by this array is about $4\,\text{kW}$. The average, year-round, is $12\,\text{kWh}$ per day. That's $20\,\text{W}$ per square metre of panel.

Figure 6.6. Two solar warriors enjoying their photovoltaic system, which powers their electric cars and home. The array of 120 panels ($300\,\text{W}$ each, $2.2\,\text{m}^2$ each) has an area of $268\,\text{m}^2$, a peak output (allowing for losses in DC–to–AC conversion) of $30.5\,\text{kW}$, and an average output – in California, near Santa Cruz – of $5\,\text{kW}$ ($19\,\text{W/m}^2$). Photo kindly provided by Kenneth Adelman.
www.solarwarrior.com

Fantasy time: solar farming

If a breakthrough of solar technology occurs and the cost of photovoltaics came down enough that we could deploy panels all over the countryside, what is the maximum conceivable production? Well, if we covered 5% of the UK with 10%-efficient panels, we'd have

$$10\% \times 100\,\text{W/m}^2 \times 200\,\text{m}^2 \text{ per person}$$
$$\simeq 50\,\text{kWh/day/person}.$$

I assumed only 10%-efficient panels, by the way, because I imagine that solar panels would be mass-produced on such a scale only if they were very cheap, and it's the lower-efficiency panels that will get cheap first. The power density (the power per unit area) of such a solar farm would be

$$10\% \times 100\,\text{W/m}^2 = 10\,\text{W/m}^2.$$

This power density is twice that of the Bavaria Solarpark (figure 6.7).

Could this flood of solar panels co-exist with the army of windmills we imagined in Chapter 4? Yes, no problem: windmills cast little shadow, and ground-level solar panels have negligible effect on the wind. How audacious is this plan? The solar power capacity required to deliver this 50 kWh per day per person in the UK is more than 100 times all the photovoltaics in the whole world. So should I include the PV farm in my sustainable production stack? I'm in two minds. At the start of this book I said I wanted to explore what the laws of physics say about the limits of sustainable energy, assuming money is no object. On those grounds, I should certainly go ahead, industrialize the countryside, and push the PV farm onto the stack. At the same time, I want to help people figure out what we should be doing between *now* and 2050. And today, electricity from solar farms would be four times as expensive as the market rate. So I feel a bit irresponsible as I include this estimate in the sustainable production stack in figure 6.9 – paving 5% of the UK with solar panels seems beyond the bounds of plausibility in so many ways. If we seriously contemplated doing such a thing, it would quite probably be better to put the panels in a two-fold sunnier country and send some of the energy home by power lines. We'll return to this idea in Chapter 25.

Mythconceptions

Manufacturing a solar panel consumes more energy than it will ever deliver.

False. The *energy yield ratio* (the ratio of energy delivered by a system over its lifetime, to the energy required to make it) of a roof-mounted, grid-connected solar system in Central Northern Europe is 4, for a system with a lifetime of 20 years (Richards and Watt, 2007); and more than 7 in

Figure 6.7. A solar photovoltaic farm: the 6.3 MW (peak) Solarpark in Mühlhausen, Bavaria. Its average power per unit land area is expected to be about $5\,\text{W/m}^2$. Photo by SunPower.

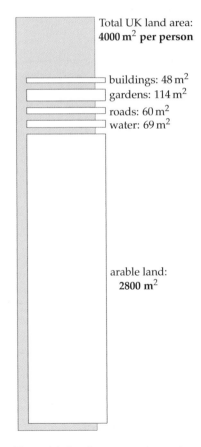

Total UK land area: **4000 m² per person**

buildings: 48 m²
gardens: 114 m²
roads: 60 m²
water: 69 m²

arable land: **2800 m²**

Figure 6.8. Land areas per person in Britain.

a sunnier spot such as Australia. (An energy yield ratio bigger than one means that a system is A Good Thing, energy-wise.) Wind turbines with a lifetime of 20 years have an energy yield ratio of 80.

Aren't photovoltaic panels going to get more and more efficient as technology improves?

I am sure that photovoltaic panels will become ever *cheaper*; I'm also sure that solar panels will become ever less energy-intensive to *manufacture*, so their energy yield ratio will improve. But this chapter's photovoltaic estimates weren't constrained by the economic cost of the panels, nor by the energy cost of their manufacture. This chapter was concerned with the maximum conceivable power delivered. Photovoltaic panels with 20% efficiency are already close to the theoretical limit (see this chapter's endnotes). I'll be surprised if this chapter's estimate for roof-based photovoltaics ever needs a significant upward revision.

Solar biomass

> *All of a sudden, you know, we may be in the energy business by being able to grow grass on the ranch! And have it harvested and converted into energy. That's what's close to happening.*
>
> George W. Bush, February 2006

All available bioenergy solutions involve first growing green stuff, and then doing something with the green stuff. How big could the energy collected by the green stuff possibly be? There are four main routes to get energy from solar-powered biological systems:

1. We can grow specially-chosen plants and burn them in a power station that produces electricity or heat or both. We'll call this "coal substitution."

2. We can grow specially-chosen plants (oil-seed rape, sugar cane, or corn, say), turn them into ethanol or biodiesel, and shove that into cars, trains, planes or other places where such chemicals are useful. Or we might cultivate genetically-engineered bacteria, cyanobacteria, or algae that directly produce hydrogen, ethanol, or butanol, or even electricity. We'll call all such approaches "petroleum substitution."

3. We can take by-products from other agricultural activities and burn them in a power station. The by-products might range from straw (a by-product of Weetabix) to chicken poo (a by-product of McNuggets). Burning by-products is coal substitution again, but using ordinary plants, not the best high-energy plants. A power station that burns agricultural by-products won't deliver as much power per unit area of farmland as an optimized biomass-growing facility, but it has the

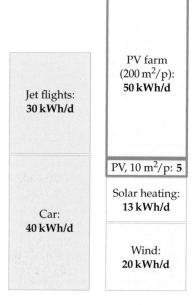

Figure 6.9. Solar photovoltaics: a $10 \, m^2$ array of building-mounted south-facing panels with 20% efficiency can deliver about 5 kWh per day of electrical energy. If 5% of the country were coated with 10%-efficient solar panels (200 m^2 of panels per person) they would deliver 50 kWh/day/person.

advantage that it doesn't monopolize the land. Burning methane gas from landfill sites is a similar way of getting energy, but it's sustainable only as long as we have a sustainable source of junk to keep putting into the landfill sites. (Most of the landfill methane comes from wasted food; people in Britain throw away about 300 g of food per day per person.) Incinerating household waste is another slightly less roundabout way of getting power from solar biomass.

4. We can grow plants and feed them directly to energy-requiring humans or other animals.

For all of these processes, the first staging post for the energy is in a chemical molecule such as a carbohydrate in a green plant. We can therefore estimate the power obtainable from any and all of these processes by estimating how much power could pass through that first staging post. All the subsequent steps involving tractors, animals, chemical facilities, landfill sites, or power stations can only lose energy. So the power at the first staging post is an upper bound on the power available from all plant-based power solutions.

So, let's simply estimate the power at the first staging post. (In Chapter D, we'll go into more detail, estimating the maximum contribution of each process.) The average harvestable power of sunlight in Britain is 100 W/m^2. The most efficient plants in Europe are about 2%-efficient at turning solar energy into carbohydrates, which would suggest that plants might deliver 2 W/m^2; however, their efficiency drops at higher light levels, and the best performance of any energy crops in Europe is closer to 0.5 W/m^2. Let's cover 75% of the country with quality green stuff. That's 3000 m^2 per person devoted to bio-energy. This is the same as the British land area

Figure 6.10. Some *Miscanthus* grass enjoying the company of Dr Emily Heaton, who is 5′4″ (163 cm) tall. In Britain, *Miscanthus* achieves a power per unit area of 0.75 W/m^2. Photo provided by the University of Illinois.

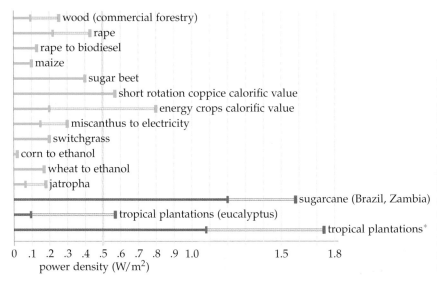

Figure 6.11. Power production, per unit area, achieved by various plants. For sources, see the end-notes. These power densities vary depending on irrigation and fertilization; ranges are indicated for some crops, for example wood has a range from 0.095–0.254 W/m^2. The bottom three power densities are for crops grown in tropical locations. The last power density (tropical plantations*) assumes genetic modification, fertilizer application, and irrigation. In the text, I use 0.5 W/m^2 as a summary figure for the best energy crops in NW Europe.

currently devoted to agriculture. So the maximum power available, ignoring all the additional costs of growing, harvesting, and processing the greenery, is

$$0.5 \, \text{W/m}^2 \times 3000 \, \text{m}^2 \text{ per person} = 36 \, \text{kWh/d per person.}$$

Wow. That's not very much, considering the outrageously generous assumptions we just made, to try to get a big number. If you wanted to get biofuels for cars or planes from the greenery, all the other steps in the chain from farm to spark plug would inevitably be inefficient. I think it'd be optimistic to hope that the overall losses along the processing chain would be as small as 33%. Even burning dried wood in a good wood boiler loses 20% of the heat up the chimney. So surely the true potential power from biomass and biofuels cannot be any bigger than 24 kWh/d per person. And don't forget, we want to use some of the greenery to make food for us and for our animal companions.

Could genetic engineering produce plants that convert solar energy to chemicals more efficiently? It's conceivable; but I haven't found any scientific publication predicting that plants in Europe could achieve net power production beyond 1 W/m².

I'll pop 24 kWh/d per person onto the green stack, emphasizing that I think this number is an over-estimate – I think the true maximum power that we could get from biomass will be smaller because of the losses in farming and processing.

I think one conclusion is clear: *biofuels can't add up* – at least, not in countries like Britain, and not as a replacement for all transport fuels. Even leaving aside biofuels' main defects – that their production competes with food, and that the additional inputs required for farming and processing often cancel out most of the delivered energy (figure 6.14) – biofuels made from plants, in a European country like Britain, can deliver so little power, I think they are scarcely worth talking about.

Notes and further reading

page no.

38   *…compensate for the tilt between the sun and the land.* The latitude of Cambridge is $\theta = 52°$; the intensity of midday sunlight is multiplied by $\cos\theta \simeq 0.6$. The precise factor depends on the time of year, and varies between $\cos(\theta + 23°) = 0.26$ and $\cos(\theta - 23°) = 0.87$.

–   *In a typical UK location the sun shines during one third of daylight hours.* The Highlands get 1100 h sunshine per year – a sunniness of 25%. The best spots in Scotland get 1400 h per year – 32%. Cambridge: 1500 ± 130 h per year – 34%. South coast of England (the sunniest part of the UK): 1700 h per year – 39%. [2rqloc] Cambridge data from [2szckw]. See also figure 6.16.

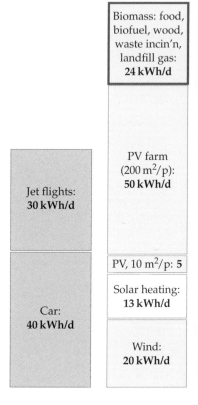

Figure 6.12. Solar biomass, including all forms of biofuel, waste incineration, and food: 24 kWh/d per person.

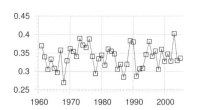

Figure 6.13. Sunniness of Cambridge: the number of hours of sunshine per year, expressed as a fraction of the total number of daylight hours.

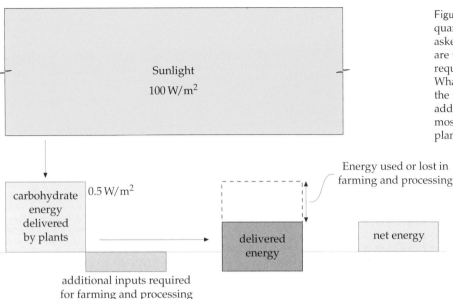

Figure 6.14. This figure illustrates the quantitative questions that must be asked of any proposed biofuel. What are the additional energy inputs required for farming and processing? What is the delivered energy? What is the *net* energy output? Often the additional inputs and losses wipe out most of the energy delivered by the plants.

38 *The average raw power of sunshine per square metre of south-facing roof in Britain is roughly 110 W/m², and of flat ground, roughly 100 W/m².* Source: NASA "Surface meteorology and Solar Energy" [5hrxls]. Surprised that there's so little difference between a tilted roof facing south and a horizontal roof? I was. The difference really is just 10% [6z9epq].

39 *…that would be about 10 m² of panels per person.* I estimated the area of south-facing roof per person by taking the area of land covered by buildings per person (48 m² in England – table I.6), multiplying by ¹/₄ to get the south-facing fraction, and bumping the area up by 40% to allow for roof tilt. This gives 16 m² per person. Panels usually come in inconvenient rectangles so some fraction of roof will be left showing; hence 10 m² of panels.

–   *The average power delivered by photovoltaic panels…*
    There's a myth going around that states that solar panels produce almost as much power in cloudy conditions as in sunshine. This is simply not true. On a bright but cloudy day, solar photovoltaic panels and plants do continue to convert some energy, but much less: photovoltaic production falls roughly ten-fold when the sun goes behind clouds (because the intensity of the incoming sunlight falls ten-fold). As figure 6.15 shows, the power delivered by photovoltaic panels is almost exactly proportional to the intensity of the sunlight – at least, if the panels are at 25 °C. To complicate things, the power delivered depends on temperature too – hotter panels have reduced power (typically 0.38% loss in power per °C) – but if you check data from real panels, e.g. at www.solarwarrior.com, you can confirm the main point: output on a cloudy day is *far less* than on a sunny day. This issue is obfuscated by some solar-panel promoters who discuss how the "efficiency" varies with sunlight. "The panels are more efficient in cloudy conditions," they say; this

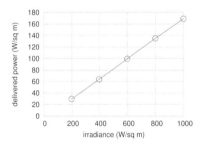

Figure 6.15. Power produced by the Sanyo HIP-210NKHE1 module as a function of light intensity (at 25 °C, assuming an output voltage of 40 V). Source: datasheet, www.sanyo-solar.eu.

average sunshine (W/m^2)

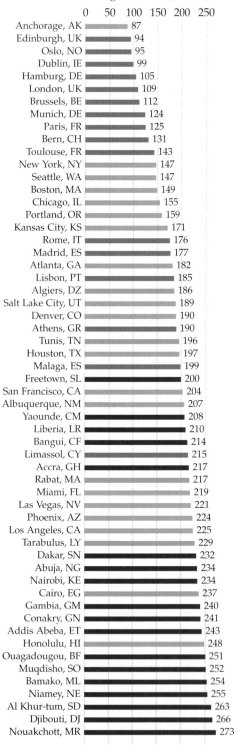

Anchorage, AK	87
Edinburgh, UK	94
Oslo, NO	95
Dublin, IE	99
Hamburg, DE	105
London, UK	109
Brussels, BE	112
Munich, DE	124
Paris, FR	125
Bern, CH	131
Toulouse, FR	143
New York, NY	147
Seattle, WA	147
Boston, MA	149
Chicago, IL	155
Portland, OR	159
Kansas City, KS	171
Rome, IT	176
Madrid, ES	177
Atlanta, GA	182
Lisbon, PT	185
Algiers, DZ	186
Salt Lake City, UT	189
Denver, CO	190
Athens, GR	190
Tunis, TN	196
Houston, TX	197
Malaga, ES	199
Freetown, SL	200
San Francisco, CA	204
Albuquerque, NM	207
Yaounde, CM	208
Liberia, LR	210
Bangui, CF	214
Limassol, CY	215
Accra, GH	217
Rabat, MA	217
Miami, FL	219
Las Vegas, NV	221
Phoenix, AZ	224
Los Angeles, CA	225
Tarabulus, LY	229
Dakar, SN	232
Abuja, NG	234
Nairobi, KE	234
Cairo, EG	237
Gambia, GM	240
Conakry, GN	241
Addis Abeba, ET	243
Honolulu, HI	248
Ouagadougou, BF	251
Muqdisho, SO	252
Bamako, ML	254
Niamey, NE	255
Al Khur-tum, SD	263
Djibouti, DJ	266
Nouakchott, MR	273

Figure 6.16. Average power of sunshine falling on a horizontal surface in selected locations in Europe, North America, and Africa.

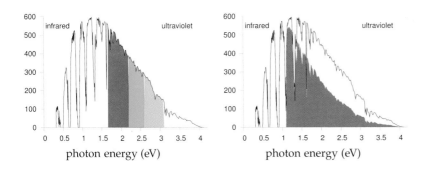

may be true, but efficiency should not be confused with delivered power.

39   *Typical solar panels have an efficiency of about 10%; expensive ones perform at 20%.* See figure 6.18. Sources: Turkenburg (2000), Sunpower www.sunpowercorp.com, Sanyo www.sanyo-solar.eu, Suntech.

–   *A device with efficiency greater than 30% would be quite remarkable.* This is a quote from Hopfield and Gollub (1978), who were writing about panels without concentrating mirrors or lenses. The theoretical limit for a standard "single-junction" solar panel without concentrators, the Shockley–Queisser limit, says that at most 31% of the energy in sunlight can be converted to electricity (Shockley and Queisser, 1961). (The main reason for this limit is that a standard solar material has a property called its band-gap, which defines a particular energy of photon that that material converts most efficiently. Sunlight contains photons with many energies; photons with energy *below* the band-gap are not used at all; photons with energy *greater* than the band-gap may be captured, but all their energy in excess of the band-gap is lost. Concentrators (lenses or mirrors) can both reduce the cost (per watt) of photovoltaic systems, and increase their efficiency. The Shockley–Queisser limit for solar panels with concentrators is 41% efficiency. The only way to beat the Shockley–Queisser limit is to make fancy photovoltaic devices that split the light into different wavelengths, processing each wavelength-range with its own personalized band-gap. These are called multiple-junction photovoltaics. Recently multiple-junction photovoltaics with optical concentrators have been reported to be about 40% efficient. [2tl7t6], www.spectrolab.com. In July 2007, the University of Delaware reported 42.8% efficiency with 20-times concentration [6hobq2], [21sx6t]. In August 2008, NREL reported 40.8% efficiency with 326-times concentration [62ccou]. Strangely, both these results were called world efficiency records. What multiple-junction devices are available on the market? Uni-solar sell a thin-film triple-junction 58 W(peak) panel with an area of 1 m^2. That implies an efficiency, in full sunlight, of only 5.8%.

40   *Figure 6.5: Solar PV data.* Data and photograph kindly provided by Jonathan Kimmitt.

–   *Heliodynamics* – www.hdsolar.com. See figure 6.19.
    A similar system is made by Arontis www.arontis.se.

Figure 6.17. Part of Shockley and Queisser's explanation for the 31% limit of the efficiency of simple photovoltaics.
Left: the spectrum of midday sunlight. The vertical axis shows the power density in W/m^2 per eV of spectral interval. The visible part of the spectrum is indicated by the coloured section.
Right: the energy captured by a photovoltaic device with a single band-gap at 1.1 eV is shown by the tomato-shaded area. Photons with energy less than the band-gap are lost. Some of the energy of photons above the band-gap is lost; for example half of the energy of every 2.2 eV photon is lost.
Further losses are incurred because of inevitable radiation from recombining charges in the photovoltaic material.

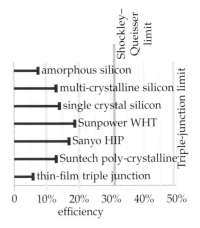

Figure 6.18. Efficiencies of solar photovoltaic modules available for sale today. In the text I assume that roof-top photovoltaics are 20% efficient, and that country-covering photovoltaics would be 10% efficient. In a location where the average power density of incoming sunlight is 100 W/m^2, 20%-efficient panels deliver 20 W/m^2.

41   *The Solarpark in Muhlhausen, Bavaria.* On average this 25-hectare farm is
     expected to deliver 0.7 MW (17 000 kWh per day).
     New York's Stillwell Avenue subway station has integrated amorphous sili-
     con thin-film photovoltaics in its roof canopy, delivering 4 W/m^2 (Fies et al.,
     2007).
     The Nellis solar power plant in Nevada was completed in December, 2007,
     on 140 acres, and is expected to generate 30 GWh per year. That's 6 W/m^2
     [5hzs5y].
     Serpa Solar Power Plant, Portugal (PV), "the world's most powerful so-
     lar power plant," [39z5m5] [2uk8q8] has sun-tracking panels occupying 60
     hectares, i.e., 600 000 m^2 or 0.6 km^2, expected to generate 20 GWh per year,
     i.e., 2.3 MW on average. That's a power per unit area of 3.8 W/m^2.

41   *The solar power capacity required to deliver 50 kWh/d per person in the UK
     is more than 100 times all the photovoltaics in the whole world.* To deliver
     50 kWh/d per person in the UK would require 125 GW average power, which
     requires 1250 GW of capacity. At the end of 2007, world installed photo-
     voltaics amounted to 10 GW peak; the build rate is roughly 2 GW per year.

–   *...paving 5% of this country with solar panels seems beyond the bounds of
     plausibility.* My main reason for feeling such a panelling of the country
     would be implausible is that Brits like using their countryside for farming
     and recreation rather than solar-panel husbandry. Another concern might be
     price. This isn't a book about economics, but here are a few figures. Going
     by the price-tag of the Bavarian solar farm, to deliver 50 kWh/d per person
     would cost €91 000 per person; if that power station lasted 20 years without
     further expenditure, the wholesale cost of the electricity would be €0.25 per
     kWh. Further reading: David Carlson, BP solar [2ahecp].

43   *People in Britain throw away about 300 g of food per day.* Source: Ventour
     (2008).

–   *Figure 6.10.* In the USA, *Miscanthus* grown without nitrogen fertilizer yields
     about 24 t/ha/y of dry matter. In Britain, yields of 12–16 t/ha/y are re-
     ported. Dry *Miscanthus* has a net calorific value of 17 MJ/kg, so the British
     yield corresponds to a power density of 0.75 W/m^2. Sources: Heaton et al.
     (2004) and [6kqq77]. The estimated yield is obtained only after three years
     of undisturbed growing.

–   *The most efficient plants are about 2% efficient; but the delivered power per
     unit area is about 0.5 W/m^2.* At low light intensities, the best British plants are
     2.4% efficient in well-fertilized fields (Monteith, 1977) but at higher light in-
     tensities, their conversion efficiency drops. According to Turkenburg (2000)
     and Schiermeier et al. (2008), the conversion efficiency of solar to biomass
     energy is less than 1%.
     Here are a few sources to back up my estimate of *0.5 W/m^2* for vegetable
     power in the UK. The Royal Commission on Environmental Pollution's esti-
     mate of the potential delivered power density from energy crops in Britain is
     0.2 W/m^2 (Royal Commission on Environmental Pollution, 2004). On page
     43 of the Royal Society's biofuels document (Royal Society working group
     on biofuels, 2008), *Miscanthus* tops the list, delivering about 0.8 W/m^2 of
     chemical power.

Figure 6.19. A
combined-heat-and-power
photovoltaic unit from
Heliodynamics. A reflector area of
32 m^2 (a bit larger than the side of a
double-decker bus) delivers up to
10 kW of heat and 1.5 kW of electrical
power. In a sun-belt country, one of
these one-ton devices could deliver
about 60 kWh/d of heat and 9 kWh/d
of electricity. These powers
correspond to average fluxes of
80 W/m^2 of heat and 12 W/m^2 of
electricity (that's per square metre of
device surface); these fluxes are
similar to the fluxes delivered by
standard solar heating panels and
solar photovoltaic panels, but
Heliodynamics's concentrating design
delivers power at a lower cost,
because most of the material is simple
flat glass. For comparison, the total
power consumption of the average
European person is 125 kWh/d.

In the World Energy Assessment published by the UNDP, Rogner (2000) writes: "Assuming a 45% conversion efficiency to electricity and yields of 15 oven dry tons per hectare per year, $2 \, km^2$ of plantation would be needed per megawatt of electricity of installed capacity running 4,000 hours a year." That is a power per unit area of $0.23 \, W(e)/m^2$. (1 W(e) means 1 watt of electrical power.)

Energy for Sustainable Development Ltd (2003) estimates that short-rotation coppices can deliver over 10 tons of dry wood per hectare per year, which corresponds to a power density of $0.57 \, W/m^2$. (Dry wood has a calorific value of 5 kWh per kg.)

According to Archer and Barber (2004), the instantaneous efficiency of a healthy leaf in optimal conditions can approach 5%, but the long-term energy-storage efficiency of modern crops is 0.5–1%. Archer and Barber suggest that by genetic modification, it might be possible to improve the storage efficiency of plants, especially *C4 plants*, which have already naturally evolved a more efficient photosynthetic pathway. C4 plants are mainly found in the tropics and thrive in high temperatures; they don't grow at temperatures below $10 \, °C$. Some examples of C4 plants are sugarcane, maize, sorghum, finger millet, and switchgrass. Zhu et al. (2008) calculate that the theoretical limit for the conversion efficiency of solar energy to biomass is 4.6% for C3 photosynthesis at $30 \, °C$ and today's 380 ppm atmospheric CO_2 concentration, and 6% for C4 photosynthesis. They say that the highest solar energy conversion efficiencies reported for C3 and C4 crops are 2.4% and 3.7% respectively; and, citing Boyer (1982), that the average conversion efficiencies of major crops in the US are 3 or 4 times lower than those record efficiencies (that is, about 1% efficient). One reason that plants don't achieve the theoretical limit is that they have insufficient capacity to use all the incoming radiation of bright sunlight. Both these papers (Zhu et al., 2008; Boyer, 1982) discuss prospects for genetic engineering of more-efficient plants.

43   *Figure 6.11.* The numbers in this figure are drawn from Rogner (2000) (net energy yields of wood, rape, sugarcane, and tropical plantations); Bayer Crop Science (2003) (rape to biodiesel); Francis et al. (2005) and Asselbergs et al. (2006) (jatropha); Mabee et al. (2006) (sugarcane, Brazil); Schmer et al. (2008) (switchgrass, marginal cropland in USA); Shapouri et al. (1995) (corn to ethanol); Royal Commission on Environmental Pollution (2004); Royal Society working group on biofuels (2008); Energy for Sustainable Development Ltd (2003); Archer and Barber (2004); Boyer (1982); Monteith (1977).

44   *Even just setting fire to dried wood in a good wood boiler loses 20% of the heat up the chimney.* Sources: Royal Society working group on biofuels (2008); Royal Commission on Environmental Pollution (2004).

7 Heating and cooling

This chapter explores how much power we spend controlling the temperature of our surroundings – at home and at work – and on warming or cooling our food, drink, laundry, and dirty dishes.

Figure 7.1. A flock of new houses.

Domestic water heating

The biggest use of hot water in a house might be baths, showers, dishwashing, or clothes-washing – it depends on your lifestyle. Let's estimate first the energy used by taking a hot bath.

The volume of bath-water is 50 cm × 15 cm × 150 cm ≃ 110 litre. Say the temperature of the bath is 50 °C (120 F) and the water coming into the house is at 10 °C. The heat capacity of water, which measures how much energy is required to heat it up, is 4200 J per litre per °C. So the energy required to heat up the water by 40 °C is

$$4200\,\text{J/litre/}^\circ\text{C} \times 110\,\text{litre} \times 40\,^\circ\text{C} \simeq 18\,\text{MJ} \simeq 5\,\text{kWh}.$$

So taking a bath uses about 5 kWh. For comparison, taking a shower (30 litres) uses about 1.4 kWh.

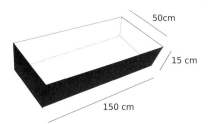

Figure 7.2. The water in a bath.

Kettles and cookers

Britain, being a civilized country, has a 230 volt domestic electricity supply. With this supply, we can use an electric kettle to boil several litres of water in a couple of minutes. Such kettles have a power of 3 kW. Why 3 kW? Because this is the biggest power that a 230 volt outlet can deliver without the current exceeding the maximum permitted, 13 amps. In countries where the voltage is 110 volts, it takes twice as long to make a pot of tea.

If a household has the kettle on for 20 minutes per day, that's an average power consumption of 1 kWh per day. (I'll work out the next few items "per household," with 2 people per household.)

One small ring on an electric cooker has the same power as a toaster: 1 kW. The higher-power hot plates deliver 2.3 kW. If you use two rings of the cooker on full power for half an hour per day, that corresponds to 1.6 kWh per day.

A microwave oven usually has its cooking power marked on the front: mine says 900 W, but it actually *consumes* about 1.4 kW. If you use the microwave for 20 minutes per day, that's 0.5 kWh per day.

A regular oven guzzles more: about 3 kW when on full. If you use the oven for one hour per day, and the oven's on full power for half of that time, that's 1.5 kWh per day.

230 V × 13 A = 3000 W

Microwave:
1400 W peak

Fridge-freezer:
100 W peak,
18 W average

Figure 7.3. Power consumption by a heating and a cooling device.

Device	power	time per day	energy per day
Cooking			
– kettle	3 kW	$^1/_3$ h	1 kWh/d
– microwave	1.4 kW	$^1/_3$ h	0.5 kWh/d
– electric cooker (rings)	3.3 kW	$^1/_2$ h	1.6 kWh/d
– electric oven	3 kW	$^1/_2$ h	1.5 kWh/d
Cleaning			
– washing machine	2.5 kW		1 kWh/d
– tumble dryer	2.5 kW	0.8 h	2 kWh/d
– airing-cupboard drying			0.5 kWh/d
– washing-line drying			0 kWh/d
– dishwasher	2.5 kW		1.5 kWh/d
Cooling			
– refrigerator	0.02 kW	24 h	0.5 kWh/d
– freezer	0.09 kW	24 h	2.3 kWh/d
– air-conditioning	0.6 kW	1 h	0.6 kWh/d

Table 7.4. Energy consumption figures for heating and cooling devices, per household.

Hot clothes and hot dishes

A clothes washer, dishwasher, and tumble dryer all use a power of about 2.5 kW when running.

A clothes washer uses about 80 litres of water per load, with an energy cost of about 1 kWh if the temperature is set to 40 °C. If we use an indoor airing-cupboard instead of a tumble dryer to dry clothes, heat is still required to evaporate the water – roughly 1.5 kWh to dry one load of clothes, instead of 3 kWh.

Totting up the estimates relating to hot water, I think it's easy to use about 12 kWh per day per person.

Hot air – at home and at work

Now, does more power go into making hot water and hot food, or into making hot air via our buildings' radiators?

One way to estimate the energy used per day for hot air is to imagine a building heated instead by electric fires, whose powers are more familiar to us. The power of a small electric bar fire or electric fan heater is 1 kW (24 kWh per day). In winter, you might need one of these per person to keep toasty. In summer, none. So we estimate that on average one modern person *needs* to use 12 kWh per day on hot air. But most people use more than they need, keeping several rooms warm simultaneously (kitchen, living room, corridor, and bathroom, say). So a plausible consumption figure for hot air is about double that: 24 kWh per day per person.

This chapter's companion Chapter E contains a more detailed account of where the heat is going in a building; this model makes it possible to

> Hot water:
> **12 kWh/d**

Figure 7.5. The hot water total at both home and work – including bathing, showering, clothes washing, cookers, kettles, microwave oven, and dishwashing – is about 12 kWh per day per person. I've given this box a light colour to indicate that this power could be delivered by low-grade thermal energy.

Figure 7.6. A big electric heater: 2 kW.

predict the heat savings from turning the thermostat down, double-glazing the windows, and so forth.

Warming the outdoors, and other luxuries

There's a growing trend of warming the outdoors with patio heaters. Typical patio heaters have a power of 15 kW. So if you use one of these for a couple of hours every evening, you are using an extra 30 kWh per day.

A more modest luxury is an electric blanket. An electric blanket for a double bed uses 140 W; switching it on for one hour uses 0.14 kWh.

Cooling

Fridge and freezer

We control the temperatures not only of the hot water and hot air with which we surround ourselves, but also of the cold cupboards we squeeze into our hothouses. My fridge-freezer, pictured in figure 7.3, consumes 18 W on average – that's roughly 0.5 kWh/d.

Air-conditioning

In countries where the temperature gets above 30 °C, air-conditioning is viewed as a necessity, and the energy cost of delivering that temperature control can be large. However, this part of the book is about British energy consumption, and Britain's temperatures provide little need for air-conditioning (figure 7.8).

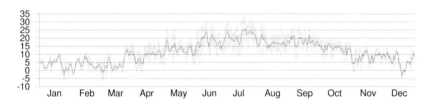

An economical way to get air-conditioning is an air-source heat pump. A window-mounted electric air-conditioning unit for a single room uses 0.6 kW of electricity and (by heat-exchanger) delivers 2.6 kW of cooling. To estimate how much energy someone might use in the UK, I assumed they might switch such an air-conditioning unit on for about 12 hours per day on 30 days of the year. On the days when it's on, the air-conditioner uses 7.2 kWh. The average consumption over the whole year is 0.6 kWh/d.

This chapter's estimate of the energy cost of cooling – 1 kWh/d per person – includes this air-conditioning and a domestic refrigerator. Society

---

> **Hot air:**
> **24 kWh/d**

Figure 7.7. Hot air total – including domestic and workplace heating – about 24 kWh per day per person.

Figure 7.8. Cambridge temperature in degrees Celsius, daily (red line), and half-hourly (blue line) during 2006.

Cooling: **1 kWh/d**

Figure 7.9. Cooling total – including a refrigerator (fridge/freezer) and a little summer air-conditioning – 1 kWh/d.

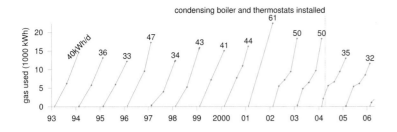

Figure 7.10. My domestic cumulative gas consumption, in kWh, each year from 1993 to 2005. The number at the top of each year's line is the average rate of energy consumption, in kWh per day. To find out what happened in 2007, keep reading!

also refrigerates food on its way from field to shopping basket. I'll estimate the power cost of the food-chain later, in Chapter 15.

Total heating and cooling

Our rough estimate of the total energy that one person might spend on heating and cooling, including home, workplace, and cooking, is 37 kWh/d per person (12 for hot water, 24 for hot air, and 1 for cooling).

Evidence that this estimate is in the right ballpark, or perhaps a little on the low side, comes from my own domestic gas consumption, which for 12 years averaged 40 kWh per day (figure 7.10). At the time I thought I was a fairly frugal user of heating, but I wasn't being attentive to my actual power consumption. Chapter 21 will reveal how much power I saved once I started paying attention.

Since heating is a big item in our consumption stack, let's check my estimates against some national statistics. Nationally, the average *domestic* consumption for space heating, water, and cooking in the year 2000 was 21 kWh per day per person, and consumption in the *service sector* for heating, cooling, catering, and hot water was 8.5 kWh/d/p. For an estimate of workplace heating, let's take the gas consumption of the University of Cambridge in 2006–7: 16 kWh/d per employee.

Totting up these three numbers, a second guess for the national spend on heating is $21 + 8.5 + 16 \simeq 45$ kWh/d per person, if Cambridge University is a normal workplace. Good, that's reassuringly close to our first guess of 37 kWh/d.

Notes and further reading

page no.

50 *An oven uses 3 kW*. Obviously there's a range of powers. Many ovens have a maximum power of 1.8 kW or 2.2 kW. Top-of-the-line ovens use as much as 6 kW. For example, the Whirlpool AGB 487/WP 4 Hotplate Electric Oven Range has a 5.9 kW oven, and four 2.3 kW hotplates.
www.kcmltd.com/electric_oven_ranges.shtml
www.1stforkitchens.co.uk/kitchenovens.html

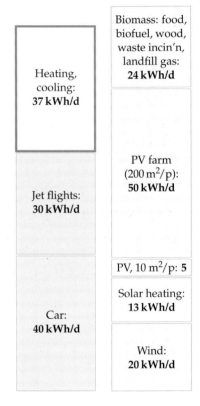

Figure 7.11. Heating and cooling – about 37 units per day per person. I've removed the shading from this box to indicate that it represents power that could be delivered by low-grade thermal energy.

51 *An airing cupboard requires roughly 1.5 kWh to dry one load of clothes.* I
    worked this out by weighing my laundry: a load of clothes, 4 kg when dry,
    emerged from my Bosch washing machine weighing 2.2 kg more (even after
    a good German spinning). The latent heat of vaporization of water at 15 °C is
    roughly 2500 kJ/kg. To obtain the daily figure in table 7.4 I assumed that one
    person has a load of laundry every three days, and that this sucks valuable
    heat from the house during the cold half of the year. (In summer, using the
    airing cupboard delivers a little bit of air-conditioning, since the evaporating
    water cools the air in the house.)

53 *Nationally, the average domestic consumption was 21 kWh/d/p; consump-*
    tion in the service sector was 8.5 kWh/d/p.   Source: Dept. of Trade and
    Industry (2002a).

–   *In 2006–7, Cambridge University's gas consumption was 16 kWh/d per em-*
    ployee. The gas and oil consumption of the University of Cambridge (not
    including the Colleges) was 76 GWh in 2006–7. I declared the University to
    be the place of work of 13 300 people (8602 staff and 4667 postgraduate re-
    searchers). Its electricity consumption, incidentally, was 99.5 GWh. Source:
    University utilities report.

8 Hydroelectricity

To make hydroelectric power, you need altitude, and you need rainfall. Let's estimate the total energy of all the rain as it runs down to sea-level.

For this hydroelectric forecast, I'll divide Britain into two: the lower, dryer bits, which I'll call "the lowlands;" and the higher, wetter bits, which I'll call "the highlands." I'll choose Bedford and Kinlochewe as my representatives of these two regions.

Let's do the lowlands first. To estimate the gravitational power of lowland rain, we multiply the rainfall in Bedford (584 mm per year) by the density of water (1000 kg/m^3), the strength of gravity (10 m/s^2) and the typical lowland altitude above the sea (say 100 m). The power per unit area works out to 0.02 W/m^2. That's the power per unit area of land on which rain falls.

When we multiply this by the area per person (2700 m^2, if the lowlands are equally shared between all 60 million Brits), we find an average raw power of about 1 kWh per day per person. This is the absolute upper limit for lowland hydroelectric power, if every river were dammed and every drop perfectly exploited. Realistically, we will only ever dam rivers with substantial height drops, with catchment areas much smaller than the whole country. Much of the water evaporates before it gets anywhere near a turbine, and no hydroelectric system exploits the full potential energy of the water. We thus arrive at a firm conclusion about lowland water power. People may enjoy making "run-of-the-river" hydro and other small-scale hydroelectric schemes, but such lowland facilities can never deliver more than 1 kWh per day per person.

Figure 8.1. Nant-y-Moch dam, part of a 55 MW hydroelectric scheme in Wales. Photo by Dave Newbould, www.origins-photography.co.uk.

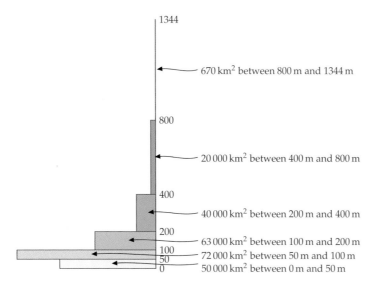

1344

670 km^2 between 800 m and 1344 m

800

20 000 km^2 between 400 m and 800 m

400

40 000 km^2 between 200 m and 400 m

200

63 000 km^2 between 100 m and 200 m

100
50

72 000 km^2 between 50 m and 100 m

0

50 000 km^2 between 0 m and 50 m

Figure 8.2. Altitudes of land in Britain. The rectangles show how much land area there is at each height.

Let's turn to the highlands. Kinlochewe is a rainier spot: it gets 2278 mm per year, four times more than Bedford. The height drops there are bigger too – large areas of land are above 300 m. So overall a twelve-fold increase in power per square metre is plausible for mountainous regions. The raw power per unit area is roughly 0.24 W/m^2. If the highlands generously share their hydro-power with the rest of the UK (at 1300 m^2 area per person), we find an upper limit of about 7 kWh per day per person. As in the lowlands, this is the upper limit on raw power if evaporation were outlawed and every drop were perfectly exploited.

What should we estimate is the plausible practical limit? Let's guess 20% of this – 1.4 kWh per day, and round it up a little to allow for production in the lowlands: 1.5 kWh per day.

The actual power from hydroelectricity in the UK today is 0.2 kWh/d per person, so this 1.5 kWh/d per person would require a seven-fold increase in hydroelectric power.

Notes and further reading

55   *Rainfall* statistics are from the BBC weather centre.

56   *The raw power per unit area [of Highland rain] is roughly 0.24 W/m^2.* We can check this estimate against the actual power density of the Loch Sloy hydro-electric scheme, completed in 1950 (Ross, 2008). The catchment area of Loch Sloy is about 83 km^2; the rainfall there is about 2900 mm per year (a bit higher than the 2278 mm/y of Kinlochewe); and the electricity output in 2006 was 142 GWh per year, which corresponds to a power density of 0.2 W per m^2 of catchment area. Loch Sloy's surface area is about 1.5 km^2, so the hydroelectric facility itself has a per unit lake area of 11 W/m^2. So the hillsides, aqueducts, and tunnels bringing water to Loch Sloy act like a 55-fold power concentrator.

–   *The actual power from hydroelectricity in the UK today is 0.2 kWh per day per person.* Source: MacLeay et al. (2007). In 2006, large-scale hydro produced 3515 GWh (from plant with a capacity of 1.37 GW); small-scale hydro, 212 GWh (0.01 kWh/d/p) (from a capacity of 153 MW).

In 1943, when the growth of hydroelectricity was in full swing, the North of Scotland Hydroelectricity Board's engineers estimated that the Highlands of Scotland could produce 6.3 TWh per year in 102 facilities – that would correspond to 0.3 kWh/d per person in the UK (Ross, 2008).

Glendoe, the first new large-scale hydroelectric project in the UK since 1957, will add capacity of 100 MW and is expected to deliver 180 GWh per year. Glendoe's catchment area is 75 km^2, so its power density works out to 0.27 W per m^2 of catchment area. Glendoe has been billed as "big enough to power Glasgow." But if we share its 180 GWh per year across the population of Glasgow (616 000 people), we get only 0.8 kWh/d per person. That is just 5% of the average electricity consumption of 17 kWh/d per person. The 20-fold exaggeration is achieved by focusing on Glendoe's *peak* output rather than its *average*, which is 5 times smaller; and by discussing "homes" rather than the total electrical power of Glasgow (see p329).

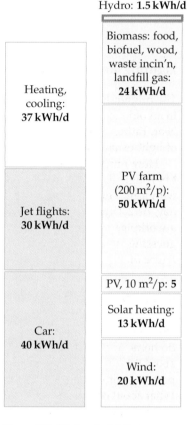

Figure 8.3. Hydroelectricity.

Figure 8.4. A 60 kW waterwheel.

9 Light

Lighting home and work

The brightest domestic lightbulbs use 250 W, and bedside lamps use 40 W. In an old-fashioned incandescent bulb, most of this power gets turned into heat, rather than light. A fluorescent tube can produce an equal amount of light using one quarter of the power of an incandescent bulb.

How much power does a moderately affluent person use for lighting? My rough estimate, based on table 9.2, is that a typical two-person home with a mix of low-energy and high-energy bulbs uses about 5.5 kWh per day, or 2.7 kWh per day per person. I assume that each person also has a workplace where they share similar illumination with their colleagues; guessing that the workplace uses 1.3 kWh/d per person, we get a round figure of 4 kWh/d per person.

Street-lights and traffic lights

Do we need to include public lighting too, to get an accurate estimate, or do home and work dominate the lighting budget? Street-lights in fact use about 0.1 kWh per day per person, and traffic lights only 0.005 kWh/d per person – both negligible, compared with our home and workplace lighting. What about other forms of public lighting – illuminated signs and bollards, for example? There are fewer of them than street-lights; and street-lights already came in well under our radar, so we don't need to modify our overall estimate of 4 kWh/d per person.

Lights on the traffic

In some countries, drivers must switch their lights on whenever their car is moving. How does the extra power required by that policy compare with the power already being used to trundle the car around? Let's say the car has four incandescent lights totalling 100 W. The electricity for those bulbs is supplied by a 25%-efficient engine powering a 55%-efficient generator, so the power required is 730 W. For comparison, a typical car going at an average speed of 50 km/h and consuming one litre per 12 km

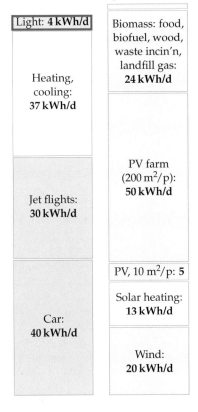

Figure 9.1. Lighting – 4 kWh per day per person.

Device	Power	Time per day	Energy per day per home
10 incandescent lights	1 kW	5 h	5 kWh
10 low-energy lights	0.1 kW	5 h	0.5 kWh

Table 9.2. Electric consumption for domestic lighting. A plausible total is 5.5 kWh per home per day; and a similar figure at work; perhaps 4 kWh per day per person.

has an average power consumption of 42 000 W. So having the lights on while driving requires 2% extra power.

What about the future's electric cars? The power consumption of a typical electric car is about 5000 W. So popping on an extra 100 W would increase its consumption by 2%. Power consumption would be smaller if we switched all car lights to light-emitting diodes, but if we pay any more attention to this topic, we will be coming down with a severe case of every-little-helps-ism.

The economics of low-energy bulbs

Generally I avoid discussing economics, but I'd like to make an exception for lightbulbs. Osram's 20 W low-energy bulb claims the same light output as a 100 W incandescent bulb. Moreover, its lifetime is said to be 15 000 hours (or "12 years," at 3 hours per day). In contrast a typical incandescent bulb might last 1000 hours. So during a 12-year period, you have this choice (figure 9.3): buy 15 incandescent bulbs and 1500 kWh of electricity (which costs roughly £150); or buy one low-energy bulb and 300 kWh of electricity (which costs roughly £30).

Should I wait until the old bulb dies before replacing it?

It feels like a waste, doesn't it? Someone put resources into making the old incandescent lightbulb; shouldn't we cash in that original investment by using the bulb until it's worn out? But the economic answer is clear: *continuing to use an old lightbulb is throwing good money after bad*. If you can find a satisfactory low-energy replacement, replace the old bulb now.

What about the mercury in compact fluorescent lights? Are LED bulbs better than fluorescents?

Researchers say that LED (light-emitting diode) bulbs will soon be even more energy-efficient than compact fluorescent lights. The efficiency of a light is measured in *lumens per watt*. I checked the numbers on my latest purchases: the Philips Genie 11 W compact fluorescent bulb (figure 9.4) has a brightness of 600 lumens, which is an efficiency of **55 lumens per watt**; regular incandescent bulbs deliver **10 lumens per watt**; the Omicron 1.3 W lamp, which has 20 white LEDs hiding inside it, has a brightness of 46 lumens, which is an efficiency of **35 lumens per watt**. So this LED bulb is almost as efficient as the fluorescent bulb. The LED industry still has a little catching up to do. In its favour, the LED bulb has a life of 50 000 hours, eight times the life of the fluorescent bulb. As I write, I see that www.cree.com is selling LEDs with a power of **100 lumens per watt**. It's projected that in the future, white LEDs will have an efficiency of over 150 lumens per watt [ynjzej]. I expect that within another couple of years, the best advice, from the point of view of both energy efficiency and avoiding mercury pollution, will be to use LED bulbs.

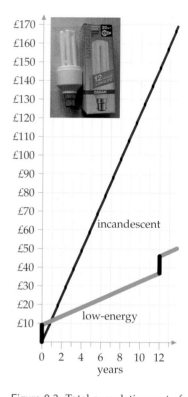

Figure 9.3. Total cumulative cost of using a traditional incandescent 100 W bulb for 3 hours per day, compared with replacing it *now* with an Osram Dulux Longlife Energy Saver (pictured). Assumptions: electricity costs 10p per kWh; replacement traditional bulbs cost 45p each; energy-saving bulbs cost £9. (I know you can find them cheaper than this, but this graph shows that even at £9, they're much more economical.)

Figure 9.4. Philips 11 W alongside Omicron 1.3 W LED bulb.

Mythconceptions

"There is no point in my switching to energy-saving lights. The "wasted" energy they put out heats my home, so it's not wasted."
    This myth is addressed in Chapter 11, p71.

Notes and further reading

page no.

57 *Street-lights use about 0.1 kWh per day per person...*   There's roughly one sodium street-light per 10 people; each light has a power of 100 W, switched on for 10 hours per day. That's 0.1 kWh per day per person.

– *...and traffic lights only 0.005 kWh/d per person.* Britain has 420 000 traffic and pedestrian signal light bulbs, consuming 100 million kWh of electricity per year. Shared between 60 million people, 100 million kWh per year is 0.005 kWh/d per person.

– *There are fewer signs and illuminated bollards than street-lights.*
    [www.highwayelectrical.org.uk]. There are 7.7 million lighting units (street lighting, illuminated signs and bollards) in the UK. Of these, roughly 7 million are street-lights and 1 million are illuminated road signs. There are 210 000 traffic signals.
    According to DUKES 2005, the total power for public lighting is 2095 GWh/y, which is 0.1 kWh/d per person.

– *55%-efficient generator* – source:
    en.wikipedia.org/wiki/Alternator. Generators in power stations are much more efficient at converting mechanical work to electricity.

Bulb type	efficiency (lumens/W)
incandescent	10
halogen	16–24
white LED	35
compact fluorescent	55
large fluorescent	94
sodium street-light	150

Table 9.5. Lighting efficiencies of commercially-available bulbs. In the future, white LEDs are expected to deliver 150 lumens per watt.

and much of it would be further than 50 km offshore. The outcome: if an area equal to a 9 km-wide strip all round the coast were filled with turbines, deep offshore wind could deliver a power of 32 kWh/d per person. A huge amount of power, yes; but still no match for our huge consumption. And we haven't spoken about the issue of wind's intermittency. We'll come back to that in Chapter 26.

I'll include this potential deep offshore contribution in the production stack, with the proviso, as I said before, that wind experts reckon deep offshore wind is prohibitively expensive.

Some comparisons and costs

So, how's our race between consumption and production coming along? Adding both shallow and deep offshore wind to the production stack, the green stack has a lead. Something I'd like you to notice about this race, though, is this contrast: how *easy* it is to toss a bigger log on the consumption fire, and how *difficult* it is to grow the production stack. As I write this paragraph, I'm feeling a little cold, so I step over to my thermostat and turn it up. It's so simple for me to consume an extra 30 kWh per day. But squeezing an extra 30 kWh per day per person from renewables requires an industrialization of the environment so large it is hard to imagine.

To create 48 kWh per day of offshore wind per person in the UK would require 60 million tons of concrete and steel – one ton per person. Annual world steel production is about 1200 million tons, which is 0.2 tons per person in the world. During the second world war, American shipyards built 2751 Liberty ships, each containing 7000 tons of steel – that's a total of 19 million tons of steel, or 0.1 tons per American. So the building of 60 million tons of wind turbines is not off the scale of achievability; but don't kid yourself into thinking that it's easy. Making this many windmills is as big a feat as building the Liberty ships.

For comparison, to make 48 kWh per day of nuclear power per person in the UK would require 8 million tons of steel and 0.14 million tons of concrete. We can also compare the 60 million tons of offshore wind hardware that we're trying to imagine with the existing fossil-fuel hardware already sitting in and around the North Sea (figure 10.4). In 1997, 200 installations and 7000 km of pipelines in the UK waters of the North Sea contained 8 million tons of steel and concrete. The newly built Langeled gas pipeline from Norway to Britain, which will convey gas with a power of 25 GW (10 kWh/d/p), used another 1 million tons of steel and 1 million tons of concrete (figure 10.5).

The UK government announced on 10th December 2007 that it would permit the creation of 33 GW of offshore wind capacity (which would deliver on average 10 GW to the UK, or 4.4 kWh per day per person), a plan branded "pie in the sky" by some in the wind industry. Let's run with a round figure of 4 kWh per day per person. This is one quarter of my

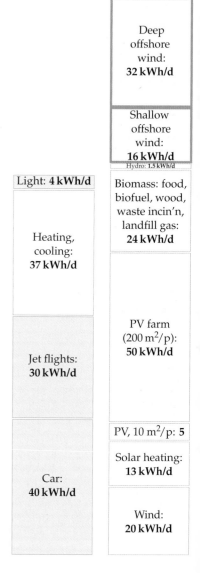

Figure 10.3. Offshore wind.

shallow 16 kWh per day per person. To obtain this average power requires roughly 10 000 "3 MW" wind turbines like those in figure 10.1. (They have a capacity of "3 MW" but on average they deliver 1 MW. I pop quotes round "3 MW" to indicate that this is a capacity, a peak power.)

What would this "33 GW"' of power cost to erect? Well, the "90 MW" Kentish Flats farm cost £105 million, so "33 GW" would cost about £33 billion. One way to clarify this £33 billion cost of offshore wind delivering 4 kWh/d per person is to share it among the UK population; that comes out to £550 per person. This is a much better deal, incidentally, than microturbines. A roof-mounted microturbine currently costs about £1500 and, even at a very optimistic windspeed of 6 m/s, delivers only 1.6 kWh/d. In reality, in a typical urban location in England, such microturbines deliver 0.2 kWh per day.

Another bottleneck constraining the planting of wind turbines is the special ships required. To erect 10 000 wind turbines ("33 GW") over a period of 10 years would require roughly 50 jack-up barges. These cost £60 million each, so an extra capital investment of £3 billion would be required. Not a show-stopper compared with the £33bn price tag already quoted, but the need for jack-up barges is certainly a detail that requires some forward planning.

Costs to birds

Do windmills kill "huge numbers" of birds? Wind farms recently got adverse publicity from Norway, where the wind turbines on Smola, a set of islands off the north-west coast, killed 9 white-tailed eagles in 10 months. I share the concern of BirdLife International for the welfare of rare birds. But I think, as always, it's important to do the numbers. It's been estimated that 30 000 birds per year are killed by wind turbines in Denmark, where windmills generate 9% of the electricity. Horror! Ban windmills! We also learn, moreover, that *traffic* kills *one million* birds per year in Denmark. Thirty-times-greater horror! Thirty-times-greater incentive to ban cars! And in Britain, 55 million birds per year are killed by *cats* (figure 10.6).

Going on emotions alone, I would like to live in a country with virtually no cars, virtually no windmills, and with plenty of cats and birds (with the cats that prey on birds perhaps being preyed upon by Norwegian white-tailed eagles, to even things up). But what I really hope is that decisions about cars and windmills are made by careful rational thought, not by emotions alone. Maybe we do need the windmills!

Figure 10.4. The Magnus platform in the northern UK sector of the North Sea contains 71 000 tons of steel. In the year 2000 this platform delivered 3.8 million tons of oil and gas – a power of 5 GW. The platform cost £1.1 billion.
Photos by Terry Cavner.

Figure 10.5. Pipes for Langeled. From Bredero–Shaw [brederoshaw.com].

Notes and further reading

page no.

60 *The Kentish Flats wind farm in the Thames Estuary...*
See www.kentishflats.co.uk. Its 30 Vestas V90 wind turbines have a total peak output of 90 MW, and the predicted average output was 32 MW (assuming a load factor of 36%). The mean wind speed at the hub height is 8.7 m/s. The turbines stand in 5 m-deep water, are spaced 700 m apart, and occupy an area of 10 km^2. The power density of this offshore wind farm was thus predicted to be 3.2 W/m^2. In fact, the average output was 26 MW, so the average load factor in 2006 was 29% [wbd8o]. This works out to a power density of 2.6 W/m^2. The North Hoyle wind farm off Prestatyn, North Wales, had a higher load factor of 36% in 2006. Its thirty 2 MW turbines occupy 8.4 km^2. They thus had an average power density of 2.6 W/m^2.

– *...shallow offshore wind, while roughly twice as costly as onshore wind, is economically feasible, given modest subsidy.* Source: Danish wind association windpower.org.

– *...deep offshore wind is at present not economically feasible.*
Source: British Wind Energy Association briefing document, September 2005, www.bwea.com. Nevertheless, a deep offshore demonstration project in 2007 put two turbines adjacent to the Beatrice oil field, 22 km off the east coast of Scotland (figure 10.8). Each turbine has a "capacity" of 5 MW and sits in a water depth of 45 m. Hub height: 107 m; diameter 126 m. All the electricity generated will be used by the oil platforms. Isn't that special! The 10 MW project cost £30 million – this price-tag of £3 per watt (peak) can be

Region	depth 5 to 30 metres		depth 30 to 50 metres	
	area (km^2)	potential resource (kWh/d/p)	area (km^2)	potential resource (kWh/d/p)
North West	3 300	6	2 000	4
Greater Wash	7 400	14	950	2
Thames Estuary	2 100	4	850	2
Other	14 000	28	45 000	87
TOTAL	27 000	52	49 000	94

Table 10.7. Potential offshore wind generation resource in proposed strategic regions, if these regions were *entirely filled* with wind turbines. From Dept. of Trade and Industry (2002b).

compared with that of Kentish Flats, £1.2 per watt (£105 million for 90 MW). www.beatricewind.co.uk

It's possible that *floating* wind turbines may change the economics of deep offshore wind.

60 *The area available for offshore wind.*
The Department of Trade and Industry's (2002) document "Future Offshore" gives a detailed breakdown of areas that are useful for offshore wind power. Table 10.7 shows the estimated resource in 76 000 km^2 of shallow and deep water. The DTI's estimated power contribution, if these areas were *entirely* filled with windmills, is 146 kWh/d per person (consisting of 52 kWh/d/p from the shallow and 94 kWh/d/p from the deep). But the DTI's estimate of the potential offshore wind generation resource is just 4.6 kWh per day per person. It might be interesting to describe how they get down from this potential resource of 146 kWh/d per person to 4.6 kWh/d per person. Why a final figure so much lower than ours? First, they imposed these limits: the water must be within 30 km of the shore and less than 40 m deep; the sea bed must not have gradient greater than 5°; shipping lanes, military zones, pipelines, fishing grounds, and wildlife reserves are excluded. Second, they assumed that only 5% of potential sites will be developed (as a result of seabed composition or planning constraints); they reduced the capacity by 50% for all sites less than 10 miles from shore, for reasons of public acceptability; they further reduced the capacity of sites with wind speed over 9 m/s by 95% to account for "development barriers presented by the hostile environment;" and other sites with average wind speed 8–9 m/s had their capacities reduced by 5%.

61 *...if we take the total coastline of Britain (length: 3000 km), and put a strip of turbines 4 km wide all the way round...* Pedants will say that "the coastline of Britain is not a well-defined length, because the coast is a fractal." Yes, yes, it's a fractal. But, dear pedant, please take a map and put a strip of turbines 4 km wide around mainland Britain, and see if it's not the case that your strip is indeed about 3000 km long.

– *Horns Reef* (Horns Rev). The difficulties with this "160 MW" Danish wind farm off Jutland [www.hornsrev.dk] are described by Halkema (2006).

When it is in working order, Horns Reef's load factor is 0.43 and its average power per unit area is 2.6 W/m^2.

62 *Liberty ships –*
www.liberty-ship.com/html/yards/introduction.html

– *...fossil fuel installations in the North Sea contained 8 million tons of steel and concrete* – Rice and Owen (1999).

– *The UK government announced on 10th December 2007 that it would permit the creation of 33 GW of offshore capacity...* [25e59w].

– *... "pie in the sky"*. Source: Guardian [2t2vjq].

63 *What would "33 GW" of offshore wind cost?* According to the DTI in November 2002, electricity from offshore wind farms costs about £50 per MWh (5p per kWh) (Dept. of Trade and Industry, 2002b, p21). Economic facts vary, however, and in April 2007 the estimated cost of offshore was up to £92 per MWh (Dept. of Trade and Industry, 2007, p7). By April 2008, the price of offshore wind evidently went even higher: Shell pulled out of their commitment to build the London Array. It's because offshore wind is so expensive that the Government is having to increase the number of ROCs (renewable obligation certificates) per unit of offshore wind energy. The ROC is the unit of subsidy given out to certain forms of renewable electricity generation. The standard value of a ROC is £45, with 1 ROC per MWh; so with a wholesale price of roughly £40/MWh, renewable generators are getting paid £85 per MWh. So 1 ROC per MWh is not enough subsidy to cover the cost of £92 per MWh. In the same document, estimates for other renewables (medium levelized costs in 2010) are as follows. Onshore wind: £65–89/MWh; co-firing of biomass: £53/MWh; large-scale hydro: £63/MWh; sewage gas: £38/MWh; solar PV: £571/MWh; wave: £196/MWh; tide: £177/MWh.
"Dale Vince, chief executive of green energy provider Ecotricity, which is engaged in building onshore wind farms, said that he supported the Government's [offshore wind] plans, but only if they are not to the detriment of onshore wind. 'It's dangerous to overlook the fantastic resource we have in this country... By our estimates, it will cost somewhere in the region of £40bn to build the 33 GW of offshore power Hutton is proposing. We could do the same job onshore for £20bn'." [57984r]

– *In a typical urban location in England, microturbines deliver 0.2 kWh per day*. Source: *Third Interim Report*, www.warwickwindtrials.org.uk/2.html. Among the best results in the Warwick Wind Trials study is a Windsave WS1000 (a 1-kW machine) in Daventry mounted at a height of 15 m above the ground, generating 0.6 kWh/d on average. But some microturbines deliver only 0.05 kWh per day – Source: Donnachadh McCarthy: "My carbon-free year," *The Independent*, December 2007 [6oc3ja]. The Windsave WS1000 wind turbine, sold across England in B&Q's shops, won an Eco-Bollocks award from *Housebuilder's Bible* author Mark Brinkley: "Come on, it's time to admit that the roof-mounted wind turbine industry is a complete fiasco. Good money is being thrown at an invention that doesn't work. This is the Sinclair C5 of the Noughties." [5soql2]. The Met Office and Carbon Trust published a report in July 2008 [6g2jm5], which estimates that, if small-scale

Figure 10.8. Construction of the Beatrice demonstrator deep offshore windfarm. Photos kindly provided by Talisman Energy (UK) Limited.

Figure 10.9. Kentish Flats. Photos © Elsam (`elsam.com`). Used with permission.

turbines were installed at all houses where economical in the UK, they would generate in total roughly $0.7\,\mathrm{kWh/d/p}$. They advise that roof-mounted turbines in towns are usually worse than useless: "in many urban situations, roof-mounted turbines may not pay back the carbon emitted during their production, installation and operation."

63 *Jack-up barges cost £60 million each.*
Source: `news.bbc.co.uk/1/hi/magazine/7206780.stm`. I estimated that we would need roughly 50 of them by assuming that there would be 60 work-friendly days each year, and that erecting a turbine would take 3 days.

Further reading: UK wind energy database [`www.bwea.com/ukwed/`].

11   Gadgets

One of the greatest dangers to society is the phone charger. The BBC News has been warning us of this since 2005:

> "The nuclear power stations will all be switched off in a few years. How can we keep Britain's lights on? ... **unplug your mobile-phone charger when it's not in use**."

Sadly, a year later, Britain hadn't got the message, and the BBC was forced to report:

> "**Britain tops energy waste league**."

And how did this come about? The BBC rams the message home:

> "65% of UK consumers leave chargers on."

From the way reporters talk about these planet-destroying black objects, it's clear that they are roughly as evil as Darth Vader. But how evil, exactly?

In this chapter we'll find out the truth about chargers. We'll also investigate their cousins in the gadget parade: computers, phones, and TVs. Digital set-top boxes. Cable modems. In this chapter we'll estimate the power used in running them and charging them, but not in manufacturing the toys in the first place – we'll address that in the later chapter on "stuff."

 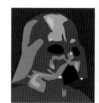

Vader   Charger

Figure 11.1. Planet destroyers. Spot the difference.

Figure 11.2. These five chargers – three for mobile phones, one for a pocket PC, and one for a laptop – registered less than one watt on my power meter.

The truth about chargers

Modern phone chargers, when left plugged in with no phone attached, use about half a watt. In our preferred units, this is a power consumption of about 0.01 kWh per day. For anyone whose consumption stack is over 100 kWh per day, the BBC's advice, *always unplug the phone charger*, could potentially reduce their energy consumption by one hundredth of one percent (if only they would do it).

Every little helps!

I don't think so. Obsessively switching off the phone-charger is like bailing the Titanic with a teaspoon. Do switch it off, but please be aware how tiny a gesture it is. Let me put it this way:

> All the energy saved in switching off your charger for one day is used up in *one second* of car-driving.

> The energy saved in switching off the charger for *one year* is equal to the energy in a single hot bath.

Admittedly, some older chargers use more than half a watt – if it's warm to the touch, it's probably using one watt or even three (figure 11.3). A three-watt-guzzling charger uses 0.07 kWh per day. I think that it's a good idea to switch off such a charger – it *will* save you three pounds per year. But don't kid yourself that you've "done your bit" by so doing. 3 W is only a tiny fraction of total energy consumption.

OK, that's enough bailing the Titanic with a teaspoon. Let's find out where the electricity is really being used.

Gadgets that really suck

Table 11.4 shows the power consumptions, in watts, of a houseful of gadgets. The first column shows the power consumption when the device is actually being used – for example, when a sound system is actually playing sound. The second column shows the consumption when the device is switched on, but sitting doing nothing. I was particularly shocked to find that a laser-printer sitting idle consumes 17 W – the same as the average consumption of a fridge-freezer! The third column shows the consumption when the gadget is explicitly asked to go to sleep or standby. The fourth shows the consumption when it is completely switched off – but still left plugged in to the mains. I'm showing all these powers in watts – to convert back to our standard units, remember that 40 W is 1 kWh/d. A nice rule of thumb, by the way, is that each watt costs about one pound per year (assuming electricity costs 10p per kWh).

The biggest guzzlers are the computer, its screen, and the television, whose consumption is in the hundreds of watts, when on. Entertainment systems such as stereos and DVD players swarm in the computer's wake, many of them consuming 10 W or so. A DVD player may cost just £20 in the shop, but if you leave it switched on all the time, it's costing you another £10 per year. Some stereos and computer peripherals consume several watts even when switched off, thanks to their mains-transformers. To be sure that a gadget is truly off, you need to switch it off at the wall.

Figure 11.3. This wasteful cordless phone and its charger use 3 W when left plugged in. That's **0.07 kWh/d**. If electricity costs 10p per kWh then a 3 W trickle costs £3 per year.

Powering the hidden tendrils of the information age

According to Jonathan Koomey (2007), the computer-servers in US data-centres and their associated plumbing (air conditioners, backup power systems, and so forth) consumed 0.4 kWh per day per person – just over 1% of US electricity consumption. That's the consumption figure for 2005, which, by the way, is twice as big as the consumption in 2000, because the number of servers grew from 5.6 million to 10 million.

Gadget	Power consumption (W)			
	on and active	on but inactive	standby	off
Computer and peripherals:				
computer box	80	55		2
cathode-ray display	110		3	0
LCD display	34		2	1
projector	150		5	
laser printer	500	17		
wireless & cable-modem	9			
Laptop computer	16	9		0.5
Portable CD player	2			
Bedside clock-radio	1.1	1		
Bedside clock-radio	1.9	1.4		
Digital radio	9.1		3	
Radio cassette-player	3	1.2		1.2
Stereo amplifier	6			6
Stereo amplifier II	13			0
Home cinema sound	7	7	4	
DVD player	7	6		
DVD player II	12	10	5	
TV	100		10	
Video recorder	13		1	
Digital TV set top box	6		5	
Clock on microwave oven	2			
Xbox	160		2.4	
Sony Playstation 3	190		2	
Nintendo Wii	18		2	
Answering machine		2		
Answering machine II		3		
Cordless telephone		1.7		
Mobile phone charger	5	0.5		
Vacuum cleaner	1600			

Table 11.4. Power consumptions of various gadgets, in watts. 40 W is 1 kWh/d.

Laptop: 16 W   Computer: 80 W

LCD   CRT   Printer: 17 W
31 W   108 W   (on, idle)

Projector: 150 W   Digital radio: 8 W

Other gadgets

A vacuum cleaner, if you use it for a couple of hours per week, is equivalent to about $0.2\,\text{kWh/d}$. Mowing the lawn uses about $0.6\,\text{kWh}$. We could go on, but I suspect that computers and entertainment systems are the big suckers on most people's electrical balance-sheet.

This chapter's summary figure: it'll depend how many gadgets you have at home and work, but a healthy houseful or officeful of gadgets left on all the time could easily use $5\,\text{kWh/d}$.

Mythconceptions

"There is no point in my switching off lights, TVs, and phone chargers during the winter. The 'wasted' energy they put out heats my home, so it's not wasted."

This myth is *True* for a few people, but only during the winter; but *False* for most.

If your house is being heated by electricity through ordinary bar fires or blower heaters then, yes, it's much the same as heating the house with any electricity-wasting appliances. But if you are in this situation, you should change the way you heat your house. Electricity is high-grade energy, and heat is low-grade energy. *It's a waste to turn electricity into heat.* To be precise, if you make only one unit of heat from a unit of electricity, that's a waste. Heaters called air-source heat pumps or ground-source heat pumps can do much better, delivering 3 or 4 units of heat for every unit of electricity consumed. They work like back-to-front refrigerators, pumping heat into your house from the outside air (see Chapter 21).

For the rest, whose homes are heated by fossil fuels or biofuels, it's a good idea to avoid using electrical gadgets as a heat source for your home – at least for as long as our increases in electricity-demand are served from fossil fuels. It's better to burn the fossil fuel at home. The point is, if you use electricity from an ordinary fossil power station, more than half of the energy from the fossil fuel goes sadly up the cooling tower. Of the energy that gets turned into electricity, about 8% is lost in the transmission system. If you burn the fossil fuel in your home, more of the energy goes directly into making hot air for you.

Notes and further reading

page no.

68 *The BBC News has been warning us ... unplug your mobile-phone charger.*
The BBC News article from 2005 said: "the nuclear power stations will all be switched off in a few years. How can we keep Britain's lights on? Here's three ways you can save energy: switch off video recorders when they're not in use; don't leave televisions on standby; and unplug your mobile-phone charger when it's not in use."

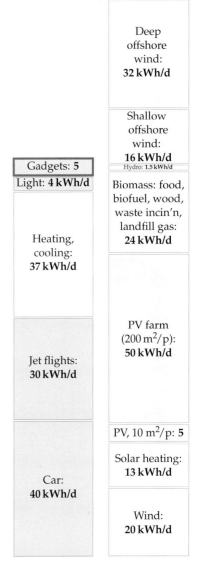

Figure 11.5. Information systems and other gadgets.

68   *Modern phone chargers, when left plugged in with no phone attached, use about half a watt.* The Maplin power meter in figure 11.2 is not accurate enough to measure this sort of power. I am grateful to Sven Weier and Richard McMahon of Cambridge University Engineering Department who measured a standard Nokia charger in an accurate calorimeter; they found that, when not connected to the mobile, it wastes 0.472 W. They made additional interesting measurements: the charger, when connected to a fully-charged mobile phone, wastes 0.845 W; and when the charger is doing what it's meant to do, charging a partly-charged Nokia mobile, it wastes 4.146 W as heat. Pedants sometimes ask "what about the *reactive power* of the charger?" This is a technical niggle, not really worth our time. For the record, I measured the reactive power (with a crummy meter) and found it to be about 2 VA per charger. Given that the power loss in the national grid is 8% of the delivered power, I reckon that the power loss associated with the reactive power is at most 0.16 W. When actually making a phone-call, the mobile uses 1 W.

Further reading: Kuehr (2003).

Figure 11.6. Advertisement from the "DIY planet repairs" campaign. The text reads "**Unplug**. If every London household unplugged their mobile-phone chargers when not in use, we could save 31,000 tonnes of CO_2 and £7.75m per year." `london.gov.uk/diy/`

12 Wave

If wave power offers hope to any country, then it must offer hope to the United Kingdom and Ireland – flanked on the one side by the Atlantic Ocean, and on the other by the North Sea.

First, let's clarify where waves come from: *sun makes wind and wind makes waves.*

Most of the sunlight that hits our planet warms the oceans. The warmed water warms the air above it, and produces water vapour. The warmed air rises; as it rises it cools, and the water eventually re-condenses, forming clouds and rain. At its highest point, the air is cooled down further by the freezing blackness of space. The cold air sinks again. This great solar-powered pump drives air round and round in great convection rolls. From our point of view on the surface, these convection rolls produce the winds. Wind is second-hand solar energy. As wind rushes across open water, it generates waves. Waves are thus third-hand solar energy. (The waves that crash on a beach are nothing to do with the tides.)

In open water, waves are generated whenever the wind speed is greater than about 0.5 m/s. The wave crests move at about the speed of the wind that creates them, and in the same direction. The *wavelength* of the waves (the distance between crests) and the *period* (the time between crests) depend on the speed of the wind. The longer the wind blows for, and the greater the expanse of water over which the wind blows, the greater the *height* of the waves stroked up by the wind. Thus since the prevailing winds over the Atlantic go from west to east, the waves arriving on the Atlantic coast of Europe are often especially big. (The waves on the east coast of the British Isles are usually much smaller, so my estimates of potential wave power will focus on the resource in the Atlantic Ocean.)

Waves have long memory and will keep going in the same direction for days after the wind stopped blowing, until they bump into something. In seas where the direction of the wind changes frequently, waves born on different days form a superposed jumble, travelling in different directions.

If waves travelling in a particular direction encounter objects that absorb energy from the waves – for example, a row of islands with sandy beaches – then the seas beyond the object are calmer. The objects cast a shadow, and there's less energy in the waves that get by. So, whereas sunlight delivers a power per unit *area*, waves deliver a power per unit *length* of coastline. You can't have your cake and eat it. You can't collect wave energy two miles off-shore *and* one mile off-shore. Or rather, you can try, but the two-mile facility will absorb energy that would have gone to the one-mile facility, and it won't be replaced. The fetch required for wind to stroke up big waves is thousands of miles.

We can find an upper bound on the maximum conceivable power that could be obtained from wave power by estimating the incoming power

Figure 12.1. A Pelamis wave energy collector is a sea snake made of four sections. It faces nose-on towards the incoming waves. The waves make the snake flex, and these motions are resisted by hydraulic generators. The peak power from one snake is 750 kW; in the best Atlantic location one snake would deliver 300 kW on average. Photo from Pelamis wave power
www.pelamiswave.com.

73

per unit length of exposed coastline, and multiplying by the length of coastline. We ignore the question of what mechanism could collect all this power, and start by working out how much power it is.

The power of Atlantic waves has been measured: it's about 40 kW per metre of exposed coastline. That sounds like a lot of power! If everyone owned a metre of coastline and could harness their whole 40 kW, that would be plenty of power to cover modern consumption. However, *our population is too big.* There is not enough Atlantic-facing coastline for everyone to have their own metre.

As the map on p73 shows, Britannia rules about 1000 km of Atlantic coastline (one million metres), which is $^1/_{60}$ m per person. So the total raw incoming power is 16 kWh per day per person. If we extracted all this power, the Atlantic, at the seaside, would be as flat as a millpond. Practical systems won't manage to extract all the power, and some of the power will inevitably be lost during conversion from mechanical energy to electricity. Let's assume that brilliant wave-machines are 50%-efficient at turning the incoming wave power into electricity, and that we are able to pack wave-machines along 500 km of Atlantic-facing coastline. That would mean we could deliver 25% of this theoretical bound. That's 4 kWh per day per person. As usual, I'm intentionally making pretty extreme assumptions to boost the green stack – I expect the assumption that we could line *half of the Atlantic coastline* with wave absorbers will sound bananas to many readers.

How do the numbers assumed in this calculation compare with today's technology? As I write, there are just three wave machines working in deep water: three Pelamis wave energy collectors (figure 12.1) built in Scotland and deployed off Portugal. No actual performance results have been published, but the makers of the Pelamis ("designed with survival as the key objective before power capture efficiency") predict that a two-kilometre-long wave-farm consisting of 40 of their sea-snakes would deliver 6 kW per metre of wave-farm. Using this number in the previous calculation, the power delivered by 500 kilometres of wave-farm is reduced to 1.2 kWh per day per person. While wave power may be useful for small communities on remote islands, I suspect it can't play a significant role in the solution to Britain's sustainable energy problem.

What's the weight of a Pelamis, and how much steel does it contain? One snake with a maximum power of 750 kW weighs 700 tons, including 350 tons of ballast. So it has about 350 tons of steel. That's a weight-to-power ratio of roughly 500 kg per kW (peak). We can compare this with the steel requirements for offshore wind: an offshore wind-turbine with a maximum power of 3 MW weighs 500 tons, including its foundation. That's a weight-to-power ratio of about 170 kg per kW, one third of the wave machine's. The Pelamis is a first prototype; presumably with further investment and development in wave technology, the weight-to-power ratio would fall.

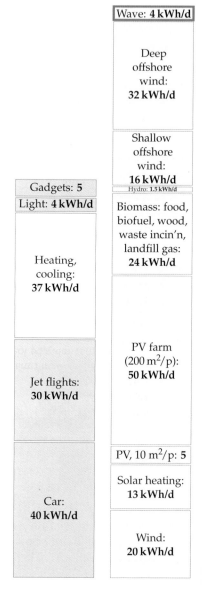

Figure 12.2. Wave.

Notes and further reading

page no.

73   *Waves are generated whenever the wind speed is greater than about 0.5 m/s.*
     The wave crests move at about the speed of the wind that creates them. The
     simplest theory of wave-production (Faber, 1995, p. 337) suggests that (for
     small waves) the wave crests move at about half the speed of the wind that
     creates them. It's found empirically however that, the longer the wind blows
     for, the longer the wavelength of the dominant waves present, and the greater
     their velocity. The characteristic speed of fully-developed seas is almost ex-
     actly equal to the wind-speed 20 metres above the sea surface (Mollison,
     1986).

Photo by Terry Cavner.

–   *The waves on the east coast of the British Isles are usually much smaller.*
     Whereas the wave power at Lewis (Atlantic) is 42 kW/m, the powers at the
     east-coast sites are: Peterhead: 4 kW/m; Scarborough: 8 kW/m; Cromer:
     5 kW/m. Source: Sinden (2005). Sinden says: "The North Sea Region expe-
     riences a very low energy wave environment."

74   *Atlantic wave power is 40 kW per metre of exposed coastline.*
     (Chapter F explains how we can estimate this power using a few facts about
     waves.) This number has a firm basis in the literature on Atlantic wave
     power (Mollison et al., 1976; Mollison, 1986, 1991). From Mollison (1986), for
     example: "the large scale resource of the NE Atlantic, from Iceland to North
     Portugal, has a net resource of 40–50 MW/km, of which 20–30 MW/km is
     potentially economically extractable." At any point in the open ocean, three
     powers per unit length can be distinguished: the total power passing through
     that point in all directions (63 kW/m on average at the Isles of Scilly and
     67 kW/m off Uist); the net power intercepted by a directional collecting de-
     vice oriented in the optimal direction (47 kW/m and 45 kW/m respectively);
     and the power per unit coastline, which takes into account the misalignment
     between the optimal orientation of a directional collector and the coastline
     (for example in Portugal the optimal orientation faces northwest and the
     coastline faces west).

Uist°

Scilly°

100 km

–   *Practical systems won't manage to extract all the power, and some of the*
     power will inevitably be lost during conversion from mechanical energy to
     electricity. The UK's first grid-connected wave machine, the Limpet on Islay,
     provides a striking example of these losses. When it was designed its con-
     version efficiency from wave power to grid power was estimated to be 48%,
     and the average power output was predicted to be 200 kW. However losses
     in the capture system, flywheels and electrical components mean the actual
     average output is 21 kW – just 5% of the predicted output (Wavegen, 2002).

13 Food and farming

Modern agriculture is the use of land to convert petroleum into food.

Albert Bartlett

We've already discussed in Chapter 6 how much sustainable power could be *produced* through greenery; in this chapter we discuss how much power is currently *consumed* in giving us our daily bread.

A moderately active person with a weight of 65 kg consumes food with a chemical energy content of about 2600 "Calories" per day. A "Calorie," in food circles, is actually 1000 chemist's calories (1 kcal). 2600 "Calories" per day is about 3 kWh per day. Most of this energy eventually escapes from the body as heat, so one function of a typical person is to act as a space heater with an output of a little over 100 W, a medium-power lightbulb. Put 10 people in a small cold room, and you can switch off the 1 kW convection heater.

How much energy do we actually consume in order to get our 3 kWh per day? If we enlarge our viewpoint to include the inevitable upstream costs of food production, then we may find that our energy footprint is substantially bigger. It depends if we are vegan, vegetarian or carnivore.

The vegan has the smallest inevitable footprint: 3 kWh per day of energy from the plants he eats.

The energy cost of drinking milk

I love milk. If I drinka-pinta-milka-day, what energy does that require? A typical dairy cow produces 16 litres of milk per day. So my one pint per day (half a litre per day) requires that I employ $^1/_{32}$ of a cow. Oh, hang on – I love cheese too. And to make 1 kg of Irish Cheddar takes about 9 kg of milk. So consuming 50 g of cheese per day requires the production of an extra 450 g of milk. OK: my milk and cheese habit requires that I employ $^1/_{16}$ of a cow. And how much power does it take to run a cow? Well, if a cow weighing 450 kg has similar energy requirements per kilogram to a human (whose 65 kg burns 3 kWh per day) then the cow must be using about 21 kWh/d. Does this extrapolation from human to cow make you uneasy? Let's check these numbers: www.dairyaustralia.com.au says that a suckling cow of weight 450 kg needs 85 MJ/d, which is 24 kWh/d. Great, our guess wasn't far off! So my $^1/_{16}$ share of a cow has an energy consumption of about 1.5 kWh per day. This figure ignores other energy costs involved in persuading the cow to make milk and the milk to turn to cheese, and of getting the milk and cheese to travel from her to me. We'll cover some of these costs when we discuss freight and supermarkets in Chapter 15.

Figure 13.1. A salad Niçoise.

Minimum: **3 kWh/d**

Figure 13.2. Minimum energy requirement of one person.

Milk, cheese: **1.5 kWh/d**

Figure 13.3. Milk and cheese.

Eggs

A "layer" (a chicken that lays eggs) eats about 110 g of chicken feed per day. Assuming that chicken feed has a metabolizable energy content of 3.3 kWh per kg, that's a power consumption of 0.4 kWh per day per chicken. Layers yield on average 290 eggs per year. So eating two eggs a day requires a power of 1 kWh per day. Each egg itself contains 80 kcal, which is about 0.1 kWh. So from an energy point of view, egg production is 20% efficient.

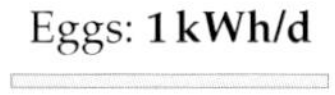

Eggs: **1 kWh/d**

Figure 13.4. Two eggs per day.

The energy cost of eating meat

Let's say an enthusiastic meat-eater eats about half a pound a day (227 g). (This is the average meat consumption of Americans.) To work out the power required to maintain the meat-eater's animals as they mature and wait for the chop, we need to know for how long the animals are around, consuming energy. Chicken, pork, or beef?

Chicken, sir? Every chicken you eat was clucking around being a chicken for roughly 50 days. So the steady consumption of half a pound a day of chicken requires about 25 pounds of chicken to be alive, preparing to be eaten. And those 25 pounds of chicken consume energy.

Pork, madam? Pigs are around for longer – maybe 400 days from birth to bacon – so the steady consumption of half a pound a day of pork requires about 200 pounds of pork to be alive, preparing to be eaten.

Cow? Beef production involves the longest lead times. It takes about 1000 days of cow-time to create a steak. So the steady consumption of half a pound a day of beef requires about 500 pounds of beef to be alive, preparing to be eaten.

To condense all these ideas down to a single number, let's assume you eat half a pound (227 g) per day of meat, made up of equal quantities of chicken, pork, and beef. This meat habit requires the perpetual sustenance of 8 pounds of chicken meat, 70 pounds of pork meat, and 170 pounds of cow meat. That's a total of 110 kg of meat, or 170 kg of animal (since about two thirds of the animal gets turned into meat). And if the 170 kg of animal has similar power requirements to a human (whose 65 kg burns 3 kWh/d) then the power required to fuel the meat habit is

Carnivory: **8 kWh/d**

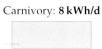

Figure 13.5. Eating meat requires extra power because we have to feed the queue of animals lining up to be eaten by the human.

$$170 \, \text{kg} \times \frac{3 \, \text{kWh/d}}{65 \, \text{kg}} \simeq 8 \, \text{kWh/d}.$$

I've again taken the physiological liberty of assuming "animals are like humans;" a more accurate estimate of the energy to make chicken is in this chapter's endnotes. No matter, I only want a ballpark estimate, and here it is. The power required to make the food for a typical consumer of vegetables, dairy, eggs, and meat is $1.5 + 1.5 + 1 + 8 = 12 \, \text{kWh}$ per day. (The daily calorific balance of this rough diet is 1.5 kWh from vegetables;

0.7 kWh from dairy; 0.2 kWh from eggs; and 0.5 kWh from meat – a total of 2.9 kWh per day.)

This number does not include any of the power costs associated with farming, fertilizing, processing, refrigerating, and transporting the food. We'll estimate some of those costs below, and some in Chapter 15.

Do these calculations give an argument in favour of vegetarianism, on the grounds of lower energy consumption? It depends on where the animals feed. Take the steep hills and mountains of Wales, for example. Could the land be used for anything other than grazing? Either these rocky pasturelands are used to sustain sheep, or they are not used to help feed humans. You can think of these natural green slopes as maintenance-free biofuel plantations, and the sheep as automated self-replicating biofuel-harvesting machines. The energy losses between sunlight and mutton are substantial, but there is probably no better way of capturing solar power in such places. (I'm not sure whether this argument for sheep-farming in Wales actually adds up: during the worst weather, Welsh sheep are moved to lower fields where their diet is supplemented with soya feed and other food grown with the help of energy-intensive fertilizers; what's the true energy cost? I don't know.) Similar arguments can be made in favour of carnivory for places such as the scrublands of Africa and the grasslands of Australia; and in favour of dairy consumption in India, where millions of cows are fed on by-products of rice and maize farming.

On the other hand, where animals are reared in cages and fed grain that humans could have eaten, there's no question that it would be more energy-efficient to cut out the middlehen or middlesow, and feed the grain directly to humans.

Figure 13.6. Will harvest energy crops for food.

Fertilizer and other energy costs in farming

The embodied energy in Europe's fertilizers is about 2 kWh per day per person. According to a report to DEFRA by the University of Warwick, farming in the UK in 2005 used an energy of 0.9 kWh per day per person for farm vehicles, machinery, heating (especially greenhouses), lighting, ventilation, and refrigeration.

The energy cost of Tiddles, Fido, and Shadowfax

Animal companions! Are you the servant of a dog, a cat, or a horse?

There are perhaps 8 million cats in Britain. Let's assume you look after one of them. The energy cost of Tiddles? If she eats 50 g of meat per day (chicken, pork, and beef), then the last section's calculation says that the power required to make Tiddles' food is just shy of 2 kWh per day. A vegetarian cat would require less.

Similarly if your dog Fido eats 200 g of meat per day, and carbohydrates

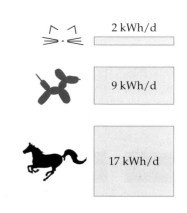

Figure 13.7. The power required for animal companions' food.

amounting to 1 kWh per day, then the power required to make his food is about 9 kWh per day.

Shadowfax the horse weighs about 400 kg and consumes 17 kWh per day.

Mythconceptions

I heard that the energy footprint of food is so big that "it's better to drive than to walk."

Whether this is true depends on your diet. It's certainly possible to find food whose fossil-fuel energy footprint is bigger than the energy delivered to the human. A bag of crisps, for example, has an embodied energy of 1.4 kWh of fossil fuel per kWh of chemical energy eaten. The embodied energy of meat is higher. According to a study from the University of Exeter, the typical diet has an embodied energy of roughly 6 kWh per kWh eaten. To figure out whether driving a car or walking uses less energy, we need to know the transport efficiency of each mode. For the typical car of Chapter 3, the energy cost was 80 kWh per 100 km. Walking uses a net energy of 3.6 kWh per 100 km – 22 times less. So if you live entirely on food whose footprint is greater than 22 kWh per kWh then, yes, the energy cost of getting you from A to B in a fossil-fuel-powered vehicle is less than if you go under your own steam. But if you have a typical diet (6 kWh per kWh) then "it's better to drive than to walk" is a myth. Walking uses one quarter as much energy.

Notes and further reading

page no.

76 *A typical dairy cow produces 16 litres of milk per day.* There are 2.3 million dairy cows in the UK, each producing around 5900 litres per year. Half of all milk produced by cows is sold as liquid milk. www.ukagriculture.com, www.vegsoc.org/info/cattle.html

77 *It takes about 1000 days of cow-time to create a steak.* 33 months from conception to slaughterhouse: 9 months' gestation and 24 months' rearing. www.shabdenparkfarm.com/farming/cattle.htm

– *Chicken.* A full-grown (20-week old) layer weighs 1.5 or 1.6 kg. Its feed has an energy content of 2850 kcal per kg, which is 3.3 kWh per kg, and its feed consumption rises to 340 g per week when 6 weeks old, and to 500 g per week when aged 20 weeks. Once laying, the typical feed required is 110 g per day.
Meat chickens' feed has an energy content of 3.7 kWh per kg. Energy consumption is 400–450 kcal per day per hen (0.5 kWh/d per hen), with 2 kg being a typical body weight. A meat chicken weighing 2.95 kg consumes a total of 5.32 kg of feed [5h69fm]. So the embodied energy of a meat chicken is about 6.7 kWh per kg of animal, or 10 kWh per kg of eaten meat.

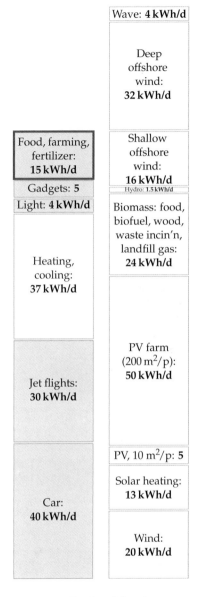

Figure 13.8. Food and farming.

If I'd used this number instead of my rough guess, the energy contribution of the chicken would have been bumped up a little. But given that the mixed-meat diet's energy footprint is dominated by the beef, it really doesn't matter that I underestimated the chickens. Sources: Subcommittee on Poultry Nutrition, National Research Council (1994), www.nap.edu/openbook.php?isbn=0309048923, MacDonald (2008), and www.statistics.gov.uk/statbase/datasets2.asp.

77   *let's assume you eat half a pound (227 g) a day of meat, made up of equal quantities of chicken, pork, and beef.* This is close to the average meat consumption in America, which is 251 g per day – made up of 108 g chicken, 81 g beef, and 62 g pork (MacDonald, 2008).

78   *The embodied energy in Europe's fertilizers is about 2 kWh per day per person.* In 1998–9, Western Europe used 17.6 Mt per year of fertilizers: 10 Mt of nitrates, 3.5 Mt of phosphate and 4.1 Mt potash. These fertilizers have energy footprints of 21.7, 4.9, and 3.8 kWh per kg respectively. Sharing this energy out between 375 million people, we find a total footprint of 1.8 kWh per day per person. Sources: Gellings and Parmenter (2004), International Fertilizer Industry Association [5pwojp].

–   *Farming in the UK in 2005 used an energy of 0.9 kWh per day per person.* Source: Warwick HRI (2007).

79   *A bag of crisps has an embodied energy of 1.4 kWh of fossil fuel per kWh of chemical energy eaten.* I estimated this energy from the carbon footprint of a bag of crisps: 75 g CO_2 for a standard 35 g bag [5bj8k3]. Of this footprint, 44% is associated with farming, 30% with processing, 15% packaging, and 11% transport and disposal. The chemical energy delivered to the consumer is 770 kJ. So this food has a carbon footprint of 350 g per kWh. Assuming that most of this carbon footprint is from fossil fuels at 250 g CO_2 per kWh, the energy footprint of the crisps is 1.4 kWh of fossil fuel per kWh of chemical energy eaten.

–   *The typical diet has an embodied energy of roughly 6 kWh per kWh eaten.* Coley (2001) estimates the embodied energy in a typical diet is 5.75 times the derived energy. Walking has a CO_2 footprint of 42 g/km; cycling, 30 g/km. For comparison, driving an average car emits 183 g/km.

–   *Walking uses 3.6 kWh per 100 km.* A walking human uses a total of 6.6 kWh per 100 km [3s576h]; we subtract off the resting energy to get the energy footprint of walking (Coley, 2001).

Further reading: Weber and Matthews (2008).

14 Tide

The moon and earth are in a whirling, pirouetting dance around the sun. Together they tour the sun once every year, at the same time whirling around each other once every 28 days. The moon also turns around once every 28 days so that she always shows the same face to her dancing partner, the earth. The prima donna earth doesn't return the compliment; she pirouettes once every day. This dance is held together by the force of gravity: every bit of the earth, moon, and sun is pulled towards every other bit of earth, moon, and sun. The sum of all these forces is *almost* exactly what's required to keep the whirling dance on course. But there are very slight imbalances between the gravitational forces and the forces required to maintain the dance. It is these imbalances that give rise to the tides.

The imbalances associated with the whirling of the moon and earth around each other are about three times as big as the imbalances associated with the earth's slower dance around the sun, so the size of the tides varies with the phase of the moon, as the moon and sun pass in and out of alignment. At full moon and new moon (when the moon and sun are in line with each other) the imbalances reinforce each other, and the resulting big tides are called *spring tides*. (Spring tides are *not* "tides that occur at spring-time;" spring tides happen every two weeks like clockwork.) At the intervening half moons, the imbalances partly cancel and the tides are smaller; these smaller tides are called *neap tides*. Spring tides have roughly twice the amplitude of neap tides: the spring high tides are twice as high above mean sea level as neap high tides, the spring low tides are twice as low as neap low tides, and the tidal currents are twice as big at springs as at neaps.

Why are there two high tides and two low tides per day? Well, if the earth were a perfect sphere, a smooth billiard ball covered by oceans, the tidal effect of the earth-moon whirling would be to deform the water slightly towards and away from the moon, making the water slightly rugby-ball shaped (figure 14.1). Someone living on the equator of this billiard-ball earth, spinning round once per day within the water cocoon, would notice the water level going up and down twice per day: up once as he passed under the nose of the rugby-ball, and up a second time as he passed under its tail. This cartoon explanation is some way from reality. In reality, the earth is not smooth, and it is not uniformly covered by water (as you may have noticed). Two humps of water cannot whoosh round the earth once per day because the continents get in the way. The true behaviour of the tides is thus more complicated. In a large body of water such as the Atlantic Ocean, tidal crests and troughs form but, unable to whoosh round the earth, they do the next best thing: they whoosh around the perimeter of the Ocean. In the North Atlantic there are two crests and two troughs, all circling the Atlantic in an anticlockwise direction once a

away from the moon — towards the moon

Figure 14.1. An ocean covering a billiard-ball earth. We're looking down on the North pole, and the moon is 60 cm off the page to the right. The earth spins once per day inside a rugby-ball-shaped shell of water. The oceans are stretched towards and away from the moon because the gravitational forces supplied by the moon don't perfectly match the required centripetal force to keep the earth and moon whirling around their common centre of gravity.
Someone standing on the equator (rotating as indicated by the arrow) will experience two high waters and two low waters per day.

day. Here in Britain we don't directly see these Atlantic crests and troughs – we are set back from the Atlantic proper, separated from it by a few hundred miles of paddling pool called the continental shelf. Each time one of the crests whooshes by in the Atlantic proper, it sends a crest up our paddling pool. Similarly each Atlantic trough sends a trough up the paddling pool. Consecutive crests and troughs are separated by six hours. Or to be more precise, by six and a quarter hours, since the time between moon-rises is about 25, not 24 hours.

The speed at which the crests and troughs travel varies with the depth of the paddling pool. The shallower the paddling pool gets, the slower the crests and troughs travel and the larger they get. Out in the ocean, the tides are just a foot or two in height. Arriving in European estuaries, the tidal range is often as big as four metres. In the northern hemisphere, the Coriolis force (a force, associated with the rotation of the earth, that acts only on moving objects) makes all tidal crests and troughs tend to hug the right-hand bank as they go. For example, the tides in the English channel are bigger on the French side. Similarly, the crests and troughs entering the North Sea around the Orkneys hug the British side, travelling down to the Thames Estuary then turning left at the Netherlands to pay their respects to Denmark.

Tidal energy is sometimes called lunar energy, since it's mainly thanks to the moon that the water sloshes around so. Much of the tidal energy, however, is really coming from the rotational energy of the spinning earth. The earth is very gradually slowing down.

So, how can we put tidal energy to use, and how much power could we extract?

Rough estimates of tidal power

When you think of tidal power, you might think of an artificial pool next to the sea, with a water-wheel that is turned as the pool fills or empties (figures 14.2 and 14.3). Chapter G shows how to estimate the power available from such tide-pools. Assuming a range of 4 m, a typical range in many European estuaries, the maximum power of an artificial tide-pool that's filled rapidly at high tide and emptied rapidly at low tide, generating power from both flow directions, is about $3 \, \text{W/m}^2$. This is the same as the power per unit area of an offshore wind farm. And we already know how big offshore wind farms need to be to make a difference. *They need to be country-sized.* So similarly, to make tide-pools capable of producing power comparable to Britain's total consumption, we'd need the total area of the tide-pools to be similar to the area of Britain.

Amazingly, Britain is already supplied with a natural tide-pool of just the required dimensions. This tide-pool is known as the North Sea (figure 14.5). If we simply insert generators in appropriate spots, significant power can be extracted. The generators might look like underwater wind-

Figure 14.2. Woodbridge tide-pool and tide-mill. Photos kindly provided by Ted Evans.

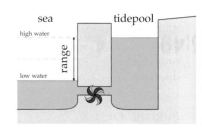

Figure 14.3. An artificial tide-pool. The pool was filled at high tide, and now it's low tide. We let the water out through the electricity generator to turn the water's potential energy into electricity.

tidal range	power density
2 m	$1 \, \text{W/m}^2$
4 m	$3 \, \text{W/m}^2$
6 m	$7 \, \text{W/m}^2$
8 m	$13 \, \text{W/m}^2$

Table 14.4. Power density (power per unit area) of tide-pools, assuming generation from both the rising and the falling tide.

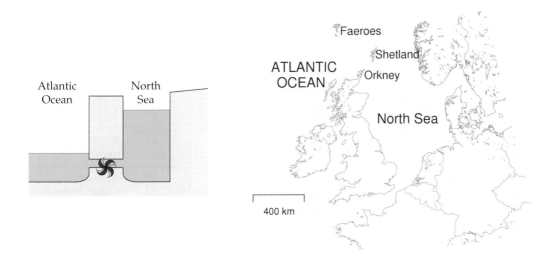

Figure 14.5. The British Isles are in a fortunate position: the North Sea forms a natural tide-pool, in and out of which great sloshes of water pour twice a day.

mills. Because the density of water is roughly 1000 times that of air, the power of water flow is 1000 times greater than the power of wind at the same speed. We'll come back to tide farms in a moment, but first let's discuss how much raw tidal energy rolls around Britain every day.

Raw incoming tidal power

The tides around Britain are genuine tidal waves – unlike tsunamis, which are called "tidal waves," but are nothing to do with tides. Follow a high tide as it rolls in from the Atlantic. The time of high tide becomes progressively later as we move east up the English channel from the Isles of Scilly to Portsmouth and on to Dover. The crest of the tidal wave progresses up the channel at about 70 km/h. (The crest of the wave moves much faster than the water itself, just as ordinary waves on the sea move faster than the water.) Similarly, a high tide moves clockwise round Scotland, rolling down the North Sea from Wick to Berwick and on to Hull at a speed of about 100 km/h. These two high tides converge on the Thames Estuary. By coincidence, the Scottish crest arrives about 12 hours later than the crest that came via Dover, so it arrives in near-synchrony with the next high tide via Dover, and London receives the normal two high tides per day.

The power we can extract from tides can never be more than the total power of these tidal waves from the Atlantic. The total power crossing the lines in figure 14.6 has been measured; on average it amounts to 100 kWh per day per person. If we imagine extracting 10% of this incident energy, and if the conversion and transmission processes are 50% efficient, the average power delivered would be 5 kWh per day per person.

This is a tentative first guess, made without specifying any technical

Figure 14.6. The average incoming power of lunar tidal waves crossing these two lines has been measured to be 250 GW. This raw power, shared between 60 million people, is 100 kWh per day per person.

pump water *out* of the low lagoon, making its level even lower than low water. The energy required to pump down the level of the low lagoon is then repaid with interest at high tide, when power is generated by letting water into the low lagoon. Similarly, extra water can be pumped into the high lagoon at high tide, using energy generated by the low lagoon. Whatever state the tide is in, one lagoon or the other would be able to generate power. Such a pair of tidal lagoons could also work as a pumped storage facility, storing excess energy from the electricity grid.

The average power per unit area of tidal lagoons in British waters could be $4.5\,\text{W/m}^2$, so if tidal lagoons with a total area of $800\,\text{km}^2$ were created (as indicated in figure 14.9), the power generated would be $1.5\,\text{kWh/d}$ per person.

Figure 14.9. Two tidal lagoons, each with an area of $400\,\text{km}^2$, one off Blackpool, and one in the Wash. The Severn estuary is also highlighted for comparison.

Beauties of tide

Totting everything up, the barrage, the lagoons, and the tidal stream farms could deliver something like $11\,\text{kWh/d}$ per person (figure 14.10).

Tide power has never been used on an industrial scale in Britain, so it's hard to know what economic and technical challenges will be raised as we build and maintain tide-turbines – corrosion, silt accumulation, entanglement with flotsam? But here are seven reasons for being excited about tidal power in the British Isles. 1. Tidal power is completely predictable; unlike wind and sun, tidal power is a renewable on which one could depend; it works day and night all year round; using tidal lagoons, energy can be stored so that power can be delivered on demand. 2. Successive high and low tides take about 12 hours to progress around the British Isles, so the strongest currents off Anglesey, Islay, Orkney and Dover occur at different times from each other; thus, together, a collection of tide farms could produce a more constant contribution to the electrical grid than one tide farm, albeit a contribution that wanders up and down with the phase of the moon. 3. Tidal power will last for millions of years. 4. It doesn't require high-cost hardware, in contrast to solar photovoltaic power. 5. Moreover, because the power density of a typical tidal flow is greater than the power density of a typical wind, a $1\,\text{MW}$ tide turbine is smaller in size than a $1\,\text{MW}$ wind turbine; perhaps tide turbines could therefore be cheaper than wind turbines. 6. Life below the waves is peaceful; there is no such thing as a freak tidal storm; so, unlike wind turbines, which require costly engineering to withstand rare windstorms, underwater tide turbines will not require big safety factors in their design. 7. Humans mostly live on the land, and they can't see under the sea, so objections to the visual impact of tide turbines should be less strong than the objections to wind turbines.

Mythconceptions

Tidal power, while clean and green, should not be called renewable. Extracting power from the tides slows down the earth's rotation. We definitely can't use tidal power long-term.

False. The natural tides already slow down the earth's rotation. The natural rotational energy loss is roughly 3 TW (Shepherd, 2003). Thanks to natural tidal friction, each century, the day gets longer by 2.3 milliseconds. Many tidal energy extraction systems are just extracting energy that would have been lost anyway in friction. But even if we *doubled* the power extracted from the earth–moon system, tidal energy would still last more than a billion years.

Notes and further reading

page no.

82   *The power of an artificial tide-pool.* The power per unit area of a tide-pool is derived in Chapter G, p311.

–   *Britain is already supplied with a natural tide-pool … known as the North Sea.* I should not give the impression that the North Sea fills and empties just like a tide-pool on the English coast. The flows in the North Sea are more complex because the time taken for a bump in water level to propagate across the Sea is similar to the time between tides. Nevertheless, there are whopping tidal currents in and out of the North Sea, and within it too.

83   *The total incoming power of lunar tidal waves crossing these lines has been measured to be 100 kWh per day per person.* Source: Cartwright et al. (1980). For readers who like back-of-envelope models, Chapter G shows how to estimate this power from first principles.

84   *La Rance* generated 16 TWh over 30 years. That's an average power of 60 MW. (Its peak power is 240 MW.) The tidal range is up to 13.5 m; the impounded area is 22 km^2; the barrage 750 m long. Average power density: 2.7 W/m^2. Source: [6xrm5q].

85   *The engineers' reports on the Severn barrage… say 17 TWh/year.* (Taylor, 2002b). This (2 GW) corresponds to 5% of current UK total electricity consumption, on average.

86   *Power per unit area of tidal lagoons could be 4.5 W/m^2.* MacKay (2007a).

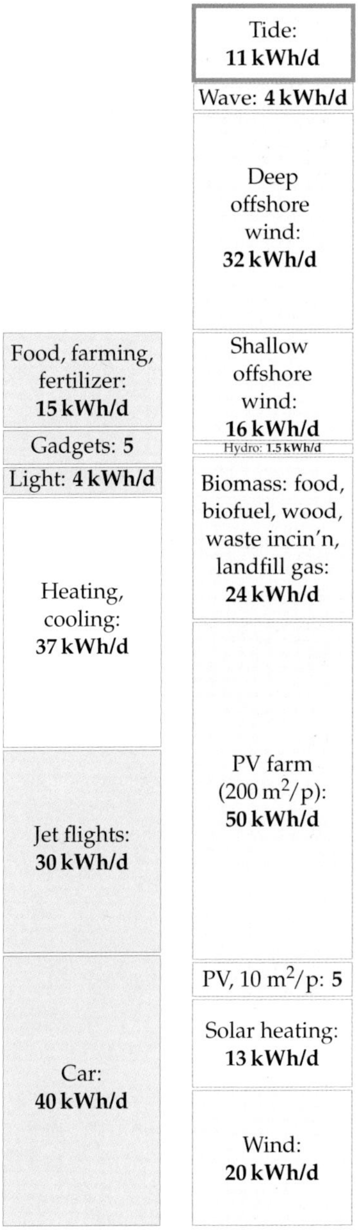

		Tide: **11 kWh/d**
		Wave: **4 kWh/d**
	Deep offshore wind: **32 kWh/d**	
Food, farming, fertilizer: **15 kWh/d**	Shallow offshore wind: **16 kWh/d**	
Gadgets: **5**	Hydro: **1.5 kWh/d**	
Light: **4 kWh/d**	Biomass: food, biofuel, wood, waste incin'n, landfill gas: **24 kWh/d**	
Heating, cooling: **37 kWh/d**		
	PV farm (200 m^2/p): **50 kWh/d**	
Jet flights: **30 kWh/d**		
	PV, 10 m^2/p: **5**	
	Solar heating: **13 kWh/d**	
Car: **40 kWh/d**	Wind: **20 kWh/d**	

Figure 14.10. Tide.

15 Stuff

One of the main sinks of energy in the "developed" world is the creation of stuff. In its natural life cycle, stuff passes through three stages. First, a new-born stuff is displayed in shiny packaging on a shelf in a shop. At this stage, stuff is called "goods." As soon as the stuff is taken home and sheds its packaging, it undergoes a transformation from "goods" to its second form, "clutter." The clutter lives with its owner for a period of months or years. During this period, the clutter is largely ignored by its owner, who is off at the shops buying more goods. Eventually, by a miracle of modern alchemy, the clutter is transformed into its final form, rubbish. To the untrained eye, it can be difficult to distinguish this "rubbish" from the highly desirable "good" that it used to be. Nonetheless, at this stage the discerning owner pays the dustman to transport the stuff away.

Let's say we want to understand the full energy-cost of a stuff, perhaps with a view to designing better stuff. This is called life-cycle analysis. It's conventional to chop the energy-cost of anything from a hair-dryer to a cruise-ship into four chunks:

Phase R: Making **raw materials**. This phase involves digging minerals out of the ground, melting them, purifying them, and modifying them into manufacturers' lego: plastics, glasses, metals, and ceramics, for example. The energy costs of this phase include the transportation costs of trundling the raw materials to their next destination.

Phase P: Production. In this phase, the raw materials are processed into a manufactured product. The factory where the hair-dryer's coils are wound, its graceful lines moulded, and its components carefully snapped together, uses heat and light. The energy costs of this phase include packaging and more transportation.

Phase U: Use. Hair-dryers and cruise-ships both guzzle energy when they're used as intended.

Phase D: Disposal. This phase includes the energy cost of putting the stuff back in a hole in the ground (landfill), or of turning the stuff back into raw materials (recycling); and of cleaning up all the pollution associated with the stuff.

To understand how much energy a stuff's life requires, we should estimate the energy costs of all four phases and add them up. Usually one of these four phases dominates the total energy cost, so to get a reasonable estimate of the total energy cost we need accurate estimates only of the cost of that dominant phase. If we wish to redesign a stuff so as to reduce its total energy cost, we should usually focus on reducing the cost of the dominant phase, while making sure that energy-savings in that phase

Figure 15.1. Selfridges' rubbish advertisement.

	embodied energy (kWh per kg)
fossil fuel	10
wood	5
paper	10
glass	7
PET plastic	30
aluminium	40
steel	6

Table 15.2. Embodied energy of materials.

aren't being undone by accompanying increases in the energy costs of the other three phases.

Rather than estimating in detail how much power the perpetual production and transport of all stuff requires, let's first cover just a few common examples: drink containers, computers, batteries, junk mail, cars, and houses. This chapter focuses on the energy costs of phases R and P. These energy costs are sometimes called the "embodied" or "embedded" energy of the stuff – slightly confusing names, since usually that energy is neither literally embodied nor embedded in the stuff.

Drink containers

Let's assume you have a coke habit: you drink five cans of multinational chemicals per day, and throw the aluminium cans away. For this stuff, it's the raw material phase that dominates. The production of metals is energy intensive, especially for aluminium. Making one aluminium drinks-can needs 0.6 kWh. So a five-a-day habit wastes energy at a rate of 3 kWh/d.

As for a 500 ml water bottle made of PET (which weighs 25 g), the embodied energy is 0.7 kWh – just as bad as an aluminium can!

Other packaging

The average Brit throws away 400 g of packaging per day – mainly food packaging. The embodied energy content of packaging ranges from 7 to 20 kWh per kg as we run through the spectrum from glass and paper to plastics and steel cans. Taking the typical embodied energy content to be 10 kWh/kg, we deduce that the energy footprint of packaging is 4 kWh/d. A little of this embodied energy is recoverable by waste incineration, as we'll discuss in Chapter 27.

Computers

Making a personal computer costs 1800 kWh of energy. So if you buy a new computer every two years, that corresponds to a power consumption of 2.5 kWh per day.

Batteries

The energy cost of making a rechargeable nickel-cadmium AA battery, storing 0.001 kWh of electrical energy and having a mass of 25 g, is 1.4 kWh (phases R and P). If the energy cost of disposable batteries is similar, throwing away two AA batteries per month uses about 0.1 kWh/d. The energy cost of batteries is thus likely to be a minor item in your stack of energy consumption.

Aluminium: **3 kWh/d**

Packaging:
4 kWh/d

Figure 15.3. Five aluminium cans per day is 3 kWh/d. The embodied energy in other packaging chucked away by the average Brit is 4 kWh/d.

Chips: **2.5 kWh/d**

Figure 15.4. She's making chips. Photo: ABB.
Making one personal computer every two years costs 2.5 kWh per day.

Newspapers, magazines, and junk mail

A 36-page newspaper, distributed for free at railway stations, weighs 90 g. The Cambridge Weekly News (56 pages) weighs 150 g. *The Independent* (56 pages) weighs 200 g. A 56-page property-advertising glossy magazine and Cambridgeshire Pride Magazine (32 pages), both delivered free at home, weigh 100 g and 125 g respectively.

This river of reading material and advertising junk pouring through our letterboxes contains energy. It also costs energy to make and deliver. Paper has an embodied energy of 10 kWh per kg. So the energy embodied in a typical personal flow of junk mail, magazines, and newspapers, amounting to 200 g of paper per day (that's equivalent to one *Independent* per day for example) is about 2 kWh per day.

Paper recycling would save about half of the energy of manufacture; waste incineration or burning the paper in a home fire may make use of some of the contained energy.

Newspapers,
junk mail,
magazines:
2 kWh/d

Bigger stuff

The largest stuff most people buy is a house.

In Chapter H, I estimate the energy cost of making a new house. Assuming we replace each house every 100 years, the estimated energy cost is 2.3 kWh/d. This is the energy cost of creating the *shell* of the house only – the foundation, bricks, tiles, and roof beams. If the average house occupancy is 2.3, the average energy expenditure on house building is thus estimated to be 1 kWh per day per person.

What about a car, and a road? Some of us own the former, but we usually share the latter. A new car's embodied energy is 76 000 kWh – so if you get one every 15 years, that's an average energy cost of 14 kWh per day. A life-cycle analysis by Treloar, Love, and Crawford estimates that building an Australian road costs 7600 kWh per metre (a continuously reinforced concrete road), and that, including maintenance costs, the total cost over 40 years was 35 000 kWh per metre. Let's turn this into a ballpark figure for the energy cost of British roads. There are 28 000 miles of trunk roads and class-1 roads in Britain (excluding motorways). Assuming 35 000 kWh per metre per 40 years, those roads cost us 2 kWh/d per person.

House-building: **1 kWh/d**

Car-making:
14 kWh/d

Road-building: **2 kWh/d**

Transporting the stuff

Up till now I've tried to make estimates of *personal* consumption. "If you chuck away five coke-cans, that's 3 kWh; if you buy *The Independent*, that's 2 kWh." From here on, however, things are going to get a bit less personal. As we estimate the energy required to transport stuff around the country and around the planet, I'm going to look at national totals and divide them by the population.

Figure 15.5. Food-miles – Pasties, hand-made in Helston, Cornwall, shipped 580 km for consumption in Cambridge.

Cambridge

Helston

100 km

Freight transport is measured in ton-kilometres (t-km). If one ton of Cornish pasties are transported 580 km (figure 15.5) then we say 580 t-km of freight transport have been achieved. The energy intensity of road transport in the UK is about 1 kWh per t-km.

When the container ship in figure 15.6 transports 50 000 tons of cargo a distance of 10 000 km, it achieves 500 million t-km of freight transport. The energy intensity of freight transport by this container ship is 0.015 kWh per t-km. Notice how much more efficient transport by container-ship is than transport by road. These energy intensities are displayed in figure 15.8.

Transport of stuff by road

In 2006, the total amount of road transport in Britain by heavy goods vehicles was 156 billion t-km. Shared between 60 million, that comes to 7 t-km per day per person, which costs 7 kWh per day per person (assuming an energy intensity of 1 kWh per ton-km). One quarter of this transport, by the way, was of food, drink, and tobacco.

Transport by water

In 2002, 560 million tons of freight passed through British ports. The Tyndall Centre calculated that Britain's share of the energy cost of international shipping is 4 kWh/d per person.

Transport of water; taking the pee

Water's not a very glamorous stuff, but we use a lot of it – about 160 litres

Figure 15.6. The container ship *Ever Uberty* at Thamesport Container Terminal. Photo by Ian Boyle www.simplonpc.co.uk.

Road freight: **7 kWh/d**

Figure 15.7. The lorry delivereth and the lorry taketh away. Energy cost of UK road freight: 7 kWh/d per person.

Shipping: **4 kWh/d**

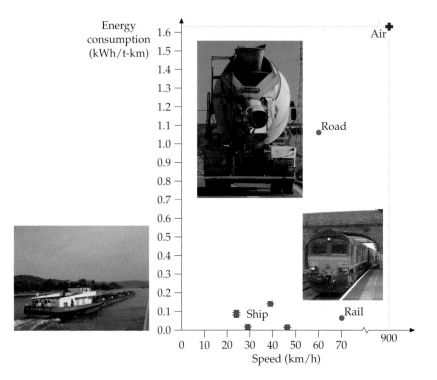

Figure 15.8. Energy requirements of
different forms of freight-transport.
The vertical coordinate shows the
energy consumed in kWh per net
ton-km, (that is, the energy per t-km
of freight moved, not including the
weight of the vehicle).
See also figure 20.23 (energy
requirements of passenger transport).

Water transport requires energy
because boats make waves.
Nevertheless, transporting freight by
ship is surprisingly energy efficient.

per day per person. In turn, we provide about 160 litres per day per person
of sewage to the water companies. The cost of pumping water around the
country and treating our sewage is about 0.4 kWh per day per person.

Desalination

At the moment the UK doesn't spend energy on water desalination. But
there's talk of creating desalination plants in London. What's the energy
cost of turning salt water into drinking water? The least energy-intensive
method is reverse osmosis. Take a membrane that lets through only water, put salt water on one side of it, and pressurize the salt water. Water
reluctantly oozes through the membrane, producing purer water – reluctantly, because pure water separated from salt has low entropy, and nature
prefers high entropy states where everything is mixed up. We must pay
high-grade energy to achieve unmixing.

   The Island of Jersey has a desalination plant that can produce 6000 m³
of pure water per day (figure 15.10). Including the pumps for bringing
the water up from the sea and through a series of filters, the whole plant
uses a power of 2 MW. That's an energy cost of 8 kWh per m³ of water
produced. At a cost of 8 kWh per m³, a daily water consumption of 160
litres would require 1.3 kWh per day.

Water delivery
and removal:
0.4 kWh/d

Figure 15.9. Water delivery:
0.3 kWh/d; sewage processing:
0.1 kWh/d.

Figure 15.10. Part of the reverse-osmosis facility at Jersey Water's desalination plant. The pump in the foreground, right, has a power of 355 kW and shoves seawater at a pressure of 65 bar into 39 spiral-wound membranes in the banks of blue horizontal tubes, left, delivering 1500 m^3 per day of clean water. The clean water from this facility has a total energy cost of 8 kWh per m^3.

Stuff retail

Supermarkets in the UK consume about 11 TWh of energy per year. Shared out equally between 60 million happy shoppers, that's a power of 0.5 kWh per day per person.

Supermarkets:
0.5 kWh/d

The significance of imported stuff

In standard accounts of "Britain's energy consumption" or "Britain's carbon footprint," imported goods are *not* counted. Britain used to make its own gizmos, and our per-capita footprint in 1910 was as big as America's is today. Now Britain doesn't manufacture so much (so our energy consumption and carbon emissions have dropped a bit), but we still love gizmos, and we get them made for us by other countries. Should we ignore the energy cost of making the gizmo, because it's imported? I don't think so. Dieter Helm and his colleagues in Oxford estimate that under a correct account, allowing for imports and exports, Britain's carbon footprint is nearly *doubled* from the official "11 tons CO_2e per person" to about 21 tons. This implies that the biggest item in the average British person's energy footprint is the energy cost of making imported stuff.

In Chapter H, I explore this idea further, by looking at the weight of Britain's imports. Leaving aside our imports of fuels, we import a little over 2 tons per person of stuff every year, of which about 1.3 tons per person are processed and manufactured stuff like vehicles, machinery, white goods, and electrical and electronic equipment. That's about 4 kg per day per person of processed stuff. Such goods are mainly made of materials whose production required at least 10 kWh of energy per kg of stuff. I thus estimate that this pile of cars, fridges, microwaves, computers, photocopiers and televisions has an embodied energy of at least 40 kWh per day per person.

To summarize all these forms of stuff and stuff-transport, I will put on the consumption stack 48 kWh per day per person for the making of stuff (made up of at least 40 for imports, 2 for a daily newspaper, 2 for road-making, 1 for house-making, and 3 for packaging); and another 12 kWh per day per person for the transport of the stuff by sea, by road, and by pipe, and the storing of food in supermarkets.

Work till you shop.

Traditional saying

Notes and further reading

page no.

89 *One aluminium drinks can costs 0.6 kWh.* The mass of one can is 15 g. Estimates of the total energy cost of aluminium manufacture vary from 60 MJ/kg to 300 MJ/kg. [yx7zm4], [r22oz], [yhrest]. The figure I used is from The Aluminum Association [y5as53]: 150 MJ per kg of aluminium (40 kWh/kg).

– *The embodied energy of a water bottle made of PET.* Source: Hammond and Jones (2006) – PET's embodied energy is 30 kWh per kg.

– *The average Brit throws away 400 g of packaging per day.* In 1995, Britain used 137 kg of packaging per person (Hird et al., 1999).

– *A personal computer costs 1800 kWh of energy.* Manufacture of a PC requires (in energy and raw materials) the equivalent of about 11 times its own weight of fossil fuels. Fridges require 1–2 times their weight. Cars require 1–2 times their weight. Williams (2004); Kuehr (2003).

– *. . . a rechargeable nickel-cadmium battery.* Source: Rydh and Karlström (2002).

– *. . . steel. . .* From Swedish Steel, "The consumption of coal and coke is 700 kg per ton of finished steel, equal to approximately 5320 kWh per ton of finished steel. The consumption of oil, LPG and electrical power is 710 kWh per ton finished product. Total [primary] energy consumption is thus approx. 6000 kWh per ton finished steel." (6 kWh per kg.) [y2ktgg]

90 *A new car's embodied energy is 76 000 kWh.* Source: Treloar et al. (2004). Burnham et al. (2007) give a lower figure: 30 500 kWh for the net life-cycle energy cost of a car. One reason for the difference may be that the latter life-cycle analysis assumes the vehicle is recycled, thus reducing the net materials cost.

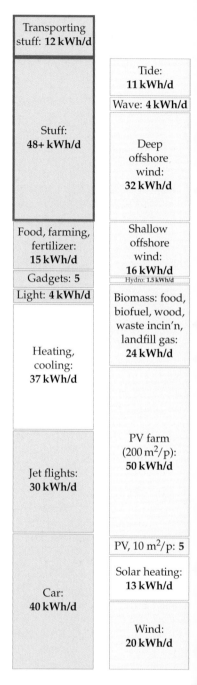

Figure 15.11. Making our stuff costs at least 48 kWh/d. Delivering the stuff costs 12 kWh/d.

90   *Paper has an embodied energy of 10 kWh per kg.* Making newspaper from virgin wood has an energy cost of about 5 kWh/kg, and the paper itself has an energy content similar to that of wood, about 5 kWh/kg. (Source: Ucuncu (1993); Erdincler and Vesilind (1993); see p284.) Energy costs vary between mills and between countries. 5 kWh/kg is the figure for a Swedish newspaper mill in 1973 from Norrström (1980), who estimated that efficiency measures could reduce the cost to about 3.2 kWh/kg. A more recent full life-cycle analysis (Denison, 1997) estimates the net energy cost of production of newsprint in the USA from virgin wood followed by a typical mix of landfilling and incineration to be 12 kWh/kg; the energy cost of producing newsprint from recycled material and recycling it is 6 kWh/kg.

91   *The energy intensity of road transport in the UK is about 1 kWh per t-km.* Source: `www.dft.gov.uk/pgr/statistics/datatablespublications/energyenvironment`.

–   *The energy intensity of freight transport by this container ship is 0.015 kWh per ton-km.* The *Ever Uberty* – length 285 m, breadth 40 m – has a capacity of 4948 TEUs, deadweight 63 000 t, and a service speed of 25 knots; its engine's normal delivered power is 44 MW. One TEU is the size of a small 20-foot container – about 40 m^3. Most containers you see today are 40-foot containers with a size of 2 TEU. A 40-foot container weighs 4 tons and can carry 26 tons of stuff. Assuming its engine is 50%-efficient, this ship's energy consumption works out to 0.015 kWh of chemical energy per ton-km. `www.mhi.co.jp/en/products/detail/container_ship_ever_uberty.html`

–   *Britain's share of international shipping...* Source: Anderson et al. (2006).

92   *Figure 15.8.* **Energy consumptions of ships.** The five points in the figure are a container ship (46 km/h), a dry cargo vessel (24 km/h), an oil tanker (29 km/h), an inland marine ship (24 km/h), and the NS Savannah (39 km/h).

Dry cargo vessel   0.08 kWh/t-km. A vessel with a grain capacity of 5200 m^3 carries 3360 deadweight tons. (Deadweight tonnage is the mass of cargo that the ship can carry.) It travels at speed 13 kn (24 km/h); its one engine with 2 MW delivered power consumes 186 g of fuel-oil per kWh of delivered energy (42% efficiency). `conoship.com/uk/vessels/detailed/page7.htm`

Oil tanker   A modern oil tanker uses 0.017 kWh/t-km [6lbrab]. Cargo weight 40 000 t. Capacity: 47 000 m^3. Main engine: 11.2 MW maximum delivered power. Speed at 8.2 MW: 15.5 kn (29 km/h). The energy contained in the oil cargo is 520 million kWh. So 1% of the energy in the oil is used in transporting the oil one-quarter of the way round the earth (10 000 km).

Roll-on, roll-off carriers   The ships of Wilh. Wilhelmsen shipping company deliver freight-transport with an energy cost between 0.028 and 0.05 kWh/t-km [5ctx4k].

92   *Water delivery and sewage treatment costs 0.4 kWh/d per person.* The total energy use of the water industry in 2005–6 was 7703 GWh. Supplying 1 m^3 of water has an energy cost of 0.59 kWh. Treating 1 m^3 of sewage has an energy cost of 0.63 kWh. For anyone interested in greenhouse-gas emissions, water supply has a footprint of 289 g CO_2 per m^3 of water delivered, and wastewater treatment, 406 g CO_2 per m^3 of wastewater.
Domestic water consumption is 151 litres per day per person. Total water consumption is 221 l/d per person. Leakage amounts to 57 litres per day per person. Sources: Parliamentary Office of Science and Technology [`www.parliament.uk/documents/upload/postpn282.pdf`], Water UK (2006).

93   *Supermarkets in the UK consume 11 TWh/y.* [yqbzl3]

–   *Helm et al. suggest that, allowing for imports and exports, Britain's carbon footprint is nearly **doubled** to about 21 tons.* Helm et al. (2007).

16 Geothermal

Geothermal energy comes from two sources: from radioactive decay in the crust of the earth, and from heat trickling through the mantle from the earth's core. The heat in the core is there because the earth used to be red-hot, and it's still cooling down and solidifying; the heat in the core is also being topped up by tidal friction: the earth flexes in response to the gravitational fields of the moon and sun, in the same way that an orange changes shape if you squeeze it and roll it between your hands.

Geothermal is an attractive renewable because it is "always on," independent of the weather; if we make geothermal power stations, we can switch them on and off so as to follow demand.

But how much geothermal power is available? We could estimate geothermal power of two types: the power available at an ordinary location on the earth's crust; and the power available in special hot spots like Iceland (figure 16.3). While the right place to first develop geothermal technology is definitely the special hot spots, I'm going to assume that the greater total resource comes from the ordinary locations, since ordinary locations are so much more numerous.

The difficulty with making *sustainable* geothermal power is that the speed at which heat travels through solid rock limits the rate at which heat can be sustainably sucked out of the red-hot interior of the earth. It's like trying to drink a crushed-ice drink through a straw. You stick in the straw, and suck, and you get a nice mouthful of cold liquid. But after a little more sucking, you find you're sucking air. You've extracted all the liquid from the ice around the tip of the straw. Your initial rate of sucking wasn't sustainable.

If you stick a straw down a 15-km hole in the earth, you'll find it's nice and hot there, easily hot enough to boil water. So, you could stick two straws down, and pump cold water down one straw and suck from the other. You'll be sucking up steam, and you can run a power station. Limitless power? No. After a while, your sucking of heat out of the rock will have reduced the temperature of the rock. You weren't sucking sustainably. You now have a long wait before the rock at the tip of your straws warms up again. A possible attitude to this problem is to treat geothermal heat the same way we currently treat fossil fuels: as a resource to be mined rather than collected sustainably. Living off geothermal heat in this way might be better for the planet than living unsustainably off fossil fuels; but perhaps it would only be another stop-gap giving us another 100 years of unsustainable living? In this book I'm most interested in *sustainable* energy, as the title hinted. Let's do the sums.

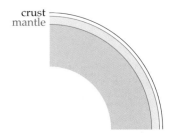

Figure 16.1. An earth in section.

Figure 16.2. Some granite.

Figure 16.3. Geothermal power in
Iceland. Average geothermal
electricity generation in Iceland
(population, 300 000) in 2006 was
300 MW (24 kWh/d per person).
More than half of Iceland's electricity
is used for aluminium production.
Photo by Gretar Ívarsson.

Geothermal power that would be sustainable forever

First imagine using geothermal energy sustainably by sticking down straws to an appropriate depth, and sucking *gently*. Sucking at such a rate that the rocks at the end of the our straws don't get colder and colder. This means sucking at the natural rate at which heat is already flowing out of the earth.

As I said before, geothermal energy comes from two sources: from radioactive decay in the crust of the earth, and from heat trickling through the mantle from the earth's core. In a typical continent, the heat flow from the centre coming through the mantle is about 10 mW/m². The heat flow at the surface is 50 mW/m². So the radioactive decay has added an extra 40 mW/m² to the heat flow from the centre.

So at a typical location, the maximum power we can get per unit area is 50 mW/m². But that power is not high-grade power, it's low-grade heat that's trickling through at the ambient temperature up here. We presumably want to make electricity, and that's why we must drill down. Heat is useful only if it comes from a source at a higher temperature than the ambient temperature. The temperature increases with depth as shown in figure 16.4, reaching a temperature of about 500 °C at a depth of 40 km. Between depths of 0 km where the heat flow is biggest but the rock temperature is too low, and 40 km, where the rocks are hottest but the heat flow is 5 times smaller (because we're missing out on all the heat generated from radioactive decay) there is an optimal depth at which we should suck. The exact optimal depth depends on what sort of sucking and power-station machinery we use. We can bound the maximum sustainable power

one milliwatt (1 mW) is 0.001 W.

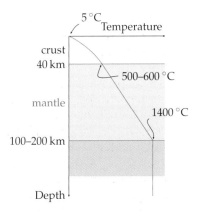

Figure 16.4. Temperature profile in a typical continent.

by finding the optimal depth assuming that we have an ideal engine for turning heat into electricity, and that drilling to any depth is free.

For the temperature profile shown in figure 16.4, I calculated that the optimal depth is about 15 km. Under these conditions, an ideal heat engine would deliver $17\,mW/m^2$. At the world population density of 43 people per square km, that's 10 kWh per person per day, if *all* land area were used. In the UK, the population density is 5 times greater, so wide-scale geothermal power of this sustainable-forever variety could offer at most 2 kWh per person per day.

This is the sustainable-forever figure, ignoring hot spots, assuming perfect power stations, assuming every square metre of continent is exploited, and assuming that drilling is free. And that it is possible to drill 15-km-deep holes.

Geothermal power as mining

The other geothermal strategy is to treat the heat as a resource to be mined. In "enhanced geothermal extraction" from hot dry rocks (figure 16.5), we first drill down to a depth of 5 or 10 km, and fracture the rocks by pumping in water. (This step may create earthquakes, which don't go down well with the locals.) Then we drill a second well into the fracture zone. Then we pump water down one well and extract superheated water or steam from the other. This steam can be used to make electricity or to deliver heat. What's the hot dry rock resource of the UK? Sadly, Britain is not well endowed. Most of the hot rocks are concentrated in Cornwall, where some geothermal experiments were carried out in 1985 in a research facility at Rosemanowes, now closed. Consultants assessing these experiments concluded that "generation of electrical power from hot dry rock was unlikely to be technically or commercially viable in Cornwall, or elsewhere in the UK, in the short or medium term." Nonetheless, what is the resource? The biggest estimate of the hot dry rock resource in the UK is a total energy of 130 000 TWh, which, according to the consultants, could conceivably contribute 1.1 kWh per day per person of electricity for about 800 years.

Other places in the world have more promising hot dry rocks, so if you want to know the geothermal answers for other countries, be sure to ask a local. But sadly for Britain, geothermal will only ever play a tiny part.

Doesn't Southampton use geothermal energy already? How much does that deliver?

Yes, Southampton Geothermal District Heating Scheme was, in 2004 at least, the only geothermal heating scheme in the UK. It provides the city with a supply of hot water. The geothermal well is part of a combined heat, power, and cooling system that delivers hot and chilled water to customers, and sells electricity to the grid. Geothermal energy contributes about 15% of the 70 GWh of heat per year delivered by this system. The population

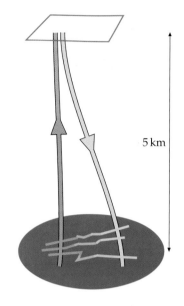

5 km

Figure 16.5. Enhanced geothermal extraction from hot dry rock. One well is drilled and pressurized to create fractures. A second well is drilled into the far side of the fracture zone. Then cold water is pumped down one well and heated water (indeed, steam) is sucked up the other.

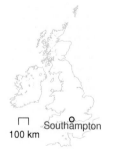

Southampton

100 km

of Southampton at the last census was 217 445, so the geothermal power being delivered there is **0.13 kWh/d** per person in Southampton.

Notes and further reading

page no.

97 *The heat flow at the surface is 50 mW/m². Massachusetts Institute of Technology (2006) says 59 mW/m² average, with a range, in the USA, from 25 mW to 150 mW. Shepherd (2003) gives 63 mW/m².*

98 *"Generation of electrical power from hot dry rock was unlikely to be technically or commercially viable in the UK".* Source: MacDonald et al. (1992). See also Richards et al. (1994).

– *The biggest estimate of the hot dry rock resource in the UK ... could conceivably contribute 1.1 kWh per day per person of electricity for about 800 years.* Source: MacDonald et al. (1992).

– *Other places in the world have more promising hot dry rocks.* There's a good study (Massachusetts Institute of Technology, 2006) describing the USA's hot dry rock resource. Another more speculative approach, researched by Sandia National Laboratories in the 1970s, is to drill all the way down to magma at temperatures of 600–1300 °C, perhaps 15 km deep, and get power there. The website www.magma-power.com reckons that the heat in pools of magma under the US would cover US energy consumption for 500 or 5000 years, and that it could be extracted economically.

– *Southampton Geothermal District Heating Scheme.* www.southampton.gov.uk.

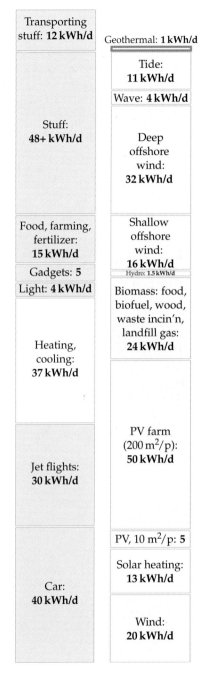

Figure 16.6. Geothermal.

17   Public services

Every gun that is made, every warship launched, every rocket fired signifies, in the final sense, a theft from those who hunger and are not fed, those who are cold and are not clothed.

This world in arms is not spending money alone. It is spending the sweat of its laborers, the genius of its scientists, the hopes of its children.

President Dwight D. Eisenhower – April, 1953

The energy cost of "defence"

Let's try to estimate how much energy we spend on our military.

In 2007–8, the fraction of British central government expenditure that went to defence was £33 billion/£587 billion = 6%. If we include the UK's spending on counter-terrorism and intelligence (£2.5 billion per year and rising), the total for defensive activities comes to £36 billion.

As a crude estimate we might guess that 6% of this £36 billion is spent on energy at a cost of 2.7p per kWh. (6% is the fraction of GDP that is spent on energy, and 2.7p is the average price of energy.) That works out to about 80 TWh per year of energy going into defence: making bullets, bombs, nuclear weapons; making devices for delivering bullets, bombs, and nuclear weapons; and roaring around keeping in trim for the next game of good-against-evil. In our favourite units, this corresponds to 4 kWh per day per person.

The cost of nuclear defence

The financial expenditure by the USA on manufacturing and deploying nuclear weapons from 1945 to 1996 was $5.5 trillion (in 1996 dollars).

Nuclear-weapons spending over this period exceeded the combined total federal spending for education; agriculture; training, employment, and social services; natural resources and the environment; general science, space, and technology; community and regional development (including disaster relief); law enforcement; and energy production and regulation.

If again we assume that 6% of this expenditure went to energy at a cost of 5¢ per kWh, we find that the energy cost of having nuclear weapons was 26 000 kWh per American, or 1.4 kWh per day per American (shared among 250 million Americans over 51 years).

What energy would have been delivered to the lucky recipients, had all those nuclear weapons been used? The energies of the biggest thermonuclear weapons developed by the USA and USSR are measured in megatons of TNT. A ton of TNT is 1200 kWh. The bomb that destroyed Hiroshima had the energy of 15 000 tons of TNT (18 million kWh). A *megaton* bomb delivers an energy of 1.2 billion kWh. If dropped on a city of one million, a megaton bomb makes an energy donation of 1200 kWh per person, equivalent to 120 litres of petrol per person. The total energy of the USA's nuclear arsenal today is 2400 megatons, contained in 10 000 warheads. In the good old days when folks really took defence seriously, the arsenal's energy was 20 000 megatons. These bombs, if used, would have delivered an energy of about 100 000 kWh per American. That's equivalent to 7 kWh per day per person for a duration of 40 years – similar to all the electrical energy supplied to America by nuclear power.

Energy cost of making nuclear materials for bombs

The main nuclear materials are plutonium, of which the USA has produced 104 t, and high-enriched uranium (HEU), of which the USA has produced 994 t. Manufacturing these materials requires energy.

The most efficient plutonium-production facilities use 24 000 kWh of heat when producing 1 gram of plutonium. So the direct energy-cost of making the USA's 104 tons of plutonium (1945–1996) was at least 2.5 trillion kWh which is 0.5 kWh per day per person (if shared between 250 million Americans).

The main energy-cost in manufacturing HEU is the cost of enrichment. Work is required to separate the ^{235}U and ^{238}U atoms in natural uranium in order to create a final product that is richer in ^{235}U. The USA's production of 994 tons of highly-enriched uranium (the USA's total, 1945–1996) had an energy cost of about 0.1 kWh per day per person.

> *"Trident creates jobs."* Well, so does relining our schools with asbestos, but that doesn't mean we should do it!
>
> Marcus Brigstocke

Universities

According to Times Higher Education Supplement (30 March 2007), UK universities use 5.2 billion kWh per year. Shared out among the whole population, that's a power of 0.24 kWh per day per person.

So higher education and research seem to have a much lower energy cost than defensive war-gaming.

There may be other energy-consuming public services we could talk about, but at this point I'd like to wrap up our race between the red and green stacks.

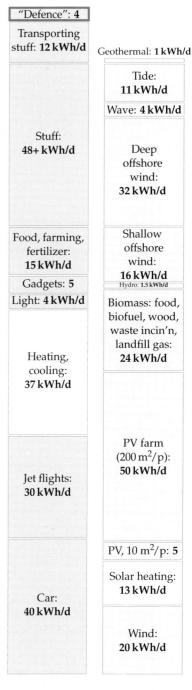

Figure 17.1. The energy cost of defence in the UK is estimated to be about 4 kWh per day per person.

Notes and further reading

page no.

100 *military energy budget.* The UK budget can be found at [yttg7p]; defence gets £33.4 billion [fcqfw] and intelligence and counter-terrorism £2.5 billion per year [2e4fcs]. According to p14 of the Government's Expenditure Plans 2007/08 [33x5kc], the "total resource budget" of the Department of Defence is a bigger sum, £39 billion, of which £33.5 billion goes for "provision of defence capability" and £6 billion for armed forces pay and pensions and war pensions. A breakdown of this budget can be found here: [35ab2c]. See also [yg5fsj], [yfgjna], and www.conscienceonline.org.uk.

The US military's energy consumption is published: "The Department of Defense is the largest single consumer of energy in the United States. In 2006, it spent $13.6 billion to buy 110 million barrels of petroleum fuel [roughly 190 billion kWh] and 3.8 billion kWh of electricity" (Dept. of Defense, 2008). This figure describes the direct use of fuel and electricity and doesn't include the embodied energy in the military's toys. Dividing by the US population of 300 million, it comes to 1.7 kWh/d per person.

– *The financial expenditure by the USA on manufacturing and deploying nuclear weapons from 1945 to 1996 was $5.5 trillion (in 1996 dollars).* Source: Schwartz (1998).

101 *Energy cost of plutonium production.* [slbae].

– *The USA's production of 994 tons of HEU…* Material enriched to between 4% and 5% ^{235}U is called low-enriched uranium (LEU). 90%-enriched uranium is called high-enriched uranium (HEU). It takes three times as much work to enrich uranium from its natural state to 5% LEU as it does to enrich LEU to 90% HEU. The nuclear power industry measures these energy requirements in a unit called the separative work unit (SWU). To produce a kilogram of ^{235}U as HEU takes 232 SWU. To make 1 kg of ^{235}U as LEU (in 22.7 kg of LEU) takes about 151 SWU. In both cases one starts from natural uranium (0.71% ^{235}U) and discards depleted uranium containing 0.25% ^{235}U.

The commercial nuclear fuel market values an SWU at about $100. It takes about 100 000 SWU of enriched uranium to fuel a typical 1000 MW commercial nuclear reactor for a year. Two uranium enrichment methods are currently in commercial use: gaseous diffusion and gas centrifuge. The gaseous diffusion process consumes about 2500 kWh per SWU, while modern gas centrifuge plants require only about 50 kWh per SWU. [yh45h8], [t2948], [2ywzee]. A modern centrifuge produces about 3 SWU per year.

The USA's production of 994 tons of highly-enriched uranium (the USA's total, 1945–1996) cost 230 million SWU, which works out to 0.1 kWh/d per person (assuming 250 million Americans, and using 2500 kWh/SWU as the cost of diffusion enrichment).

18 Can we live on renewables?

The red stack in figure 18.1 adds up to 195 kWh per day per person. The green stack adds up to about 180 kWh/d/p. A close race! But please remember: in calculating our production stack we threw all economic, social, and environmental constraints to the wind. Also, some of our green contributors are probably incompatible with each other: our photovoltaic panels and hot-water panels would clash with each other on roofs; and our solar photovoltaic farms using 5% of the country might compete with the energy crops with which we covered 75% of the country. If we were to lose just one of our bigger green contributors – for example, if we decided that deep offshore wind is not an option, or that panelling 5% of the country with photovoltaics at a cost of £200 000 per person is not on – then the production stack would no longer match the consumption stack.

Furthermore, even if our red consumption stack were lower than our green production stack, it would not necessarily mean our energy sums are adding up. You can't power a TV with cat food, nor can you feed a cat from a wind turbine. Energy exists in different forms – chemical, electrical, kinetic, and heat, for example. For a sustainable energy plan to add up, we need both the forms and amounts of energy consumption and production to match up. Converting energy from one form to another – from chemical to electrical, as at a fossil-fuel power station, or from electrical to chemical, as in a factory making hydrogen from water – usually involves substantial losses of useful energy. We will come back to this important detail in Chapter 27, which will describe some energy plans that do add up.

Here we'll reflect on our estimates of consumption and production, compare them with official averages and with other people's estimates, and discuss how much power renewables could plausibly deliver in a country like Britain.

The questions we'll address in this chapter are:

1. Is the size of the red stack roughly correct? What is the *average* consumption of Britain? We'll look at the official energy-consumption numbers for Britain and a few other countries.

2. Have I been unfair to renewables, underestimating their potential? We'll compare the estimates in the green stack with estimates published by organizations such as the Sustainable Development Commission, the Institution of Electrical Engineers, and the Centre for Alternative Technology.

3. What happens to the green stack when we take into account social and economic constraints?

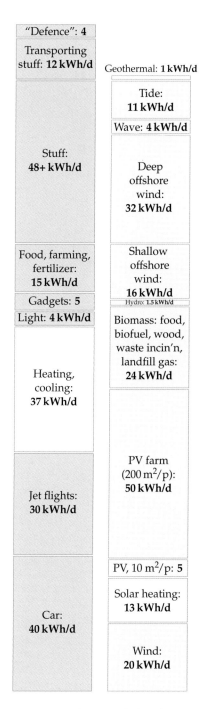

Figure 18.1. The state of play after we added up all the traditional renewables.

Red reflections

Our estimate of a typical affluent person's consumption (figure 18.1) has reached 195 kWh per day. It is indeed true that many people use this much energy, and that many more aspire to such levels of consumption. The *average* American consumes about **250 kWh per day**. If we all raised our standard of consumption to an average American level, the green production stack would definitely be dwarfed by the red consumption stack.

What about the average European and the average Brit? Average European consumption of "primary energy" (which means the energy contained in raw fuels, plus wind and hydroelectricity) is about 125 kWh per day per person. The UK average is also 125 kWh per day per person.

These official averages do not include two energy flows. First, the "embedded energy" in *imported* stuff (the energy expended in making the stuff) is not included at all. We estimated in Chapter 15 that the embedded energy in imported stuff is at least 40 kWh/d per person. Second, the official estimates of "primary energy consumption" include only industrial energy flows – things like fossil fuels and hydroelectricity – and don't keep track of the natural embedded energy in food: energy that was originally harnessed by photosynthesis.

Another difference between the red stack we slapped together and the national total is that in most of the consumption chapters so far we tended to ignore the energy lost in converting energy from one form to another, and in transporting energy around. For example, the "car" estimate in Part I covered only the energy in the petrol, not the energy used at the oil refinery that makes the petrol, nor the energy used in trundling the oil and petrol from A to B. The national total accounts for all the energy, before any conversion losses. Conversion losses in fact account for about 22% of total national energy consumption. Most of these conversion losses happen at power stations. Losses in the electricity transmission network chuck away 1% of total national energy consumption.

When building our red stack, we tried to imagine how much energy a typical affluent person uses. Has this approach biased our perception of the importance of different activities? Let's look at some official numbers. Figure 18.2 shows the breakdown of energy consumption by end use. The top two categories are transport and heating (hot air and hot water). Those two categories also dominated the red stack in Part I. Good.

Figure 18.2. Energy consumption, broken down by end use, according to the Department for Trade and Industry.

Road transport	Petroleum	22.5
Railways	Petroleum	0.4
Water transport	Petroleum	1.0
Aviation	Petroleum	7.4
All modes	Electricity	0.4
All energy used by transport		31.6

Table 18.3. 2006 breakdown of energy consumption by transport mode, in kWh/d per person.
Source: Dept. for Transport (2007).

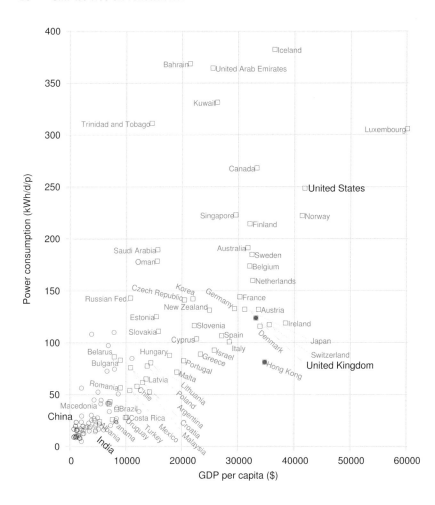

Figure 18.4. Power consumption per capita, versus GDP per capita, in purchasing-power-parity US dollars. Squares show countries having "high human development;" circles, "medium" or "low." Figure 30.1 (p231) shows the same data on logarithmic scales.

Let's look more closely at transport. In our red stack, we found that the energy footprints of driving a car 50 km per day and of flying to Cape Town once per year are roughly equal. Table 18.3 shows the relative importances of the different transport modes in the national balance-sheet. In the national averages, aviation is smaller than road transport.

How do Britain's official consumption figures compare with those of other countries? Figure 18.4 shows the power consumptions of lots of countries or regions, versus their gross domestic products (GDPs). There's an evident correlation between power consumption and GDP: the higher a country's GDP (per capita), the more power it consumes per capita. The UK is a fairly typical high-GDP country, surrounded by Germany, France, Japan, Austria, Ireland, Switzerland, and Denmark. The only notable exception to the rule "big GDP implies big power consumption" is Hong Kong. Hong Kong's GDP per capita is about the same as Britain's, but

Figure 18.5. Hong Kong. Photo by Samuel Louie and Carol Spears.

Hong Kong's power consumption is about $80\,\text{kWh/d/p}$.

The message I take from these country comparisons is that the UK is a fairly typical European country, and therefore provides a good case study for asking the question "How can a country with a high quality of life get its energy sustainably?"

Green reflections

People often say that Britain has plenty of renewables. Have I been mean to green? Are my numbers a load of rubbish? Have I underestimated sustainable production? Let's compare my green numbers first with several estimates found in the Sustainable Development Commission's publication *The role of nuclear power in a low carbon economy. Reducing CO$_2$ emissions – nuclear and the alternatives.* Remarkably, even though the Sustainable Development Commission's take on sustainable resources is very positive ("We have huge tidal, wave, biomass and solar resources"), *all the estimates in the Sustainable Development Commission's document are smaller than mine!* (To be precise, all the estimates of the renewables total are smaller than my total.) The Sustainable Development Commission's publication gives estimates from four sources detailed below (IEE, Tyndall, IAG, and PIU). Figure 18.6 shows my estimates alongside numbers from these four sources and numbers from the Centre for Alternative Technology (CAT). Here's a description of each source.

IEE The Institute of Electrical Engineers published a report on renewable energy in 2002 – a summary of possible contributions from renewables in the UK. The second column of figure 18.6 shows the "technical potential" of a variety of renewable technologies for UK electricity generation – "an upper limit that is unlikely ever to be exceeded even with quite dramatic changes in the structure of our society and economy." According to the IEE, the total of all renewables' technical potential is about $27\,\text{kWh/d}$ per person.

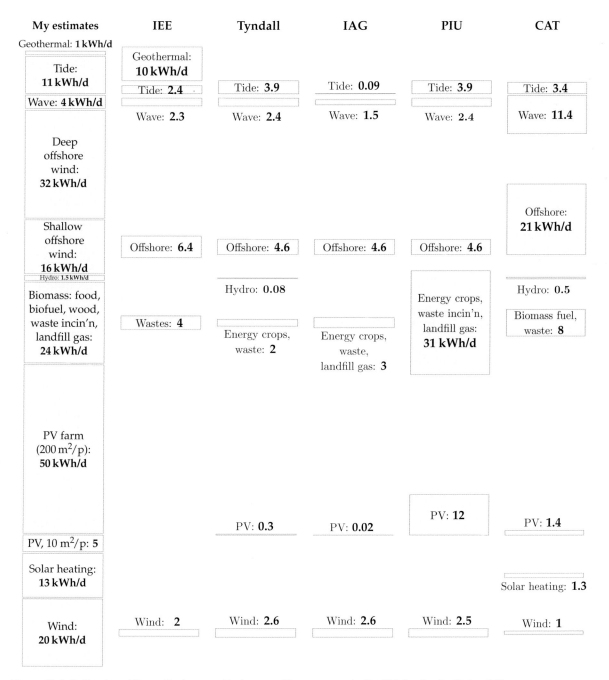

Figure 18.6. Estimates of theoretical or practical renewable resources in the UK, by the Institute of Electrical Engineers, the Tyndall Centre, the Interdepartmental Analysts Group, and the Performance and Innovation Unit; and the proposals from the Centre for Alternative Technology's "Island Britain" plan for 2027.

Figure 18.8. Where the wild things are. One of the grounds for objecting to wind farms is the noise they produce. I've chopped out of this map of the British mainland a 2-km-radius exclusion zone surrounding every hamlet, village, and town. These white areas would presumably be excluded from wind-farm development. The remaining black areas would perhaps also be largely excluded because of the need to protect tranquil places from industrialization. Settlement data from www.openstreetmap.org.

10 km

More forestry?	"No, it ruins the countryside."
Waste incineration?	"No, I'm worried about health risks, traffic congestion, dust and noise."
Hydroelectricity?	"Yes, but not *big* hydro – that harms the environment."
Offshore wind?	"No, I'm more worried about the ugly powerlines coming ashore than I was about a Nazi invasion."

Wave or geothermal power?   "No, far too expensive."

After all these objections, I fear that the maximum Britain would ever get from renewables would be something like what's shown in the bottom right of figure 18.7.

Figure 18.8 offers guidance to anyone trying to erect wind farms in Britain. On a map of the British mainland I've shown in white a 2-km-radius exclusion zone surrounding every hamlet, village, and town. These white areas would presumably be excluded from wind-farm development because they are too close to the humans. I've coloured in black all regions

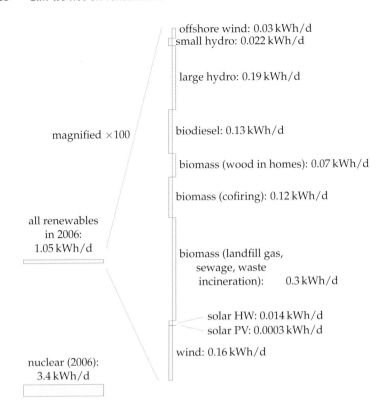

offshore wind: 0.03 kWh/d
small hydro: 0.022 kWh/d

large hydro: 0.19 kWh/d

biodiesel: 0.13 kWh/d

biomass (wood in homes): 0.07 kWh/d

biomass (cofiring): 0.12 kWh/d

biomass (landfill gas,
sewage, waste
incineration):   0.3 kWh/d

solar HW: 0.014 kWh/d
solar PV: 0.0003 kWh/d

wind: 0.16 kWh/d

magnified ×100

all renewables
in 2006:
1.05 kWh/d

nuclear (2006):
3.4 kWh/d

Figure 18.9. Production of renewables and nuclear energy in the UK in 2006. All powers are expressed per-person, as usual. The breakdown of the renewables on the right hand side is scaled up 100-fold vertically.

that are *more than 2 km* from any human settlement. These areas are largely excluded from wind-farm development because they are *tranquil*, and it's essential to protect tranquil places from industrialization. If you want to avoid objections to your wind farm, pick any piece of land that is not coloured black or white.

> *Some of these environmentalists who have good hearts but confused minds are almost a barrier to tackling climate change.*
>
> Malcolm Wicks, Minister of State for Energy

We are drawing to the close of Part I. The assumption was that we want to get off fossil fuels, for one or more of the reasons listed in Chapter 1 – climate change, security of supply, and so forth. Figure 18.9 shows how much power we currently get from renewables and nuclear. They amount to just 4% of our total power consumption.

The two conclusions we can draw from Part I are:

1. *To make a difference, renewable facilities have to be country-sized.*

   For any renewable facility to make a contribution comparable to our current consumption, *it has to be country-sized.* To get a big contribution from wind, we used wind farms with the area of Wales. To get a big contribution from solar photovoltaics, we required half the area of Wales. To get a big contribution from waves, we imagined wave farms covering 500 km of coastline. To make energy crops with a big contribution, we took 75% of the whole country.

Renewable facilities have to be country-sized because all renewables are so diffuse. Table 18.10 summarizes most of the powers-per-unit-area that we encountered in Part I.

To sustain Britain's lifestyle on its renewables alone would be very difficult. A renewable-based energy solution will necessarily be large and intrusive.

2. *It's not going to be easy* to make a plan that adds up using renewables alone. If we are serious about getting off fossil fuels, Brits are going to have to learn to start saying "yes" to something. Indeed to several somethings.

   In Part II I'll ask, "assuming that we can't get production from renewables to add up to our current consumption, what are the other options?"

Power per unit land or water area	
Wind	2 W/m^2
Offshore wind	3 W/m^2
Tidal pools	3 W/m^2
Tidal stream	6 W/m^2
Solar PV panels	5–20 W/m^2
Plants	0.5 W/m^2
Rain-water (highlands)	0.24 W/m^2
Hydroelectric facility	11 W/m^2
Geothermal	0.017 W/m^2

Table 18.10. Renewable facilities have to be country-sized because all renewables are so diffuse.

Notes and further reading

page no.

104 *UK average energy consumption is 125 kWh per day per person.* I took this number from the UNDP Human Development Report, 2007.
The DTI (now known as DBERR) publishes a Digest of United Kingdom Energy Statistics every year. [uzek2]. In 2006, according to DUKES, total primary energy demand was 244 million tons of oil equivalent, which corresponds to 130 kWh per day per person.
I don't know the reason for the small difference between the UNDP number and the DUKES number, but I can explain why I chose the slightly lower number. As I mentioned on p27, DUKES uses the same energy-summing convention as me, declaring one kWh of chemical energy to be equal to one kWh of electricity. But there's one minor exception: DUKES defines the "primary energy" produced in nuclear power stations to be the thermal energy, which in 2006 was 9 kWh/d/p; this was converted (with 38% efficiency) to 3.4 kWh/d/p of supplied electricity; in my accounts, I've focused on the electricity produced by hydroelectricity, other renewables, and nuclear power; this small switch in convention reduces the nuclear contribution by about 5 kWh/d/p.

– *Losses in the electricity transmission network chuck away 1% of total national energy consumption.* To put it another way, the losses are 8% of the electricity generated. This 8% loss can be broken down: roughly 1.5% is lost in the long-distance high-voltage system, and 6% in the local public supply system. Source: MacLeay et al. (2007).

105 *Figure 18.4.* Data from UNDP Human Development Report, 2007. [3av4s9]

108 *In the Middle Ages, the average person's lifestyle consumed a power of 20 kWh per day.* Source: Malanima (2006).

110 *"I'm more worried about the ugly powerlines coming ashore than I was about a Nazi invasion."* Source: [6frj55].

Part II

Making a difference

19 Every BIG helps

We've established that the UK's present lifestyle can't be sustained on the UK's own renewables (except with the industrialization of country-sized areas of land and sea). So, what are our options, if we wish to get off fossil fuels and live sustainably? We can balance the energy budget either by reducing demand, or by increasing supply, or, of course, by doing both.

Have no illusions. To achieve our goal of getting off fossil fuels, these reductions in demand and increases in supply must be *big*. Don't be distracted by the myth that "every little helps." *If everyone does a little, we'll achieve only a little.* We must do a lot. What's required are *big* changes in demand and in supply.

"But surely, if 60 million people all do a little, it'll add up to a lot?" No. This "if-everyone" multiplying machine is just a way of making something small *sound* big. The "if-everyone" multiplying machine churns out inspirational statements of the form "if *everyone* did X, then it would provide enough energy/water/gas to do Y," where Y sounds impressive. Is it surprising that Y sounds big? Of course not. We got Y by multiplying X by the number of people involved – 60 million or so! Here's an example from the Conservative Party's otherwise straight-talking *Blueprint for a Green Economy*:

> "The mobile phone charger averages around . . . 1 W consumption, but if every one of the country's 25 million mobile phones chargers were left plugged in and switched on they would consume enough electricity (219 GWh) to power 66 000 homes for one year."

66 000? Wow, what a lot of homes! Switch off the chargers! 66 000 sounds a lot, but the sensible thing to compare it with is the total number of homes that we're imagining would participate in this feat of conservation, namely *25 million* homes. 66 000 is just *one quarter of one percent* of 25 million. So while the statement quoted above is true, I think a calmer way to put it is:

> If you leave your mobile phone charger plugged in, it uses one quarter of one percent of your home's electricity.

And if everyone does it?

> If *everyone* leaves their mobile phone charger plugged in, those chargers will use one quarter of one percent of their homes' electricity.

The "if-everyone" multiplying machine is a bad thing because it deflects people's attention towards 25 million minnows instead of 25 million sharks. The mantra *"Little changes can make a big difference"* is bunkum, when applied to climate change and power. It may be true that "many people doing

"We were going to have a wind turbine but they're not very efficient"

Figure 19.1. Reproduced by kind permission of PRIVATE EYE / Robert Thompson www.private-eye.co.uk.

a little adds up to a lot," if all those "littles" are somehow focused into a single "lot" – for example, if one million people donate £10 to *one* accident-victim, then the victim receives £10 million. That's a lot. But power is a very different thing. We all use power. So to achieve a "big difference" in total power consumption, you need almost everyone to make a "big" difference to their own power consumption.

So, what's required are *big* changes in demand and in supply. Demand for power could be reduced in three ways:

1. by reducing our population (figure 19.2);

2. by changing our lifestyle;

3. by keeping our lifestyle, but reducing its energy intensity through "efficiency" and "technology."

Supply could be increased in three ways:

1. We could get off fossil fuels by investing in "clean coal" technology. Oops! Coal is a fossil fuel. Well, never mind – let's take a look at this idea. If we used coal "sustainably" (a notion we'll define in a moment), how much power could it offer? If we don't care about sustainability and just want "security of supply," could coal offer that?

2. We could invest in nuclear fission. Is current nuclear technology "sustainable"? Is it at least a stop-gap that might last for 100 years?

3. We could buy, beg, or steal renewable energy from other countries – bearing in mind that most countries will be in the same boat as Britain and will have no renewable energy to spare; and also bearing in mind that sourcing renewable energy from another country doesn't magically shrink the renewable power facilities required. If we import renewable energy from other countries in order to avoid building renewable facilities the size of Wales in *our* country, someone will have to build facilities roughly the size of Wales in those other countries.

The next seven chapters discuss first how to reduce demand substantially, and second how to increase supply to meet that reduced, but still "huge," demand. In these chapters, I won't mention *all* the good ideas. I'll discuss just the *big* ideas.

Cartoon Britain

To simplify and streamline our discussion of demand reduction, I propose to work with a cartoon of British energy consumption, omitting lots of details in order to focus on the big picture. My cartoon-Britain consumes

While the footprint of each individual cannot be reduced to zero, the absence of an individual does do so.

Chris Rapley, former Director of the British Antarctic Survey

We need fewer people, not greener ones.

Daily Telegraph, 24 July 2007

Democracy cannot survive overpopulation. Human dignity cannot survive overpopulation.

Isaac Asimov

Figure 19.2. Population growth and emissions... Cartoon courtesy of Colin Wheeler.

energy in just three forms: heating, transport, and electricity. The heating consumption of cartoon-Britain is 40 kWh per day per person (currently all supplied by fossil fuels); the transport consumption is also 40 kWh per day per person (currently all supplied by fossil fuels); and the electricity consumption is 18 kWh(e) per day per person; the electricity is currently almost all generated from fossil fuels; the conversion of fossil-fuel energy to electricity is 40% efficient, so supplying 18 kWh(e) of electricity in today's cartoon-Britain requires a fossil-fuel input of 45 kWh per day per person. This simplification ignores some fairly sizeable details, such as agriculture and industry, and the embodied energy of imported goods! But I'd like to be able to have a *quick* conversation about the main things we need to do to get off fossil fuels. Heating, transport, and electricity account for more than half of our energy consumption, so if we can come up with a plan that delivers heating, transport, and electricity sustainably, then we have made a good step on the way to a more detailed plan that adds up.

Having adopted this cartoon of Britain, our discussions of demand reduction will have just three bits. First, how can we reduce transport's energy-demand and eliminate all fossil fuel use for transport? This is the topic of Chapter 20. Second, how can we reduce heating's energy-demand and eliminate all fossil fuel use for heating? This is the topic of Chapter 21. Third, what about electricity? Chapter 22 discusses efficiency in electricity consumption.

Three supply options – clean coal, nuclear, and other people's renewables – are then discussed in Chapters 23, 24, and 25. Finally, Chapter 26 discusses how to cope with fluctuations in demand and fluctuations in renewable power production.

Having laid out the demand-reducing and supply-increasing options, Chapters 27 and 28 discuss various ways to put these options together to make plans that add up, in order to supply cartoon-Britain's transport, heating, and electricity.

I could spend many pages discussing "50 things you can do to make a difference," but I think this cartoon approach, chasing the three biggest fish, should lead to more effective policies.

But what about "stuff"? According to Part I, the embodied energy in imported stuff might be the biggest fish of all! Yes, perhaps that fish is the mammoth in the room. But let's leave defossilizing that mammoth to one side, and focus on the animals over which we have direct control.

So, here we go: let's talk about transport, heating, and electricity.

For the impatient reader

Are you eager to know the end of the story right away? Here is a quick summary, a sneak preview of Part II.

First, we electrify transport. Electrification both gets transport off fossil fuels, and makes transport more energy-efficient. (Of course, electrification

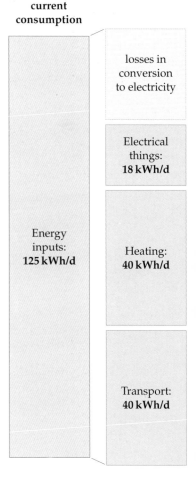

Figure 19.3. Current consumption in "cartoon-Britain 2008."

increases our demand for green electricity.)

Second, to supplement solar-thermal heating, we electrify most heating of air and water in buildings using *heat pumps*, which are four times more efficient than ordinary electrical heaters. This electrification of heating further increases the amount of green electricity required.

Third, we get all the green electricity from a mix of four sources: from our own renewables; perhaps from "clean coal;" perhaps from nuclear; and finally, and with great politeness, from other countries' renewables.

Among other countries' renewables, solar power in deserts is the most plentiful option. As long as we can build peaceful international collaborations, solar power in other people's deserts certainly has the technical potential to provide us, them, and everyone with 125 kWh per day per person.

Questions? Read on.

20 Better transport

Modern vehicle technology can reduce climate change emissions without changing the look, feel or performance that owners have come to expect.

California Air Resources Board

Roughly one third of our energy goes into transportation. Can *technology* deliver a reduction in consumption? In this chapter we explore options for achieving two goals: to deliver the biggest possible reduction in transport's energy use, *and* to eliminate fossil fuel use in transport.

Transport featured in three of our consumption chapters: Chapter 3 (cars), Chapter 5 (planes), and Chapter 15 (road freight and sea freight). So there are two sorts of transport to address: passenger transport, and freight. Our unit of passenger transport is the passenger-kilometre (p-km). If a car carries one person a distance of 100 km, it delivers 100 p-km of transportation. If it carries four people the same distance, it has delivered 400 p-km. Similarly our unit of freight transport is the ton-km (t-km). If a truck carries 5 t of cargo a distance of 100 km then it has delivered 500 t-km of freight-transport. We'll measure the energy consumption of passenger transport in "kWh per 100 passenger-kilometres," and the energy consumption of freight in "kWh per ton-km." Notice that these measures are the other way up compared to "miles per gallon": whereas we like vehicles to deliver *many* miles per gallon, we want energy-consumption to be *few* kWh per 100 p-km.

We'll start this chapter by discussing how to reduce the energy consumption of surface transport. To understand how to reduce energy consumption, we need to understand where the energy is going in surface transport. Here are the three key concepts, which are explained in more detail in Technical Chapter A.

1. In *short-distance travel* with lots of starting and stopping, the energy mainly goes into speeding up the vehicle and its contents. Key strategies for consuming less in this sort of transportation are therefore to *weigh less*, and to *go further between stops*. Regenerative braking, which captures energy when slowing down, may help too. In addition, it helps to *move slower*, and to *move less*.

2. In *long-distance travel* at steady speed, by train or automobile, most of the energy goes into making air swirl around, because you only have to accelerate the vehicle once. The key strategies for consuming less in this sort of transportation are therefore to *move slower*, and to *move less*, and to *use long, thin vehicles*.

3. In all forms of travel, there's an energy-conversion chain, which takes energy in some sort of fuel and uses some of it to push the vehicle

Figure 20.1. This chapter's starting point: an urban luxury tractor. The average UK car has a fuel consumption of 33 miles per gallon, which corresponds to an energy consumption of 80 kWh per 100 km. Can we do better?

forwards. Inevitably this energy chain has inefficiencies. In a standard fossil-fuel car, for example, only 25% is used for pushing, and roughly 75% of the energy is lost in making the engine and radiator hot. So a final strategy for consuming less energy is to make the energy-conversion chain more efficient.

These observations lead us to six principles of vehicle design and vehicle use for more-efficient surface transport: *a*) reduce the frontal area per person; *b*) reduce the vehicle's weight per person; *c*) when travelling, go at a steady speed and avoid using brakes; *d*) travel more slowly; *e*) travel less; and *f*) make the energy chain more efficient. We'll now discuss a variety of ways to apply these principles.

How to roll better

A widely quoted statistic says something along the lines of "only *1 percent* of the energy used by a car goes into moving the driver" – the implication being that, surely, by being a bit smarter, we could make cars *100* times more efficient? The answer is yes, almost, but only by applying the principles of vehicle design and vehicle use, listed above, to *extreme* degrees.

One illustration of extreme vehicle design is an eco-car, which has small frontal area and low weight, and – if any records are to be broken – is carefully driven at a low and steady speed. The *Team Crocodile* eco-car (figure 20.2) does 2184 miles per gallon (1.3 kWh per 100 km) at a speed of 15 mph (24 km/h). Weighing 50 kg and shorter in height than a traffic cone, it comfortably accommodates one teenage driver.

Hmm. I think that the driver of the urban tractor in figure 20.1 might detect a change in "look, feel and performance" if we switched them to the eco-car and instructed them to keep their speed below 15 miles per hour. So, the idea that cars could easily be 100 times more energy efficient is a myth. We'll come back to the challenge of making energy-efficient cars in a moment. But first, let's see some other ways of satisfying the principles of more-efficient surface transport.

Figure 20.3 shows a multi-passenger vehicle that is at least 25 times more energy-efficient than a standard petrol car: a bicycle. The bicycle's performance (in terms of energy per distance) is about the same as the eco-car's. Its speed is the same, its mass is lower than the eco-car's (because the human replaces the fuel tank and engine), and its effective frontal area is higher, because the cyclist is not so well streamlined as the eco-car.

Figure 20.4 shows another possible replacement for the petrol car: a train, with an energy-cost, if full, of 1.6 kWh per 100 passenger-km. In contrast to the eco-car and the bicycle, trains manage to achieve outstanding efficiency without travelling slowly, and without having a low weight per person. Trains make up for their high speed and heavy frame by exploiting the principle of small frontal area per person. Whereas a cyclist

Figure 20.2. Team Crocodile's eco-car uses 1.3 kWh per 100 km. Photo kindly provided by Team Crocodile. www.teamcrocodile.com

Figure 20.3. "Babies on board." This mode of transportation has an energy cost of 1 kWh per 100 person-km.

Figure 20.4. This 8-carriage train, at its maximum speed of 100 mph (161 km/h), consumes 1.6 kWh per 100 passenger-km, if full.

and a regular car have effective frontal areas of about 0.8 m² and 0.5 m² respectively, a full commuter train from Cambridge to London has a frontal area per passenger of 0.02 m².

But whoops, now we've broached an ugly topic – the prospect of sharing a vehicle with "all those horrible people." Well, squish aboard, and let's ask: How much could consumption be reduced by a switch from personal gas-guzzlers to excellent integrated public transport?

4.4 kWh per 100 p-km, if full

3–9 kWh per 100 seat-km, if full

7 kWh per 100 p-km, if full

21 kWh per 100 p-km, if full

Figure 20.5. Some public transports, and their energy-efficiencies, when on best behaviour.
Tubes, outer and inner.
Two high-speed trains. The electric one uses 3 kWh per 100 seat-km; the diesel, 9 kWh.
Trolleybuses in San Francisco.
Vancouver SeaBus. Photo by Larry.

Public transport

At its best, shared public transport is far more energy-efficient than individual car-driving. A diesel-powered **coach**, carrying 49 passengers and doing 10 miles per gallon at 65 miles per hour, uses 6 kWh per 100 p-km – 13 times better than the single-person car. Vancouver's **trolleybuses** consume 270 kWh per vehicle-km, and have an average speed of 15 km/h. If the trolleybus has 40 passengers on board, then its passenger transport cost is 7 kWh per 100 p-km. The Vancouver **SeaBus** has a transport cost of 83 kWh per vehicle-km at a speed of 13.5 km/h. It can seat 400 people, so its passenger transport cost when full is 21 kWh per 100 p-km. London **underground trains**, at peak times, use 4.4 kWh per 100 p-km – 18 times better than individual cars. Even **high-speed trains**, which violate two of our energy-saving principles by going twice as fast as the car and weighing a lot, are much more energy efficient: if the electric high-speed train is full, its energy cost is 3 kWh per 100 p-km – that's 27 times smaller than the car's!

However, we must be realistic in our planning. Some trains, coaches, and buses are not full (figure 20.6). So the *average* energy cost of public transport is bigger than the best-case figures just mentioned. What's the *average* energy-consumption of public transport systems, and what's a realistic appraisal of how good they could be?

In 2006–7, the total energy cost of all London's underground trains, including lighting, lifts, depots, and workshops, was 15 kWh per 100 p-km – five times better than our baseline car. In 2006–7 the energy cost of all London buses was 32 kWh per 100 p-km. Energy cost is not the only thing that matters, of course. Passengers care about speed: and the underground trains delivered higher speeds (an average of 33 km/h) than buses (18 km/h). Managers care about financial costs: the staff costs, per passenger-km, of underground trains are less than those of buses.

Figure 20.6. Some trains aren't full. Three men and a cello – the sole occupants of this carriage of the 10.30 high-speed train from Edinburgh to Kings Cross.

32 kWh per 100 p-km

9 kWh per 100 p-km

Figure 20.7. Some public transports, and their *average* energy consumptions. Left: Some red buses. Right: Croydon Tramlink. Photo by Stephen Parascandolo.

The total energy consumption of the Croydon Tramlink system (figure 20.7) in 2006–7 (including the tram depot and facilities at tram-stops) was 9 kWh per 100 p-km, with an average speed of 25 km/h.

How good could public transport be? Perhaps we can get a rough indication by looking at the data from Japan in table 20.8. At 19 kWh per 100 p-km and 6 kWh per 100 p-km, bus and rail both look promising. Rail has the nice advantage that it can solve both of our goals – reduction in energy consumption, and independence from fossil fuels. Buses and coaches have obvious advantages of simplicity and flexibility, but keeping this flexibility at the same time as getting buses and coaches to work without fossil fuels may be a challenge.

To summarise, public transport (especially electric trains, trams, and buses) seems a promising way to deliver passenger transportation – better in terms of energy per passenger-km, perhaps five or ten times better than cars. However, if people demand the flexibility of a private vehicle, what are our other options?

Energy consumption (kWh per 100 p-km)	
Car	68
Bus	19
Rail	6
Air	51
Sea	57

Table 20.8. Overall transport efficiencies of transport modes in Japan (1999).

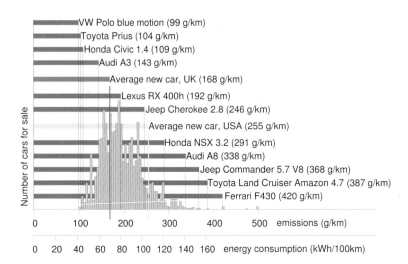

Figure 20.9. Carbon pollution, in grams CO_2 per km, of a selection of cars for sale in the UK. The horizontal axis shows the emission rate, and the height of the blue histogram indicates the number of models on sale with those emissions in 2006. Source: www.newcarnet.co.uk.
The second horizontal scale indicates approximate energy consumptions, assuming that $240\,g\,CO_2$ is associated with $1\,kWh$ of chemical energy.

Private vehicles: technology, legislation, and incentives

The energy consumption of individual cars *can* be reduced. The wide range of energy efficiencies of cars for sale proves this. In a single showroom in 2006 you could buy a Honda Civic 1.4 that uses roughly 44 kWh per 100 km, or a Honda NSX 3.2 that uses 116 kWh per 100 km (figure 20.9). The fact that people merrily *buy* from this wide range is also proof that we need extra incentives and legislation to encourage the blithe consumer to *choose* more energy-efficient cars. There are various ways to help consumers prefer the Honda Civic over the Honda NSX 3.2 gas-guzzler: raising the price of fuel; cranking up the showroom tax (the tax on new cars) in proportion to the predicted lifetime consumption of the vehicle; cranking up the road-tax on gas guzzlers; parking privileges for economical cars (figure 20.10); or fuel rationing. All such measures are unpopular with at least some voters. Perhaps a better legislative tactic would be to *enforce* reasonable energy-efficiency, rather than continuing to allow unconstrained choice; for example, we could simply *ban*, from a certain date, the sale of *any* car whose energy consumption is more than 80 kWh per 100 km; and then, over time, reduce this ceiling to 60 kWh per 100 km, then 40 kWh per 100 km, and beyond. Alternatively, to give the consumer more choice, regulations could force car manufacturers to reduce the *average* energy consumption of all the cars they sell. Additional legislation limiting the weight and frontal area of vehicles would simultaneously reduce fuel consumption and improve safety for other road-users (figure 20.11). People today choose their cars to make fashion statements. With strong efficiency legislation, there could still be a wide choice of fashions; they'd all just happen to be energy-efficient. You could choose any colour, as long as it was green.

Figure 20.10. Special parking privileges for electric cars in Ann Arbor, Michigan.

Figure 20.11. Monstercars are just tall enough to completely obscure the view and the visibility of pedestrians.

While we wait for the voters and politicians to agree to legislate for efficient cars, what other options are available?

Figure 20.12. A roundabout in Enschede, Netherlands.

Bikes

My favourite suggestion is the provision of excellent cycle facilities, along with appropriate legislation (lower speed-limits, and collision regulations that favour cyclists, for example). Figure 20.12 shows a roundabout in Enschede, Netherlands. There are two circles: the one for cars lies inside the one for bikes, with a comfortable car's length separating the two. The priority rules are the same as those of a British roundabout, except that cars exiting the central circle must give way to circulating cyclists (just as British cars give way to pedestrians on zebra crossings). Where excellent cycling facilities are provided, people will use them, as evidenced by the infinite number of cycles sitting outside the Enschede railway station (figure 20.13).

Figure 20.13. A few Dutch bikes.

Somehow, British cycle provision (figure 20.14) doesn't live up to the Dutch standard.

Figure 20.14. Meanwhile, back in Britain...
Photo on right by Mike Armstrong.

In the French city of Lyon, a privately-run public bicycle network, Vélo'v, was introduced in 2005 and has proved popular. Lyon's population of 470 000 inhabitants is served by 2000 bikes distributed around 175 cycle-stations in an area of 50 km² (figure 20.15). In the city centre, you're usually within 400 metres of a cycle-station. Users join the scheme by paying a subscription fee of €10 per year and may then hire bicycles free for all trips lasting less than 30 minutes. For longer hire periods, users pay up to €1 per hour. Short-term visitors to Lyon can buy one-week subscriptions for €1.

Other legislative opportunities

Speed limits are a simple knob that could be twiddled. As a rule, cars that travel slower use less energy (see Chapter A). With practice, drivers can learn to drive more economically: using the accelerator and brake less and always driving in the highest possible gear can give a 20% reduction in fuel consumption.

Another way to reduce fuel consumption is to reduce congestion. Stopping and starting, speeding up and slowing down, is a much less efficient way to get around than driving smoothly. Idling in stationary traffic is an especially poor deliverer of miles per gallon!

Congestion occurs when there are too many vehicles on the roads. So one simple way to reduce congestion is to group travellers into fewer vehicles. A striking way to think about a switch from cars to coaches is to calculate the road area required by the two modes. Take a trunk road on the verge of congestion, where the desired speed is 60 mph. The safe distance from one car to the next at 60 mph is 77 m. If we assume there's one car every 80 m and that each car contains 1.6 people, then vacuuming up 40 people into a single coach frees up *two kilometres* of road!

Congestion can be reduced by providing good alternatives (cycle lanes, public transport), and by charging road users extra if they contribute to congestion. In this chapter's notes I describe a fair and simple method for handling congestion-charging.

Enhancing cars

Assuming that the developed world's love-affair with the car is not about to be broken off, what are the technologies that can deliver significant energy savings? Savings of 10% or 20% are easy – we've already discussed some ways to achieve them, such as making cars smaller and lighter. Another option is to switch from petrol to diesel. Diesel engines are more expensive to make, but they tend to be more fuel-efficient. But are there technologies that can radically increase the efficiency of the energy-conversion chain? (Recall that in a standard petrol car, 75% of the energy is turned

Figure 20.15. A Vélo'v station in Lyon.

Figure 20.16. With congestion like this, it's faster to walk.

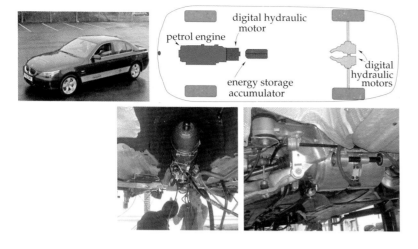

Figure 20.17. A BMW 530i modified by Artemis Intelligent Power to use digital hydraulics. Lower left: A 6-litre accumulator (the red canister), capable of storing about 0.05 kWh of energy in compressed nitrogen. Lower right: Two 200 kW hydraulic motors, one for each rear wheel, which both accelerate and decelerate the car. The car is still powered by its standard 190 kW petrol engine, but thanks to the digital hydraulic transmission and regenerative braking, it uses 30% less fuel.

into heat and blown out of the radiator!) And what about the goal of getting off fossil fuels?

In this section, we'll discuss five technologies: regenerative braking; hybrid cars; electric cars; hydrogen-powered cars; and compressed-air cars.

Regenerative braking

There are four ways to capture energy as a vehicle slows down.

1. An electric generator coupled to the wheels can charge up an electric battery or supercapacitor.

2. Hydraulic motors driven by the wheels can make compressed air, stored in a small canister.

3. Energy can be stored in a flywheel.

4. Braking energy can be stored as gravitational energy by driving the vehicle up a ramp whenever you want to slow down. This gravitational energy storage option is rather inflexible, since there must be a ramp in the right place. It's an option that's most useful for trains, and it is illustrated by the London Underground's Victoria line, which has hump-back stations. Each station is at the top of a hill in the track. Arriving trains are automatically slowed down by the hill, and departing trains are accelerated as they go down the far side of the hill. The hump-back-station design provides an energy saving of 5% and makes the trains run 9% faster.

Electric regenerative braking (using a battery to store the energy) salvages roughly 50% of the car's energy in a braking event, leading to perhaps a 20% reduction in the energy cost of city driving.

Regenerative systems using flywheels and hydraulics seem to work a little better than battery-based systems, salvaging at least 70% of the braking energy. Figure 20.17 describes a hybrid car with a petrol engine powering digitally-controlled hydraulics. On a standard driving cycle, this car uses 30% less fuel than the original petrol car. In urban driving, its energy consumption is halved, from 131 kWh per 100 km to 62 kWh per 100 km (20 mpg to 43 mpg). (Credit for this performance improvement must be shared between regenerative braking and the use of hybrid technology.) Hydraulics and flywheels are both promising ways to handle regenerative braking because small systems can handle large powers. A flywheel system weighing just 24 kg (figure 20.18), designed for energy storage in a racing car, can store 400 kJ (0.1 kWh) of energy – enough energy to accelerate an ordinary car up to 60 miles per hour (97 km/h); and it can accept or deliver 60 kW of power. Electric batteries capable of delivering that much power would weigh about 200 kg. So, unless you're already carrying that much battery on board, an electrical regenerative-braking system should probably use capacitors to store braking energy. Super-capacitors have similar energy-storage and power-delivery parameters to the flywheel's.

Figure 20.18. A flywheel regenerative-braking system. Photos courtesy of Flybrid Systems.

Hybrid cars

Hybrid cars such as the Toyota Prius (figure 20.19) have more-efficient engines and electric regenerative braking, but to be honest, today's hybrid vehicles don't really stand out from the crowd (figure 20.9).

The horizontal bars in figure 20.9 highlight a few cars including two hybrids. Whereas the average new car in the UK emits 168 g, the hybrid Prius emits about 100 g of CO_2 per km, as do several other non-hybrid vehicles – the VW Polo blue motion emits 99 g/km, and there's a Smart car that emits 88 g/km.

The Lexus RX 400h is the second hybrid, advertised with the slogan "LOW POLLUTION. ZERO GUILT." But its CO_2 emissions are 192 g/km – worse than the average UK car! The advertising standards authority ruled that this advertisement breached the advertising codes on Truthfulness, Comparisons and Environmental claims. "We considered that . . . readers were likely to understand that the car caused little or no harm to the environment, which was not the case, and had low emissions in comparison with all cars, which was also not the case."

In practice, hybrid technologies seem to give fuel savings of 20 or 30%. So neither these petrol/electric hybrids, nor the petrol/hydraulic hybrid featured in figure 20.17 seems to me to have really cracked the transport challenge. A 30% reduction in fossil-fuel consumption is impressive, but it's not enough by this book's standards. Our opening assumption was that we want to get off fossil fuels, or at least to reduce fossil fuel use by 90%. Can this goal be achieved without reverting to bicycles?

Figure 20.19. Toyota Prius – according to Jeremy Clarkson, "a very expensive, very complex, not terribly green, slow, cheaply made, and pointless way of moving around."

Figure 20.20. Electric vehicles. From left to right: the G-Wiz; the rotting corpse of a Sinclair C5; a Citroën Berlingo; and an Elettrica.

Electric vehicles

The REVA electric car was launched in June 2001 in Bangalore and is exported to the UK as the G-Wiz. The G-Wiz's electric motor has a peak power of 13 kW, and can produce a sustained power of 4.8 kW. The motor provides regenerative braking. It is powered by eight 6-volt lead acid batteries, which when fully charged give a range of "up to 77 km." A full charge consumes 9.7 kWh of electricity. These figures imply a transport cost of 13 kWh per 100 km.

Manufacturers always quote the best possible performance of their products. What happens in real life? The real-life performance of a G-Wiz in London is shown in figure 20.21. Over the course of 19 recharges, the average transport cost of this G-Wiz is 21 kWh per 100 km – about four times better than an average fossil fuel car. The best result was 16 kWh per 100 km, and the worst was 33 kWh per 100 km. If you are interested in carbon emissions, 21 kWh per 100 km is equivalent to 105 g CO_2 per km, assuming that electricity has a footprint of 500 g CO_2 per kWh.

Now, the G-Wiz sits at one end of the performance spectrum. What if we demand more – more acceleration, more speed, and more range? At the other end of the spectrum is the Tesla Roadster. The Tesla Roadster 2008 has a range of 220 miles (354 km); its lithium-ion battery pack stores 53 kWh and weighs 450 kg (120 Wh/kg). The vehicle weighs 1220 kg and its motor's maximum power is 185 kW. What is the energy-consumption of this muscle car? Remarkably, it's better than the G-Wiz: 15 kWh per 100 km. Evidence that a range of 354 km should be enough for most people most of the time comes from the fact that only 8.3% of commuters travel more than 30 km to their workplace.

I've looked up the performance figures for lots of electric vehicles – they're listed in this chapter's end-notes – and they seem to be consistent with this summary: electric vehicles can deliver transport at an energy cost of roughly 15 kWh per 100 km. That's five times better than our baseline fossil-car, and significantly better than any hybrid cars. Hurray! To achieve economical transport, we don't have to huddle together in public transport – we can still hurtle around, enjoying all the pleasures and freedoms of solo travel, thanks to electric vehicles.

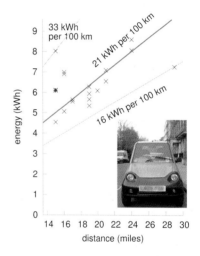

Figure 20.21. Electricity required to recharge a G-Wiz versus distance driven. Measurements were made at the socket.

Figure 20.22. Tesla Roadster: 15 kWh per 100 km. www.teslamotors.com.

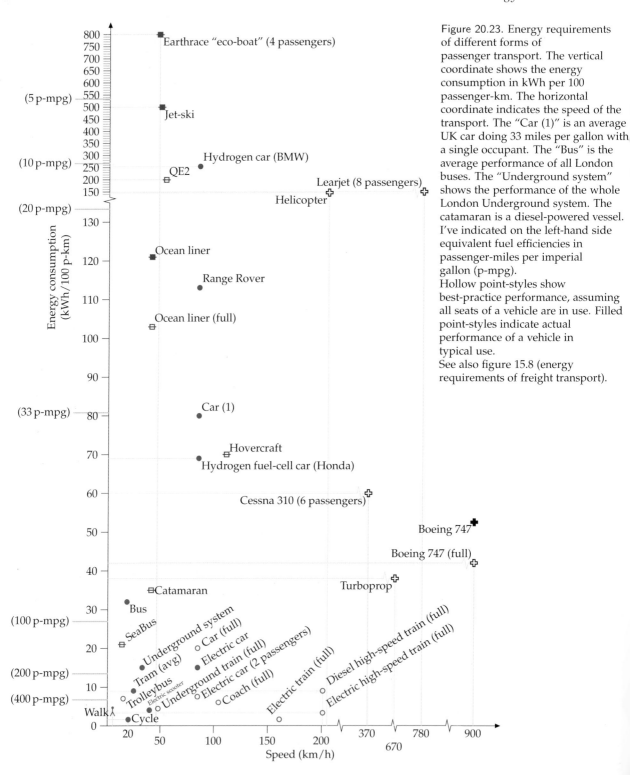

Figure 20.23. Energy requirements of different forms of passenger transport. The vertical coordinate shows the energy consumption in kWh per 100 passenger-km. The horizontal coordinate indicates the speed of the transport. The "Car (1)" is an average UK car doing 33 miles per gallon with a single occupant. The "Bus" is the average performance of all London buses. The "Underground system" shows the performance of the whole London Underground system. The catamaran is a diesel-powered vessel. I've indicated on the left-hand side equivalent fuel efficiencies in passenger-miles per imperial gallon (p-mpg).

Hollow point-styles show best-practice performance, assuming all seats of a vehicle are in use. Filled point-styles indicate actual performance of a vehicle in typical use.

See also figure 15.8 (energy requirements of freight transport).

This moment of celebration feels like a good time to unveil this chapter's big summary diagram, figure 20.23, which shows the energy requirements of all the forms of passenger-transport we have discussed and a couple that are still to come.

OK, the race is over, and I've announced two winners – public transport, and electric vehicles. But are there any other options crossing the finishing line? We have yet to hear about the compressed-air-powered car and the hydrogen car. If either of these turns out to be better than electric car, it won't affect the long-term picture very much: whichever of these three technologies we went for, the vehicles would be charged up using energy generated from a "green" source.

Compressed-air cars

Air-powered vehicles are not a new idea. Hundreds of trams powered by compressed air and hot water plied the streets of Nantes and Paris from 1879 to 1911. Figure 20.24 shows a German pneumatic locomotive from 1958. I think that in terms of energy efficiency the compressed-air technique for storing energy isn't as good as electric batteries. The problem is that compressing the air generates *heat* that's unlikely to be used efficiently; and expanding the air generates *cold*, another by-product that is unlikely to be used efficiently. But compressed air may be a superior technology to electric batteries in other ways. For example, air can be compressed thousands of times and doesn't wear out! It's interesting to note, however, that the first product sold by the Aircar company is actually an *electric* scooter. [www.theaircar.com/acf]

There's talk of Tata Motors in India manufacturing air-cars, but it's hard to be sure whether the compressed-air vehicle is going to see a revival, because no-one has published the specifications of any modern prototypes. Here's the fundamental limitation: the energy-density of compressed-air energy-stores is only about 11–28 Wh per kg, which is similar to lead-acid batteries, and roughly five times smaller than lithium-ion batteries. (See figure 26.13, p199, for details of other storage technologies.) So the range of a compressed-air car will only ever be as good as the range of the earliest electric cars. Compressed-air storage systems do have three advantages over batteries: longer life, cheaper construction, and fewer nasty chemicals.

Hydrogen cars – blimp your ride

I think hydrogen is a hyped-up bandwagon. I'll be delighted to be proved wrong, but I don't see how hydrogen is going to help us with our energy problems. Hydrogen is not a miraculous *source* of energy; it's just an energy *carrier*, like a rechargeable battery. And it is a rather inefficient energy carrier, with a whole bunch of practical defects.

The "hydrogen economy" received support from *Nature* magazine in

Figure 20.24. Top: A compressed-air tram taking on air and steam in Nantes. Powering the trams of Nantes used 4.4 kg of coal (36 kWh) per vehicle-km, or 115 kWh per 100 p-km, if the trams were full. [5qhvcb] Bottom: A compressed-air locomotive; weight 9.2 t, pressure 175 bar, power 26 kW; photo courtesy of Rüdiger Fach, Rolf-Dieter Reichert, and Frankfurter Feldbahnmuseum.

Figure 20.25. The Hummer H2H: embracing the green revolution, the American way. Photo courtesy of General Motors.

a column praising California Governor Arnold Schwarzenegger for filling up a hydrogen-powered Hummer (figure 20.25). Nature's article lauded Arnold's vision of hydrogen-powered cars replacing "polluting models" with the quote "the governor is a real-life climate action hero." But the critical question that needs to be asked when such hydrogen heroism is on display is "where is the *energy* to come from to *make* the hydrogen?" Moreover, converting energy to and from hydrogen can only be done inefficiently – at least, with today's technology.

Here are some numbers.

Figure 20.26. BMW Hydrogen 7. Energy consumption: 254 kWh per 100 km. Photo from BMW.

- In the CUTE (Clean Urban Transport for Europe) project, which was intended to demonstrate the feasibility and reliability of fuel-cell buses and hydrogen technology, fuelling the hydrogen buses required between 80% and 200% *more* energy than the baseline diesel bus.

- Fuelling the *Hydrogen 7*, the hydrogen-powered car made by BMW, requires 254 kWh per 100 km – *220% more* energy than an average European car.

If our task were "please stop using fossil fuels for transport, allowing yourself the assumption that *infinite* quantities of green electricity are available for free," then of course an energy-profligate transport solution like hydrogen might be a contender (though hydrogen faces other problems). But *green electricity is not free*. Indeed, getting green electricity on the scale of our current consumption is going to be very challenging. The fossil fuel challenge is an energy challenge. The climate-change problem is an energy problem. We need to focus on solutions that use less energy, not "solutions" that use more! *I know of no form of land transport whose energy consumption is worse than this hydrogen car.* (The only transport methods I know that are worse are jet-skis – using about 500 kWh per 100 km – and the *Earthrace* biodiesel-powered speed-boat, absurdly called an eco-boat, which uses 800 kWh per 100 p-km.)

Figure 20.27. The Earthrace "eco-boat." Photo by David Castor.

Hydrogen advocates may say "the BMW Hydrogen 7 is just an early prototype, and it's a luxury car with lots of muscle – the technology is going to get more efficient." Well, I hope so, because it has a lot of catching up to do. The Tesla Roadster (figure 20.22) is an early prototype too, and it's also a luxury car with lots of muscle. And it's more than ten times more energy-efficient than the Hydrogen 7! Feel free to put your money on the hydrogen horse if you want, and if it wins in the end, fine. But it seems daft to back the horse that's so far behind in the race. Just look at figure 20.23 – if I hadn't squished the top of the vertical axis, the hydrogen car would not have fitted on the page!

Yes, the Honda fuel-cell car, the FCX Clarity, does better – it rolls in at 69 kWh per 100 km – but my prediction is that after all the "zero-emissions" trumpeting is over, we'll find that hydrogen cars use just as much energy as the average fossil car of today.

Figure 20.28. The Honda FCX Clarity hydrogen-powered fuel-cell sedan, with a Jamie Lee Curtis for scale. Photo courtesy of automobiles.honda.com.

Here are some other problems with hydrogen. Hydrogen is a less convenient energy storage medium than most liquid fuels, because of its bulk, whether stored as a high pressure gas or as a liquid (which requires a temperature of $-253\,°C$). Even at a pressure of 700 bar (which requires a hefty pressure vessel) its energy density (energy per unit volume) is 22% of gasoline's. The cryogenic tank of the BMW Hydrogen 7 weighs 120 kg and stores 8 kg of hydrogen. Furthermore, hydrogen gradually leaks out of any practical container. If you park your hydrogen car at the railway station with a full tank and come back a week later, you should expect to find most of the hydrogen has gone.

Some questions about electric vehicles

You've shown that electric cars are more energy-efficient than fossil cars. But are they better if our objective is to reduce CO_2 emissions, and the electricity is still generated by fossil power-stations?

This is quite an easy calculation to do. Assume the electric vehicle's energy cost is 20 kWh(e) per 100 km. (I think 15 kWh(e) per 100 km is perfectly possible, but let's play sceptical in this calculation.) If grid electricity has a carbon footprint of 500 g per kWh(e) then the effective emissions of this vehicle are **100 g CO_2 per km**, which is as good as the best fossil cars (figure 20.9). So I conclude that switching to electric cars is *already* a good idea, even before we green our electricity supply.

Electric cars, like fossil cars, have costs of both manufacture and use. Electric cars may cost less to use, but if the batteries don't last very long, shouldn't you pay more attention to the manufacturing cost?

Yes, that's a good point. My transport diagram shows only the use cost. If electric cars require new batteries every few years, my numbers may be underestimates. The batteries in a Prius are expected to last just 10 years, and a new set would cost £3500. Will anyone want to own a 10-year old Prius and pay that cost? It could be predicted that most Priuses will be junked at age 10 years. This is certainly a concern for all electric vehicles that have batteries. I guess I'm optimistic that, as we switch to electric vehicles, battery technology is going to improve.

I live in a hot place. How could I drive an electric car? I demand power-hungry air-conditioning!

There's an elegant fix for this demand: fit 4 m² of photovoltaic panels in the upward-facing surfaces of the electric car. If the air-conditioning is needed, the sun must surely be shining. 20%-efficient panels will generate up to 800 W, which is enough to power a car's air-conditioning. The panels might even make a useful contribution to charging the car when it's parked, too. Solar-powered vehicle cooling was included in a Mazda in 1993; the solar cells were embedded in the glass sunroof.

I live in a cold place. How could I drive an electric car? I demand power-hungry heating!

The motor of an electric vehicle, when it's running, will on average use something like 10 kW, with an efficiency of 90–95%. Some of the lost power, the other 5–10%, will be dissipated as heat in the motor. Perhaps electric cars that are going to be used in cold places can be carefully designed so that this motor-generated heat, which might amount to 250 or 500 W, can be piped from the motor into the car. That much power would provide some significant windscreen demisting or body-warming.

Are lithium-ion batteries safe in an accident?

Some lithium-ion batteries are unsafe when short-circuited or over-heated, but the battery industry is now producing safer batteries such as lithium phosphate. There's a fun safety video at www.valence.com.

Is there enough lithium to make all the batteries for a huge fleet of electric cars?

World lithium reserves are estimated to be 9.5 million tons in ore deposits (p175). A lithium-ion battery is 3% lithium. If we assume each vehicle has a 200 kg battery, then we need 6 kg of lithium per vehicle. So the estimated reserves in ore deposits are enough to make the batteries for 1.6 billion vehicles. That's more than the number of cars in the world today (roughly 1 billion) – but not much more, so the amount of lithium may be a concern, especially when we take into account the competing ambitions of the nuclear fusion posse (Chapter 24) to guzzle lithium in their reactors. There's many thousands times more lithium in sea water, so perhaps the oceans will provide a useful backup. However, lithium specialist R. Keith Evans says "concerns regarding lithium availability for hybrid or electric vehicle batteries or other foreseeable applications are unfounded." And anyway, other lithium-free battery technologies such as zinc-air recharge-ables are being developed [www.revolttechnology.com]. I think the electric car is a goer!

The future of flying?

The superjumbo A380 is said by Airbus to be "a highly fuel-efficient air-craft." In fact, it burns just 12% less fuel per passenger than a 747.

Boeing has announced similar breakthroughs: their new 747–8 Inter-continental, trumpeted for its planet-saving properties, is (according to Boeing's advertisements) only 15% more fuel-efficient than a 747–400.

This slender rate of progress (contrasted with cars, where changes in technology deliver two-fold or even ten-fold improvements in efficiency) is explained in Technical Chapter C. Planes are up against a fundamental limit imposed by the laws of physics. Any plane, whatever its size, *has to* expend an energy of about 0.4 kWh per ton-km on keeping up and keeping

Figure 20.29. Airbus A380.

129 *the energy-density of compressed-air energy-stores is only about 11–28 Wh per kg.* The theoretical limit, assuming perfect isothermal compression: if $1\,m^3$ of ambient air is slowly compressed into a 5-litre container at 200 bar, the potential energy stored is 0.16 kWh in 1.2 kg of air. In practice, a 5-litre container appropriate for this sort of pressure weighs about 7.5 kg if made from steel or 2 kg using kevlar or carbon fibre, and the overall energy density achieved would be about 11–28 Wh per kg. The theoretical energy density is the same, whatever the volume of the container.

130 *Arnold Schwarzenegger ... filling up a hydrogen-powered Hummer.* Nature **438**, 24 November 2005. I'm not saying that hydrogen will *never* be useful for transportation; but I would hope that such a distinguished journal as *Nature* would address the hydrogen bandwagon with some critical thought, not only euphoria.

> *Hydrogen and fuel cells are not the way to go. The decision by the Bush administration and the State of California to follow the hydrogen highway is the single worst decision of the past few years.*
>   James Woolsey, Chairman of the Advisory Board of the US Clean Fuels Foundation, 27th November 2007.

In September 2008, *The Economist* wrote "Almost nobody disputes that ... eventually most cars will be powered by batteries alone."
On the other hand, to hear more from advocates of hydrogen-based transport, see the Rocky Mountain Institute's pages about the "HyperCar" www.rmi.org/hypercar/.

– *In the Clean Urban Transport for Europe project the overall energy required to power the hydrogen buses was between 80% and 200% greater than that of the baseline diesel bus.* Source: CUTE (2006); Binder et al. (2006).

– *Fuelling the hydrogen-powered car made by BMW requires three times more energy than an average car.* Half of the boot of the BMW "Hydrogen 7" car is taken up by its 170-litre hydrogen tank, which holds 8 kg of hydrogen, giving a range of 200 km on hydrogen [news.bbc.co.uk/1/hi/business/6154212.stm]. The calorific value of hydrogen is 39 kWh per kg, and the best-practice energy cost of making hydrogen is 63 kWh per kg (made up of 52 kWh of natural gas and 11 kWh of electricity) (CUTE, 2006). So filling up the 8 kg tank has an energy cost of at least 508 kWh; and if that tank indeed delivers 200 km, then the energy cost is 254 kWh per 100 km.

> *The Hydrogen 7 and its hydrogen-fuel-cell cousins are, in many ways, simply flashy distractions.*
>   David Talbot, MIT Technology Review
>   www.technologyreview.com/Energy/18301/

Honda's fuel-cell car, the FCX Clarity, weighs 1625 kg, stores 4.1 kg of hydrogen at a pressure of 345 bar, and is said to have a range of 280 miles, consuming 57 miles of road per kg of hydrogen (91 km per kg) in a standard mix of driving conditions [czjjo], [5a3ryx]. Using the cost for creating hydrogen mentioned above, assuming natural gas is used as the main energy source, this car has a transport cost of 69 kWh per 100 km.

> *Honda might be able to kid journalists into thinking that hydrogen cars are "zero emission" but unfortunately they can't fool the climate.*
>   Merrick Godhaven

132 *A lithium-ion battery is 3% lithium.* Source: Fisher et al. (2006).

– *Lithium specialist R. Keith Evans says "concerns regarding lithium availability ... are unfounded."* – Evans (2008).

133 *Two Dutch-built liners known as "The Economy Twins."* www.ssmaritime.com/rijndam-maasdam.htm.
     QE2: www.qe2.org.uk.

134 *Transrapid magnetic levitation train.* www.transrapid.de.

21 Smarter heating

In the last chapter, we learned that electrification could shrink transport's energy consumption to one fifth of its current levels; and that public transport and cycling can be about 40 times more energy-efficient than car-driving. How about heating? What sort of energy-savings can technology or lifestyle-change offer?

The power used to heat a building is given by multiplying together three quantities:

$$\text{power used} = \frac{\text{average temperature difference} \times \text{leakiness of building}}{\text{efficiency of heating system}}.$$

Figure 21.1. My house.

Let me explain this formula (which is discussed in detail in Chapter E) with an example. My house is a three-bedroom semi-detached house built about 1940 (figure 21.1). The average temperature difference between the inside and outside of the house depends on the setting of the thermostat and on the weather. If the thermostat is permanently at $20\,°C$, the average temperature difference might be $9\,°C$. The *leakiness* of the building describes how quickly heat gets out through walls, windows, and cracks, in response to a temperature difference. The leakiness is sometimes called the *heat-loss coefficient* of the building. It is measured in kWh per day per degree of temperature difference. In Chapter E, I calculate that the leakiness of my house in 2006 was $7.7\,\text{kWh/d/°C}$. The product

$$\text{average temperature difference} \times \text{leakiness of building}$$

is the rate at which heat flows out of the house by conduction and ventilation. For example, if the average temperature difference is $9\,°C$ then the heat loss is

$$9\,°C \times 7.7\,\text{kWh/d/°C} \simeq 70\,\text{kWh/d}.$$

Finally, to calculate the power required, we divide this heat loss by the efficiency of the heating system. In my house, the condensing gas boiler has an efficiency of 90%, so we find:

$$\text{power used} = \frac{9\,°C \times 7.7\,\text{kWh/d/°C}}{0.9} = 77\,\text{kWh/d}.$$

That's bigger than the space-heating requirement we estimated in Chapter 7. It's bigger for two reasons: first, this formula assumes that all the heat is supplied by the boiler, whereas in fact some heat is supplied by incidental heat gains from occupants, gadgets, and the sun; second, in Chapter 7 we assumed that a person kept just two rooms at $20\,°C$ all the time; keeping an entire house at this temperature all the time would require more.

OK, how can we reduce the power used by heating? Well, obviously, there are three lines of attack.

1. Reduce the average temperature difference. This can be achieved by turning thermostats down (or, if you have friends in high places, by changing the weather).

2. Reduce the leakiness of the building. This can be done by improving the building's insulation – think triple glazing, draught-proofing, and fluffy blankets in the loft – or, more radically, by demolishing the building and replacing it with a better insulated building; or perhaps by living in a building of smaller size per person. (Leakiness tends to be bigger, the larger a building's floor area, because the areas of external wall, window, and roof tend to be bigger too.)

3. Increase the efficiency of the heating system. You might think that 90% sounds hard to beat, but actually we can do much better.

Cool technology: the thermostat

The thermostat (accompanied by woolly jumpers) is hard to beat, when it comes to value-for-money technology. You turn it down, and your building uses less energy. Magic! In Britain, for every degree that you turn the thermostat down, the heat loss decreases by about 10%. Turning the thermostat down from 20 °C to 15 °C would nearly halve the heat loss. Thanks to incidental heat gains by the building, the savings in heating power will be even bigger than these reductions in heat loss.

Unfortunately, however, this remarkable energy-saving technology has side-effects. Some humans call turning the thermostat down a lifestyle change, and are not happy with it. I'll make some suggestions later about how to sidestep this lifestyle issue. Meanwhile, as proof that "the most important smart component in a building with smart heating is the occupant," figure 21.2 shows data from a Carbon Trust study, observing the heat consumption in twelve identical modern houses. This study permits us to gawp at the family at number 1, whose heat consumption is twice as big as that of Mr. and Mrs. Woolly at number 12. However, we should pay attention to the numbers: the family at number 1 are using 43 kWh per day. But if this is shocking, hang on – a moment ago, didn't I estimate that *my* house might use more than that? Indeed, my average gas consumption from 1993 to 2003 was a little more than 43 kWh per day (figure 7.10, p53), and I thought I was a frugal person! The problem is the *house*. All the modern houses in the Carbon Trust study had a leakiness of 2.7 kWh/d/°C, but my house had a leakiness of 7.7 kWh/d/°C! People who live in leaky houses...

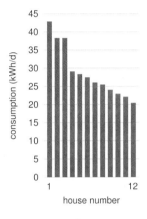

Figure 21.2. Actual heat consumption in 12 identical houses with identical heating systems. All houses had floor area 86 m^2 and were designed to have a leakiness of 2.7 kWh/d/°C. Source: Carbon Trust (2007).

The war on leakiness

What can be done with leaky old houses, apart from calling in the bulldozers? Figure 21.3 shows estimates of the space heating required in old

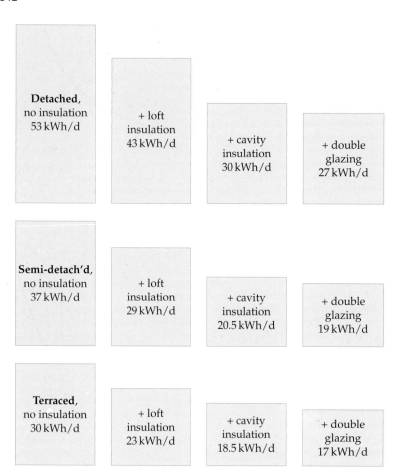

Figure 21.3. Estimates of the space heating required in a range of UK houses. From Eden and Bending (1985).

detached, semi-detached, and terraced houses as progressively more effort is put into patching them up. Adding loft insulation and cavity-wall insulation reduces heat loss in a typical old house by about 25%. Thanks to incidental heat gains, this 25% reduction in heat loss translates into roughly a 40% reduction in heating consumption.

Let's put these ideas to the test.

A case study

I introduced you to my house on page 53. Let's pick up the story. In 2004 I had a condensing boiler installed, replacing the old gas boiler. (Condensing boilers use a heat-exchanger to transfer heat from the exhaust gases to incoming air.) At the same time I removed the house's hot-water tank (so hot water is now made only on demand), and I put thermostats on all the bedroom radiators. Along with the new condensing boiler came a

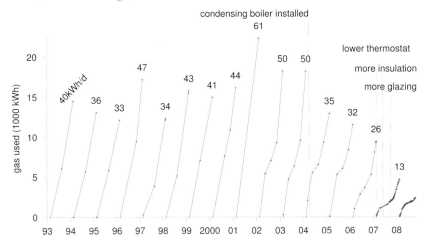

Figure 21.4. My domestic gas consumption, each year from 1993 to 2007. Each line shows the cumulative consumption during one year in kWh. The number at the end of each year is the average rate of consumption for that year, in kWh per day. Meter-readings are indicated by the blue points. Evidently, the more frequently I read my meter, the less gas I use!

new heating controller that allows me to set different target temperatures for different times of day. With these changes, my consumption decreased from an average of 50 kWh/d to about 32 kWh/d.

This reduction from 50 to 32 kWh/d is quite satisfying, but it's not enough, if the aim is to reduce one's fossil fuel footprint below one ton of CO_2 per year. 32 kWh/d of gas corresponds to over 2 tons CO_2 per year.

In 2007, I started paying more careful attention to my energy meters. I had cavity-wall insulation installed (figure 21.5) and improved my loft insulation. I replaced the single-glazed back door by a double-glazed door, and added an extra double-glazed door to the front porch (figure 21.6). Most important of all, I paid more attention to my thermostat settings. This attentiveness has led to a further halving in gas consumption. The latest year's consumption was 13 kWh/d!

Because this case study is such a hodge-podge of building modifications and behaviour changes, it's hard to be sure which changes were the most important. According to my calculations (in Chapter E), the improvements in insulation reduced the leakiness by 25%, from 7.7 kWh/d/°C to 5.8 kWh/d/°C. This is still much leakier than any modern house. It's frustratingly difficult to reduce the leakiness of an already-built house!

So, my main tip is cunning thermostat management. What's a reasonable thermostat setting to aim for? Nowadays many people seem to think that 17 °C is unbearably cold. However, the average winter-time temperature in British houses in 1970 was 13 °C! A human's perception of whether they feel warm depends on what they are doing, and what they've been doing for the last hour or so. My suggestion is, *don't think in terms of a thermostat setting*. Rather than fixing the thermostat to a single value, try just leaving it at a really low value most of the time (say 13 or 15 °C), and turn it up temporarily whenever you feel cold. It's like the lights in a library. If you allow yourself to ask the question "what is the right light level in the bookshelves?" then you'll no doubt answer "bright enough to read the

Figure 21.5. Cavity-wall insulation going in.

Figure 21.6. A new front door.

book titles," and you'll have bright lights on all the time. But that question presumes that we have to fix the light level; and we don't have to. We can fit light switches that the reader can turn on, and that switch themselves off again after an appropriate time. Similarly, thermostats don't need to be left up at 20 °C all the time.

Before leaving the topic of thermostat settings, I should mention air-conditioning. Doesn't it drive you crazy to go into a building in summer where the thermostat of the air-conditioning is set to 18 °C? These loony building managers are subjecting everyone to temperatures that in winter-time they would whinge are too cold! In Japan, the government's "Cool-Biz" guidelines recommend that air-conditioning be set to 28 °C (82 F).

Better buildings

If you get the chance to build a new building then there are lots of ways to ensure its heating consumption is much smaller than that of an old build-ing. Figure 21.2 gave evidence that modern houses are built to much better insulation standards than those of the 1940s. But the building standards in Britain could be still better, as Chapter E discusses. The three key ideas for the best results are: (1) have really thick insulation in floors, walls, and roofs; (2) ensure the building is completely sealed and use active venti-lation to introduce fresh air and remove stale and humid air, with heat exchangers passively recovering much of the heat from the removed air; (3) design the building to exploit sunshine as much as possible.

The energy cost of heat

So far, this chapter has focused on temperature control and leakiness. Now we turn to the third factor in the equation:

$$\text{power used} = \frac{\text{average temperature difference} \times \text{leakiness of building}}{\text{efficiency of heating system}}.$$

How efficiently can heat be produced? Can we obtain heat on the cheap? Today, building-heating in Britain is primarily delivered by burning a fossil fuel, natural gas, in boilers with efficiencies of 78%–90%. Can we get off fossil fuels at the same time as making building-heating more efficient?

One technology that is held up as an answer to Britain's heating prob-lem is called "combined heat and power" (CHP), or its cousin, "micro-CHP." I will explain combined heat and power now, but I've come to the conclusion that it's a bad idea, because there's a better technology for heat-ing, called heat pumps, which I'll describe in a few pages.

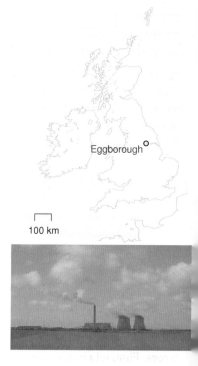

Figure 21.7. Eggborough. Not a power station participating in smart heating.

Eggborough

100 km

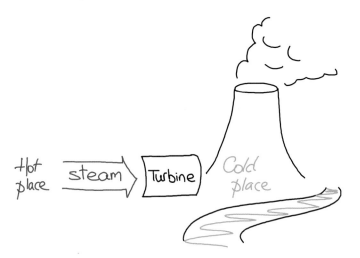

Figure 21.8. How a power station works. There has to be a cold place to condense the steam to make the turbine go round. The cold place is usually a cooling tower or river.

Combined heat and power

The standard view of conventional big centralised power stations is that they are terribly inefficient, chucking heat willy-nilly up chimneys and cooling towers. A more sophisticated view recognizes that to turn thermal energy into electricity, we inevitably have to dump heat in a cold place (figure 21.8). That is how heat engines work. There *has* to be a cold place. But surely, it's argued, we could use *buildings* as the dumping place for this "waste" heat instead of cooling towers or sea water? This idea is called "combined heat and power" (CHP) or cogeneration, and it's been widely used in continental Europe for decades – in many cities, a big power station is integrated with a district heating system. Proponents of the modern incarnation of combined heat and power, "micro-CHP," suggest that tiny power stations should be created within single buildings or small collections of buildings, delivering heat and electricity to those buildings, and exporting some electricity to the grid.

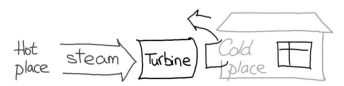

Figure 21.9. Combined heat and power. District heating absorbs heat that would have been chucked up a cooling tower.

There's certainly some truth in the view that Britain is rather backward when it comes to district heating and combined heat and power, but discussion is hampered by a general lack of numbers, and by two particular errors. First, when comparing different ways of using fuel, the wrong measure of "efficiency" is used, namely one that weights electricity as having equal value to heat. The truth is, electricity is more valuable than heat. Second, it's widely assumed that the "waste" heat in a traditional power

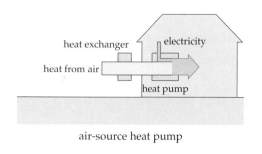

air-source heat pump

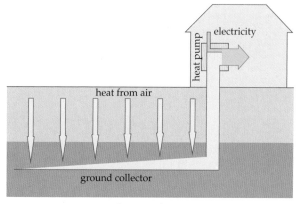

ground-source heat pump

Figure 21.10. Heat pumps.

station could be captured for a useful purpose *without impairing the power station's electricity production*. This sadly is not true, as the numbers will show. Delivering useful heat to a customer always reduces the electricity produced to some degree. The true net gains from combined heat and power are often much smaller than the hype would lead you to believe.

A final impediment to rational discussion of combined heat and power is a myth that has grown up recently, that decentralizing a technology somehow makes it greener. So whereas big centralized fossil fuel power stations are "bad," flocks of local micro-power stations are imbued with goodness. But if decentralization is actually a good idea then "small is beautiful" should be evident in the numbers. Decentralization should be able to stand on its own two feet. And what the numbers actually show is that *centralized* electricity generation has many benefits in both economic and energy terms. Only in large buildings is there any benefit to local generation, and usually that benefit is only about 10% or 20%.

The government has a target for growth of combined heat and power to 10 GW of electrical capacity by 2010, but I think that growth of gas-powered combined heat and power would be a mistake. Such combined heat and power is not green: it uses fossil fuel, and it locks us into continued use of fossil fuel. Given that heat pumps are a better technology, I believe we should leapfrog over gas-powered combined heat and power and go directly for heat pumps.

Heat pumps

Like district heating and combined heat and power, heat pumps are already widely used in continental Europe, but strangely rare in Britain. Heat pumps are back-to-front refrigerators. Feel the back of your refrigerator: it's *warm*. A refrigerator moves heat from one place (its inside) to

another (its back panel). So one way to heat a building is to turn a refrigerator inside-out – put the *inside* of the refrigerator in the garden, thus
cooling the garden down; and leave the back panel of the refrigerator in
your kitchen, thus warming the house up. What isn't obvious about this
whacky idea is that it is a really efficient way to warm your house. For
every kilowatt of power drawn from the electricity grid, the back-to-front
refrigerator can pump three kilowatts of heat from the garden, so that a
total of four kilowatts of heat gets into your house. So heat pumps are
roughly four times as efficient as a standard electrical bar-fire. Whereas
the bar-fire's efficiency is 100%, the heat pump's is 400%. The efficiency of
a heat pump is usually called its *coefficient of performance* or CoP. If the
efficiency is 400%, the coefficient of performance is 4.

Heat pumps can be configured in various ways (figure 21.10). A heat
pump can cool down the *air* in your garden using a heat-exchanger (typically a 1-metre tall white box, figure 21.11), in which case it's called an
air-source heat pump. Alternatively, the pump may cool down the *ground*
using big loops of underground plumbing (many tens of metres long),
in which case it's called a ground-source heat pump. Heat can also be
pumped from rivers and lakes.

Some heat pumps can pump heat in either direction. When an air-
source heat pump runs in reverse, it uses electricity to warm up the *outside* air and cool down the air *inside* your building. This is called air-
conditioning. Many air-conditioners are indeed heat-pumps working in
precisely this way. Ground-source heat pumps can also work as air-conditioners. So a single piece of hardware can be used to provide winter
heating and summer cooling.

People sometimes say that ground-source heat pumps use "geothermal energy," but that's not the right name. As we saw in Chapter 16,
geothermal energy offers only a tiny trickle of power per unit area (about
$50\,mW/m^2$), in most parts of the world; heat pumps have nothing to do
with this trickle, and they can be used both for heating and for cooling.
Heat pumps simply use the ground as a place to suck heat from, or to
dump heat into. When they steadily suck heat, that heat is actually being
replenished by warmth from the sun.

There's two things left to do in this chapter. We need to compare heat
pumps with combined heat and power. Then we need to discuss what are
the limits to ground-source heat pumps.

Heat pumps, compared with combined heat and power

I used to think that combined heat and power was a no-brainer. "Obviously, we should use the discarded heat from power stations to heat buildings rather than just chucking it up a cooling tower!" However, looking
carefully at the numbers describing the performance of real CHP systems,
I've come to the conclusion that there are better ways of providing electric-

Figure 21.11. The inner and outer bits of an air-source heat pump that has a coefficient of performance of 4. The inner bit is accompanied by a ball-point pen, for scale. One of these Fujitsu units can deliver 3.6 kW of heating when using just 0.845 kW of electricity. It can also run in reverse, delivering 2.6 kW of cooling when using 0.655 kW of electricity.

ity and building-heating.

I'm going to build up a diagram in three steps. The diagram shows how much electrical energy or heat energy can be delivered from chemical energy. The horizontal axis shows the electrical efficiency and the vertical axis shows the heat efficiency.

The standard solution with no CHP

In the first step, we show simple power stations and heating systems that deliver pure electricity or pure heat.

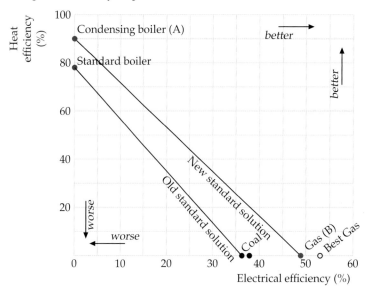

Condensing boilers (the top-left dot, A) are 90% efficient because 10% of the heat goes up the chimney. Britain's gas power stations (the bottom-right dot, B) are currently 49% efficient at turning the chemical energy of gas into electricity. If you want any mix of electricity and heat from natural gas, you can obtain it by burning appropriate quantities of gas in the electricity power station and in the boiler. Thus the new standard solution can deliver any electrical efficiency and heat efficiency on the line A–B by making the electricity and heat using two separate pieces of hardware.

To give historical perspective, the diagram also shows the old standard heating solution (an ordinary non-condensing boiler, with an efficiency of 79%) and the standard way of making electricity a few decades ago (a coal power station with an electrical efficiency of 37% or so).

Combined heat and power

Next we add combined heat and power systems to the diagram. These simultaneously deliver, from chemical energy, both electricity and heat.

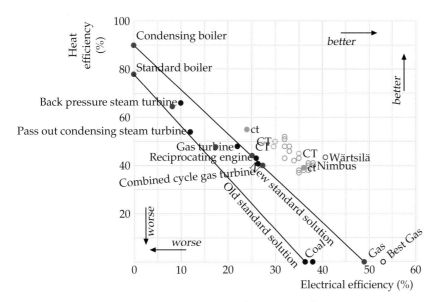

Each of the filled dots shows actual average performances of CHP systems in the UK, grouped by type. The hollow dots marked "CT" show the performances of ideal CHP systems quoted by the Carbon Trust; the hollow dots marked "Nimbus" are from a manufacturer's product specifications. The dots marked "ct" are the performances quoted by the Carbon Trust for two real systems (at Freeman Hospital and Elizabeth House).

The main thing to notice in this diagram is that the electrical efficiencies of the CHP systems are significantly smaller than the 49% efficiency delivered by single-minded electricity-only gas power stations. So the heat is not a "free by-product." Increasing the heat production hurts the electricity production.

It's common practice to lump together the two numbers (the efficiency of electricity production and heat production) into a single "total efficiency" – for example, the back pressure steam turbines delivering 10% electricity and 66% heat would be called "76% efficient," but I think this is a misleading summary of performance. After all, by this measure, the 90%-efficient condensing boiler is "more efficient" than all the CHP systems! The fact is, electrical energy is more valuable than heat.

Many of the CHP points in this figure are superior to the "old standard way of doing things" (getting electricity from coal and heat from standard boilers). And the ideal CHP systems are slightly superior to the "new standard way of doing things" (getting electricity from gas and heat from condensing boilers). But we must bear in mind that this slight superiority comes with some drawbacks – a CHP system delivers heat only to the places it's connected to, whereas condensing boilers can be planted anywhere with a gas main; and compared to the standard way of doing things, CHP systems are not so flexible in the mix of electricity and heat

they deliver; a CHP system will work best only when delivering a particular mix; this inflexibility leads to inefficiencies at times when, for example,
excess heat is produced; in a typical house, much of the electricity demand
comes in relatively brief spikes, bearing little relation to heating demand.
A final problem with some micro-CHP systems is that when they have excess electricity to share, they may do a poor job of delivering power to the
network.

Finally we add in heat pumps, which use electricity from the grid to
pump ambient heat into buildings.

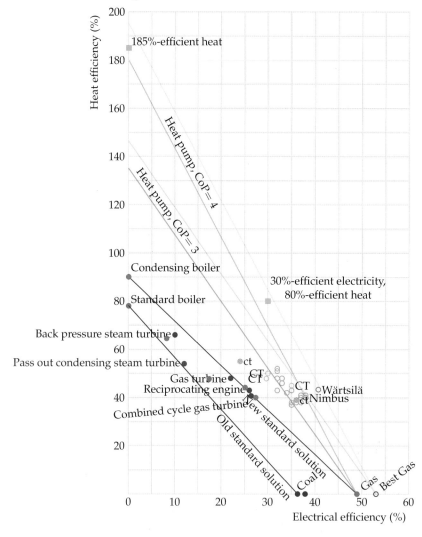

The steep green lines show the combinations of electricity and heat
that you can obtain assuming that heat pumps have a coefficient of per-

formance of 3 or 4, assuming that the extra electricity for the heat pumps is generated by an average gas power station or by a top-of-the-line gas power station, and allowing for 8% loss in the national electricity network between the power station and the building where the heat pumps pump heat. The top-of-the-line gas power station's efficiency is 53%, assuming it's running optimally. (I imagine the Carbon Trust and Nimbus made a similar assumption when providing the numbers used in this diagram for CHP systems.) In the future, heat pumps will probably get even better than I assumed here. In Japan, thanks to strong legislation favouring efficiency improvements, heat pumps are now available with a coefficient of performance of 4.9.

Notice that heat pumps offer a system that can be "better than 100%-efficient." For example the "best gas" power station, feeding electricity to heat pumps can deliver a combination of 30%-efficient electricity and 80%-efficient heat, a "total efficiency" of 110%. No plain CHP system could ever match this performance.

Let me spell this out. Heat pumps are superior in efficiency to condensing boilers, even if the heat pumps are powered by electricity from a power station burning natural gas. If you want to heat lots of buildings using natural gas, you could install condensing boilers, which are "90% efficient," or you could send the same gas to a new gas power station making electricity and install electricity-powered heat pumps in all the buildings; the second solution's efficiency would be somewhere between 140% and 185%. It's not necessary to dig big holes in the garden and install under-floor heating to get the benefits of heat pumps; the best air-source heat pumps (which require just a small external box, like an air-conditioner's) can deliver hot water to normal radiators with a coefficient of performance above 3. The air-source heat pump in figure 21.11 (p147) directly delivers warm air to an office.

I thus conclude that combined heat and power, even though it sounds a good idea, is probably not the best way to heat buildings and make electricity using natural gas, assuming that air-source or ground-source heat pumps can be installed in the buildings. The heat-pump solution has further advantages that should be emphasized: heat pumps can be located in any buildings where there is an electricity supply; they can be driven by any electricity source, so they keep on working when the gas runs out or the gas price goes through the roof; and heat pumps are flexible: they can be turned on and off to suit the demand of the building occupants.

I emphasize that this critical comparison does not mean that CHP is always a bad idea. What I'm comparing here are methods for heating ordinary buildings, which requires only very low-grade heat. CHP can also be used to deliver higher-grade heat to industrial users (at 200 °C, for example). In such industrial settings, heat pumps are unlikely to compete so well because their coefficient of performance would be lower.

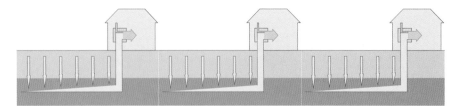

Figure 21.12. How close together can ground-source heat pumps be packed?

Limits to growth (of heat pumps)

Because the temperature of the ground, a few metres down, stays sluggishly close to 11 °C, whether it's summer or winter, the ground is theoretically a better place for a heat pump to grab its heat than the air, which in midwinter may be 10 or 15 °C colder than the ground. So heat-pump advisors encourage the choice of ground-source over air-source heat pumps, where possible. (Heat pumps work less efficiently when there's a big temperature difference between the inside and outside.)

However, the ground is not a limitless source of heat. The heat has to come from somewhere, and ground is not a very good thermal conductor. If we suck heat too fast from the ground, the ground will become as cold as ice, and the advantage of the ground-source heat pump will be diminished.

In Britain, the main purpose of heat pumps would be to get heat into buildings in the winter. The ultimate source of this heat is the sun, which replenishes heat in the ground by direct radiation and by conduction through the air. The rate at which heat is sucked from the ground must satisfy two constraints: it must not cause the ground's temperature to drop too low during the winter; and the heat sucked in the winter must be replenished somehow during the summer. If there's any risk that the *natural* trickling of heat in the summer won't make up for the heat removed in the winter, then the replenishment must be driven *actively* – for example by running the system in reverse in summer, putting heat down into the ground (and thus providing air-conditioning up top).

Let's put some numbers into this discussion. How big a piece of ground does a ground-source heat pump need? Assume that we have a neighbourhood with quite a high population density – say 6200 people per km^2 (160 m^2 per person), the density of a typical British suburb. Can *everyone* use ground-source heat pumps, without using active summer replenishment? A calculation in Chapter E (p303) gives a tentative answer of *no*: if we wanted everyone in the neighbourhood to be able to pull from the ground a heat flow of about 48 kWh/d per person (my estimate of our typical winter heat demand), we'd end up freezing the ground in the winter. Avoiding unreasonable cooling of the ground requires that the sucking rate be less than 12 kWh/d per person. So if we switch to ground-source heat pumps, we should plan to include substantial summer heat-dumping in the design, so as to refill the ground with heat for use in the winter. This summer heat-dumping could use heat from air-conditioning, or heat from

	area per person (m^2)
Bangalore	37
Manhattan	39
Paris	40
Chelsea	66
Tokyo	72
Moscow	97
Taipei	104
The Hague	152
San Francisco	156
Singapore	156
Cambridge MA	164
Sydney	174
Portsmouth	213

Table 21.13. Some urban areas per person.

roof-mounted solar water-heating panels. (Summer solar heat is stored in the ground for subsequent use in winter by Drake Landing Solar Community in Canada [www.dlsc.ca].) Alternatively, we should expect to need to use some air-source heat pumps too, and then we'll be able to get all the heat we want – as long as we have the electricity to pump it. In the UK, air temperatures don't go very far below freezing, so concerns about poor winter-time performance of air-source pumps, which might apply in North America and Scandanavia, probably do not apply in Britain.

My conclusion: can we reduce the energy we consume for heating? Yes. Can we get off fossil fuels at the same time? Yes. Not forgetting the low-hanging fruit – building-insulation and thermostat shenanigans – we should replace all our fossil-fuel heaters with electric-powered heat pumps; we can reduce the energy required to 25% of today's levels. Of course this plan for electrification would require more electricity. But even if the extra electricity came from gas-fired power stations, that would still be a much better way to get heating than what we do today, simply setting fire to the gas. Heat pumps are future-proof, allowing us to heat buildings efficiently with electricity from any source.

Nay-sayers object that the coefficient of performance of air-source heat pumps is lousy – just 2 or 3. But their information is out of date. If we are careful to buy top-of-the-line heat pumps, we can do much better. The Japanese government legislated a decade-long efficiency drive that has greatly improved the performance of air-conditioners; thanks to this drive, there are now air-source heat pumps with a coefficient of performance of 4.9; these heat pumps can make hot water as well as hot air.

Another objection to heat pumps is "oh, we can't approve of people fitting efficient air-source heaters, because they might use them for air-conditioning in the summer." Come on – I hate gratuitous air-conditioning as much as anyone, but these heat pumps are four times more efficient than any other winter heating method! Show me a better choice. Wood pellets? Sure, a few wood-scavengers can burn wood. But there is not enough wood for everyone to do so. For forest-dwellers, there's wood. For everyone else, there's heat pumps.

Notes and further reading

page no.

142 *Loft and cavity insulation reduces heat loss in a typical old house by about a quarter.* Eden and Bending (1985).

143 *The average internal temperature in British houses in 1970 was 13 °C!* Source: Dept. of Trade and Industry (2002a, para 3.11)

145 *Britain is rather backward when it comes to district heating and combined heat and power.* The rejected heat from UK power stations could meet the

heating needs of the entire country (Wood, 1985). In Denmark in 1985, district heating systems supplied 42% of space heating, with heat being transmitted 20 km or more in hot pressurized water. In West Germany in 1985, 4 million dwellings received 7 kW per dwelling from district heating. Two thirds of the heat supplied was from power stations. In Vasteras, Sweden in 1985, 98% of the city's heat was supplied from power stations.

147 *Heat pumps are roughly four times as efficient as a standard electrical barfire.* See www.gshp.org.uk.
Some heat pumps available in the UK already have a coefficient of peformance bigger than 4.0 [yok2nw]. Indeed there is a government subsidy for water-source heat pumps that applies only to pumps with a coefficient of peformance better than 4.4 [2dtx8z].
Commercial ground-source heat pumps are available with a coefficient of performance of 5.4 for cooling and 4.9 for heating [2fd8ar].

153 *Air-source heat pumps with a coefficient of performance of 4.9...* According to HPTCJ (2007), heat pumps with a coefficient of performance of 6.6 have been available in Japan since 2006. The performance of heat pumps in Japan improved from 3 to 6 within a decade thanks to government regulations. HPTCJ (2007) describe an air-source-heat-pump water-heater called Eco Cute with a coefficient of performance of 4.9. The Eco Cute came on the market in 2001. www.ecosystem-japan.com.

Further reading on heat pumps: European Heat Pump Network
    ehpn.fiz-karlsruhe.de/en/,
    www.kensaengineering.com,
    www.heatking.co.uk,
    www.iceenergy.co.uk.

Figure 21.14. Advertisement from the Mayor of London's "DIY planet repairs" campaign of 2007. The text reads "**Turn down**. If every London household turned down their thermostat by one degree, we could save 837 000 tons of CO_2 and £110m per year." [london.gov.uk/diy] Expressed in savings per person, that's 0.12 t CO_2 per year per person. That's about 1% of one person's total (11 t), so this *is* good advice. Well done, Ken!

22 Efficient electricity use

Can we cut electricity use? Yes, switching off gadgets when they're not in use is an easy way to make a difference. Energy-efficient light bulbs will save you electricity too.

We already examined gadgets in Chapter 11. Some gadgets are unimportant, but some are astonishing guzzlers. The laser-printer in my office, sitting there doing nothing, is slurping 17 W – nearly 0.5 kWh per day! A friend bought a lamp from IKEA. Its awful adaptor (figure 22.1) guzzles 10 W (0.25 kWh per day) whether or not the lamp is on. If you add up a few stereos, DVD players, cable modems, and wireless devices, you may even find that half of your home electricity consumption can be saved.

According to the International Energy Agency, standby power consumption accounts for roughly 8% of residential electricity demand. In the UK and France, the average standby power is about 0.75 kWh/d per household. The problem isn't standby itself – it's the shoddy way in which standby is implemented. It's perfectly possible to make standby systems that draw less than 0.01 W; but manufacturers, saving themselves a penny in the manufacturing costs, are saddling the consumer with an annual cost of pounds.

Figure 22.1. An awful AC lamp-adaptor from IKEA – the adaptor uses nearly 10 W even when the lamp is switched off!

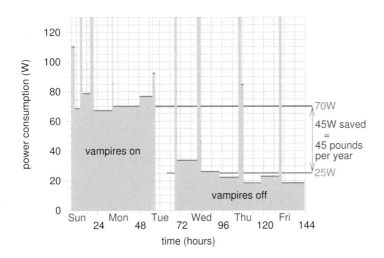

Figure 22.2. Efficiency in the offing. I measured the electricity savings from switching off vampires during a week when I was away at work most of each day, so both days and nights were almost devoid of useful activity, except for the fridge. The brief little blips of consumption are caused by the microwave, toaster, washing machine, or vacuum cleaner. On the Tuesday I switched off most of my vampires: two stereos, a DVD player, a cable modem, a wireless router, and an answering machine. The red line shows the trend of "nobody-at-home" consumption before, and the green line shows the "nobody-at-home" consumption after this change. Consumption fell by 45 W, or 1.1 kWh per day.

A vampire-killing experiment

Figure 22.2 shows an experiment I did at home. First, for two days, I measured the power consumption when I was out or asleep. Then, switching off all the gadgets that I normally left on, I measured again for three more days. I found that the power saved was 45 W – which is worth £45 per year if electricity costs 11p per unit.

Since I started paying attention to my meter readings, my total electricity consumption has halved (figure 22.3). I've cemented this saving in place by making a habit of reading my meters every week, so as to check that the electricity-sucking vampires have been banished. If this magic trick could be repeated in all homes and all workplaces, we could obviously make substantial savings. So a bunch of us in Cambridge are putting together a website devoted to making regular meter-reading fun and informative. The website, `ReadYourMeter.org`, aims to help people carry out similar experiments to mine, make sense of the resulting numbers, and get a warm fuzzy feeling from using less.

I do hope that this sort of smart-metering activity will make a difference. In the future cartoon-Britain of 2050, however, I've assumed that all such electricity savings are cancelled out by the miracle of growth. Growth is one of the tenets of our society: people are going to be wealthier, and thus able to play with more gadgets. The demand for ever-more-superlative computer games forces computers' power consumption to increase. Last decade's computers used to be thought pretty neat, but now they are found useless, and must be replaced by faster, hotter machines.

Notes and further reading

page no.

155   *Standby power consumption accounts for roughly 8% of residential electricity.* Source: International Energy Agency (2001).
For further reading on standby-power policies, see:
`www.iea.org/textbase/subjectqueries/standby.asp`.

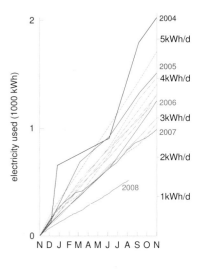

Figure 22.3. My cumulative domestic electricity consumption, in kWh, each year from 1993 to 2008. The grey lines show years from 1993 to 2003. (I haven't labelled these with their years, to avoid clutter.) The coloured lines show the years 2004 onwards. The scale on the right shows the average rate of energy consumption, in kWh per day. The vampire experiment took place on 2nd October 2007. The combination of vampire-banishment with energy-saving-lightbulb installation reduced my electricity consumption from 4 kWh/d to 2 kWh/d.

23 Sustainable fossil fuels?

It is an inescapable reality that fossil fuels will continue to be an important part of the energy mix for decades to come.

UK government spokesperson, April 2008

Our present happy progressive condition is a thing of limited duration.

William Stanley Jevons, 1865

We explored in the last three chapters the main technologies and lifestyle changes for reducing power consumption. We found that we could halve the power consumption of transport (and de-fossilize it) by switching to electric vehicles. We found that we could shrink the power consumption of heating even more (and de-fossilize it) by insulating all buildings better and using electric heat pumps instead of fossil fuels. So yes, we can reduce consumption. But still, matching even this reduced consumption with power from Britain's own renewables looks very challenging (figure 18.7, p109). It's time to discuss non-renewable options for power production.

Take the known reserves of fossil fuels, which are overwhelmingly coal: 1600 Gt of coal. Share them equally between six billion people, and burn them "sustainably." What do we mean if we talk about using up a finite resource "sustainably"? Here's the arbitrary definition I'll use: the burn-rate is "sustainable" if the resources would last 1000 years. A ton of coal delivers 8000 kWh of chemical energy, so 1600 Gt of coal shared between 6 billion people over 1000 years works out to a power of 6 kWh per day per person. A standard coal power station would turn this chemical power into electricity with an efficiency of about 37% – that means about 2.2 kWh(e) per day per person. If we care about the climate, however, then presumably we would not use a standard power station. Rather, we would go for "clean coal," also known as "coal with carbon capture and storage" – an as-yet scarcely-implemented technology that sucks most of the carbon dioxide out of the chimney-flue gases and then shoves it down a hole in the ground. Cleaning up power station emissions in this way has a significant energy cost – it would reduce the delivered electricity by about 25%. So a "sustainable" use of known coal reserves would deliver only about 1.6 kWh(e) per day per person.

We can compare this "sustainable" coal-burning rate – 1.6 Gt per year – with the current global rate of coal consumption: 6.3 Gt per year, and rising.

What about the UK alone? Britain is estimated to have 7 Gt of coal left. OK, if we share 7 Gt between 60 million people, we get 100 tons per person. If we want a 1000-year solution, this corresponds to 2.5 kWh per

Figure 23.1. Coal being delivered to Kingsnorth power station (capacity 1940 MW) in 2005. Photos by Ian Boyle www.simplonpc.co.uk.

Coal: **6 kWh/d**

Figure 23.2. "Sustainable fossil fuels."

day per person. In a power station performing carbon capture and storage, this sustainable approach to UK coal would yield $0.7\,kWh(e)$ per day per person.

Our conclusion is clear:

Clean coal is only a stop-gap.

If we do develop "clean coal" technology in order to reduce greenhouse gas emissions, we must be careful, while patting ourselves on the back, to do the accounting honestly. The coal-burning process releases greenhouse gases not only at the power station but also at the coal mine. Coal-mining tends to release methane, carbon monoxide, and carbon dioxide, both directly from the coal seams as they are exposed, and subsequently from discarded shales and mudstones; for an ordinary coal power station, these coal-mine emissions bump up the greenhouse gas footprint by about 2%, so for a "clean" coal power station, these emissions may have some impact on the accounts. There's a similar accounting problem with natural gas: if, say, 5% of the natural gas leaks out along the journey from hole in the ground to power station, then this accidental methane pollution is equivalent (in greenhouse effect) to a 40% boost in the carbon dioxide released at the power station.

Figure 23.3. A caterpillar grazing on old leaves. Photo by Peter Gunn.

New coal technologies

Stanford-based company `directcarbon.com` are developing the *Direct Carbon Fuel Cell*, which converts fuel and air directly to electricity and CO_2, without involving any water or steam turbines. They claim that this way of generating electricity from coal is twice as efficient as the standard power station.

When's the end of business as usual?

The economist Jevons did a simple calculation in 1865. People were discussing how long British coal would last. They tended to answer this question by dividing the estimated coal remaining by the rate of coal consumption, getting answers like "1000 years." But, Jevons said, consumption is *not* constant. It's been doubling every 20 years, and "progress" would have it continue to do so. So "reserves divided by consumption-rate" gives the wrong answer.

Instead, Jevons extrapolated the exponentially-growing consumption, calculating the time by which the total amount consumed would exceed the estimated reserves. This was a much shorter time. Jevons was not assuming that consumption would actually continue to grow at the same rate; rather he was making the point that growth was not sustainable. His calculation estimated for his British readership the inevitable limits to their growth, and the short time remaining before those limits would become evident. Jevons made the bold prediction that the end of British "progress" would come within 100 years of 1865. Jevons was right. British coal production peaked in 1910, and by 1965 Britain was no longer a world superpower.

Let's repeat his calculation for the world as a whole. In 2006, the coal consumption rate was 6.3 Gt per year. Comparing this with reserves of 1600 Gt of coal, people often say "there's 250 years of coal left." But if we assume "business as usual" implies a growing consumption, we get a different answer. If the growth rate of coal consumption were to continue at 2% per year (which gives a reasonable fit to the data from 1930 to 2000), then all the coal would be gone in 2096. If the growth rate is 3.4% per year (the growth rate over the last decade), the end of business-as-usual is coming before 2072. Not 250 years, but 60!

If Jevons were here today, I am sure he would firmly predict that unless we steer ourselves on a course different from business as usual, there will, by 2050 or 2060, be an end to our happy progressive condition.

Notes and further reading

page no.

157 *1000 years – my arbitrary definition of "sustainable."* As precedent for this sort of choice, Hansen et al. (2007) equate "more than 500 years" with "forever."

– *1 ton of coal equivalent = 29.3 GJ = 8000 kWh* of chemical energy. This figure does not include the energy costs of mining, transport, and carbon sequestration.

– *Carbon capture and storage* (CCS). There are several CCS technologies. Sucking the CO_2 from the flue gases is one; others gasify the coal and separate the CO_2 before combustion. See Metz et al. (2005). The first prototype coal plant with CCS was opened on 9th September 2008 by the Swedish company Vattenfall [5kpjk8].

– *UK coal.* In December 2005, the reserves and resources at *existing mines* were estimated to be 350 million tons. In November 2005, potential opencast reserves were estimated to be 620 million tons; and the underground coal gasification potential was estimated to be at least 7 billion tons. [yebuk8]

158 *Coal-mining tends to release greenhouse gases.* For information about methane release from coal-mining see www.epa.gov/cmop/, Jackson and Kershaw (1996), Thakur et al. (1996). Global emissions of methane from coal mining are about 400 Mt CO_2e per year. This corresponds to roughly 2% of the greenhouse gas emissions from burning the coal.

The average methane content in British coal seams is 4.7 m^3 per ton of coal (Jackson and Kershaw, 1996); this methane, if released to the atmosphere, has a global warming potential about 5% of that of the CO_2 from burning the coal.

158 *If 5% of the natural gas leaks, it's equivalent to a 40% boost in carbon dioxide.* Accidental methane pollution has nearly eight times as big a global-warming effect as the CO_2 pollution that would arise from burning the methane; eight times, not the standard "23 times," because "23 times" is the warming ratio between equal *masses* of methane and CO_2. Each ton of CH_4 turns into 2.75 tons of CO_2 if burned; if it leaks, it's equivalent to 23 tons of CO_2. And 23/2.75 is 8.4.

Further reading: World Energy Council [yhxf8b]

Further reading about underground coal gasification: [e2m9n]

24 Nuclear?

We made the mistake of lumping nuclear energy in with nuclear weapons, as if all things nuclear were evil. I think that's as big a mistake as if you lumped nuclear medicine in with nuclear weapons.

Patrick Moore,
former Director of Greenpeace International

Nuclear power comes in two flavours. Nuclear *fission* is the flavour that we know how to use in power stations; fission uses uranium, an exceptionally heavy element, as fuel. Nuclear *fusion* is the flavour that we don't yet know how to implement in power stations; fusion would use light elements, especially hydrogen, as its fuel. Fission reactions split up heavy nuclei into medium-sized nuclei, releasing energy. Fusion reactions fuse light nuclei into medium-sized nuclei, releasing energy.

Both forms of nuclear power, fission and fusion, have an important property: the nuclear energy available per atom is roughly one million times bigger than the chemical energy per atom of typical fuels. This means that the amounts of fuel and waste that must be dealt with at a nuclear reactor can be up to one million times smaller than the amounts of fuel and waste at an equivalent fossil-fuel power station.

Let's try to personalize these ideas. The mass of the fossil fuels consumed by "the average British person" is about 16 kg per day (4 kg of coal, 4 kg of oil, and 8 kg of gas). That means that every single day, an amount of fossil fuels with the same weight as 28 pints of milk is extracted from a hole in the ground, transported, processed, and burned somewhere on your behalf. The average Brit's fossil fuel habit creates 11 tons per year of waste carbon dioxide; that's 30 kg per day. In the previous chapter we raised the idea of capturing waste carbon dioxide, compressing it into solid or liquid form, and transporting it somewhere for disposal. Imagine that one person was responsible for capturing and dealing with all their own carbon dioxide waste. 30 kg per day of carbon dioxide is a substantial rucksack-full every day – the same weight as 53 pints of milk!

In contrast, the amount of natural uranium required to provide the same amount of energy as 16 kg of fossil fuels, in a standard fission reactor, is 2 grams; and the resulting waste weighs one quarter of a gram. (This 2 g of uranium is not as small as one millionth of 16 kg per day, by the way, because today's reactors burn up less than 1% of the uranium.) To deliver 2 grams of uranium per day, the miners at the uranium mine would have to deal with perhaps 200 g of ore per day.

So the material streams flowing into and out of nuclear reactors are small, relative to fossil-fuel streams. "Small is beautiful," but the fact that the nuclear waste stream is small doesn't mean that it's not a problem; it's just a "beautifully small" problem.

kWh/d per person

Argentina: 0.5
Armenia: 2.2
Belgium: 12.2
Brazil: 0.17
Bulgaria: 5.0
Canada: 7.4
China: 0.12
Czech Rep.: 6.6
Finland: 11.8
France: 19.0
Germany: 4.4
Hungary: 3.8
India: 0.04
Japan: 5.7
South Korea: 7.7
Lithuania: 6.9
Mexico: 0.26
Netherlands: 0.7
Pakistan: 0.04
Romania: 0.9
Russia: 2.8
Slovakia: 7.2
Slovenia: 7.4
South Africa: 0.8
Spain: 3.6
Sweden: 19.6
Switzerland: 9.7
Taiwan: 4.7
Ukraine: 5.0
UK: 2.6
USA: 7.5

Figure 24.1. Electricity generated per capita from nuclear fission in 2007, in kWh per day per person, in each of the countries with nuclear power.

"Sustainable" power from nuclear fission

Figure 24.1 shows how much electricity was generated globally by nuclear power in 2007, broken down by country.

Could nuclear power be "sustainable"? Leaving aside for a moment the usual questions about safety and waste-disposal, a key question is whether humanity could live for generations on fission. How great are the world-wide supplies of uranium, and other fissionable fuels? Do we have only a few decades' worth of uranium, or do we have enough for millennia?

To estimate a "sustainable" power from uranium, I took the total recoverable uranium in the ground and in seawater, divided it fairly between 6 billion humans, and asked "how fast can we use this if it has to last 1000 years?"

Almost all the recoverable uranium is in the oceans, not in the ground: seawater contains 3.3 mg of uranium per m^3 of water, which adds up to 4.5 billion tons worldwide. I called the uranium in the ocean "recoverable" but this is a bit inaccurate – most ocean waters are quite inaccessible, and the ocean conveyor belt rolls round only once every 1000 years or so; and no-one has yet demonstrated uranium-extraction from seawater on an industrial scale. So we'll make separate estimates for two cases: first using only mined uranium, and second using ocean uranium too.

The uranium ore in the ground that's extractable at prices below $130 per kg of uranium is about one thousandth of this. If prices went above $130 per kg, phosphate deposits that contain uranium at low concentrations would become economic to mine. Recovery of uranium from phosphates is perfectly possible, and was done in America and Belgium before 1998. For the estimate of mined uranium, I'll add both the conventional uranium ore and the phosphates, to give a total resource of 27 million tons of uranium (table 24.2).

We'll consider two ways to use uranium in a reactor: (a) the widely-used *once-through method* gets energy mainly from the ^{235}U (which makes up just 0.7% of uranium), and discards the remaining ^{238}U; (b) *fast breeder reactors*, which are more expensive to build, convert the ^{238}U to fissionable plutonium-239 and obtain roughly 60 times as much energy from the uranium.

Once-through reactors, using uranium from the ground

A once-through **one-gigawatt** nuclear power station uses **162 tons per year of uranium**. So the known mineable resources of uranium, shared between 6 billion people, would last for 1000 years if we produced nuclear power at a rate of 0.55 kWh per day per person. This sustainable rate is the output of just 136 nuclear power stations, and is half of today's nuclear power production. It's very possible this is an underestimate of uranium's potential, since, as there is not yet a uranium shortage, there is no incentive for

	million tons uranium
Australia	1.14
Kazakhstan	0.82
Canada	0.44
USA	0.34
South Africa	0.34
Namibia	0.28
Brazil	0.28
Russian Federation	0.17
Uzbekistan	0.12
World total (conventional reserves in the ground)	4.7
Phosphate deposits	22
Seawater	4 500

Table 24.2. Known recoverable resources of uranium. The top part of the table shows the "reasonable assured resources" and "inferred resources," at cost less than $130 per kg of uranium, as of 1 Jan 2005. These are the estimated resources in areas where exploration has taken place. There's also 1.3 million tons of depleted uranium sitting around in stockpiles, a by-product of previous uranium activities.

Figure 24.3. Workers push uranium slugs into the X-10 Graphite Reactor.

exploration and little uranium exploration has been undertaken since the 1980s; so maybe more mineable uranium will be discovered. Indeed, one paper published in 1980 estimated that the low-grade uranium resource is more than 1000 times greater than the 27 million tons we just assumed.

Could our current once-through use of mined uranium be sustainable? It's hard to say, since there is such uncertainty about the result of future exploration. Certainly at today's rate of consumption, once-through reactors could keep going for hundreds of years. But if we wanted to crank up nuclear power 40-fold worldwide, in order to get off fossil fuels and to allow standards of living to rise, we might worry that once-through reactors are not a sustainable technology.

Figure 24.4. Three Mile Island nuclear power plant.

Fast breeder reactors, using uranium from the ground

Uranium can be used 60 times more efficiently in fast breeder reactors, which burn up all the uranium – both the ^{238}U and the ^{235}U (in contrast to the once-through reactors, which burn mainly ^{235}U). As long as we don't chuck away the spent fuel that is spat out by once-through reactors, this source of depleted uranium could be used too, so uranium that is put in once-through reactors need not be wasted. If we used all the mineable uranium (plus the depleted uranium stockpiles) in 60-times-more-efficient fast breeder reactors, the power would be 33 kWh per day per person. Attitudes to fast breeder reactors range from "this is a dangerous failed experimental technology whereof one should not speak" to "we can and should start building breeder reactors right away." I am not competent to comment on the risks of breeder technology, and I don't want to mix ethical assertions with factual assertions. My aim is just to help understand the numbers. The one ethical position I wish to push is "we should have a plan that adds up."

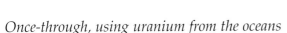

Figure 24.5. Dounreay Nuclear Power Development Establishment, whose primary purpose was the development of fast breeder reactor technology. Photo by John Mullen.

Once-through, using uranium from the oceans

The oceans' uranium, if completely extracted and used in once-through reactors, corresponds to a total energy of

$$\frac{4.5 \text{ billion tons per planet}}{162 \text{ tons uranium per GW-year}} = 28 \text{ million GW-years per planet.}$$

How fast could uranium be extracted from the oceans? The oceans circulate slowly: half of the water is in the Pacific Ocean, and deep Pacific waters circulate to the surface on the great ocean conveyor only every 1600 years. Let's imagine that 10% of the uranium is extracted over such a 1600-year period. That's an extraction rate of 280 000 tons per year. In once-through reactors, this would deliver power at a rate of

$$2.8 \text{ million GW-years } / 1600 \text{ years } = 1750 \text{ GW,}$$

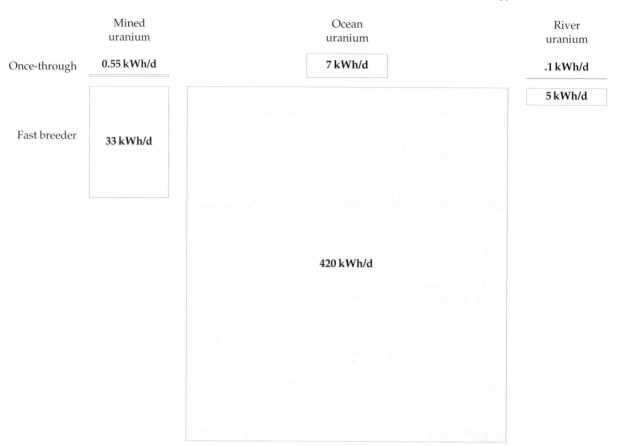

	Mined uranium	Ocean uranium	River uranium
Once-through	0.55 kWh/d	7 kWh/d	.1 kWh/d
			5 kWh/d
Fast breeder	33 kWh/d	420 kWh/d	

Figure 24.6. "Sustainable" power from uranium. For comparison, world nuclear power production today is 1.2 kWh/d per person. British nuclear power production used to be 4 kWh/d per person and is declining.

which, shared between 6 billion people, is 7 kWh per day per person. (There's currently 369 GW of nuclear reactors, so this figure corresponds to a 4-fold increase in nuclear power over today's levels.) I conclude that ocean extraction of uranium would turn today's once-through reactors into a "sustainable" option – assuming that the uranium reactors can cover the energy cost of the ocean extraction process.

Fast breeder reactors, using uranium from the oceans

If fast reactors are 60 times more efficient, the same extraction of ocean uranium could deliver 420 kWh per day per person. At last, a sustainable figure that beats current consumption! – but only with the joint help of two technologies that are respectively scarcely-developed and unfashionable: ocean extraction of uranium, and fast breeder reactors.

Using uranium from rivers

The uranium in the oceans is being topped up by rivers, which deliver uranium at a rate of 32 000 tons per year. If 10% of this influx were captured, it would provide enough fuel for 20 GW of once-through reactors, or 1200 GW of fast breeder reactors. The fast breeder reactors would deliver 5 kWh per day per person.

All these numbers are summarized in figure 24.6.

What about costs?

As usual in this book, my main calculations have paid little attention to economics. However, since the potential contribution of ocean-uranium-based power is one of the biggest in our "sustainable" production list, it seems appropriate to discuss whether this uranium-power figure is at all economically plausible.

Japanese researchers have found a technique for extracting uranium from seawater at a cost of \$100–300 per kilogram of uranium, in comparison with a current cost of about \$20/kg for uranium from ore. Because uranium contains so much more energy per ton than traditional fuels, this 5-fold or 15-fold increase in the cost of uranium would have little effect on the cost of nuclear power: nuclear power's price is dominated by the cost of power-station construction and decommissioning, not by the cost of the fuel. Even a price of \$300/kg would increase the cost of nuclear energy by only about 0.3 p per kWh. The expense of uranium extraction could be reduced by combining it with another use of seawater – for example, power-station cooling.

We're not home yet: does the Japanese technique scale up? What is the energy cost of processing all the seawater? In the Japanese experiment, three cages full of adsorbent uranium-attracting material weighing 350 kg collected "more than 1 kg of yellow cake in 240 days;" this figure corresponds to about 1.6 kg per year. The cages had a cross-sectional area of 48 m^2. To power a once-through 1 GW nuclear power station, we need 160 000 kg per year, which is a production rate 100 000 times greater than the Japanese experiment's. If we simply scaled up the Japanese technique, which accumulated uranium passively from the sea, a power of 1 GW would thus need cages having a collecting area of 4.8 km^2 and containing a weight of 350 000 tons of adsorbent material – more than the weight of the steel in the reactor itself. To put these large numbers in human terms, if uranium were delivering, say, 22 kWh per day per person, each 1 GW reactor would be shared between 1 million people, each of whom needs 0.16 kg of uranium per year. So each person would require one tenth of the Japanese experimental facility, with a weight of 35 kg per person, and an area of 5 m^2 per person. The proposal that such uranium-extraction facilities should be created is thus similar in scale to proposals such as "every person should have 10 m^2 of solar panels" and "every person should have a one-ton car and a dedicated parking place for it." A large investment, yes, but not absurdly off scale. And that was the calculation for once-through reactors. For fast breeder reactors, 60 times less uranium is required, so the mass per person of the uranium collector would be 0.5 kg.

Thorium

Thorium is a radioactive element similar to uranium. Formerly used to make gas mantles, it is about three times as abundant in the earth's crust as uranium. Soil commonly contains around 6 parts per million of thorium, and some minerals contain 12% thorium oxide. Seawater contains little thorium, because thorium oxide is insoluble. Thorium can be completely burned up in simple reactors (in contrast to standard uranium reactors which use only about 1% of natural uranium). Thorium is used in nuclear reactors in India. If uranium ore runs low, thorium will probably become the dominant nuclear fuel.

Thorium reactors deliver 3.6 billion kWh of heat per ton of thorium, which implies that a 1 GW reactor requires about 6 tons of thorium per year, assuming its generators are 40% efficient. Worldwide thorium resources are estimated to total about 6 million tons, four times more than the known reserves shown in table 24.7. As with the uranium resources, it seems plausible that these thorium resources are an underestimate, since thorium prospecting is not highly valued today. If we assume, as with uranium, that these resources are used up over 1000 years and shared equally among 6 billion people, we find that the "sustainable" power thus generated is 4 kWh/d per person.

An alternative nuclear reactor for thorium, the "energy amplifier" or "accelerator-driven system" proposed by Nobel laureate Carlo Rubbia and his colleagues would, they estimated, convert 6 million tons of thorium to 15 000 TWy of energy, or 60 kWh/d per person over 1000 years. Assuming conversion to electricity at 40% efficiency, this would deliver 24 kWh/d per person for 1000 years. And the waste from the energy amplifier would be much less radioactive too. They argue that, in due course, many times more thorium would be economically extractable than the current 6 million tons. If their suggestion – 300 times more – is correct, then thorium and the energy amplifier could offer 120 kWh/d per person for 60 000 years.

Land use

Let's imagine that Britain decides it is serious about getting off fossil fuels, and creates a lot of new nuclear reactors, even though this may not be "sustainable." If we build enough reactors to make possible a significant decarbonization of transport and heating, can we fit the required nuclear reactors into Britain? The number we need to know is the power

Country	Reserves (1000 tons)
Turkey	380
Australia	300
India	290
Norway	170
USA	160
Canada	100
South Africa	35
Brazil	16
Other countries	95
World total	1 580

Table 24.7. Known world thorium resources in monazite (economically extractable).

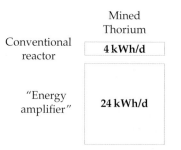

Figure 24.8. Thorium options.

per unit area of nuclear power stations, which is about $1000\,W/m^2$ (figure 24.10). Let's imagine generating 22 kWh per day per person of nuclear power – equivalent to 55 GW (roughly the same as France's nuclear power), which could be delivered by 55 nuclear power stations, each occupying one square kilometre. That's about 0.02% of the area of the country. Wind farms delivering the same average power would require 500 times as much land: 10% of the country. If the nuclear power stations were placed in pairs around the coast (length about 3000 km, at 5 km resolution), then there'd be two every 100 km. Thus while the area required is modest, the fraction of coastline gobbled by these power stations would be about 2% (2 kilometres in every 100).

Economics of cleanup

What's the cost of cleaning up nuclear power sites? The nuclear decommissioning authority has an annual budget of £2 billion for the next 25 years. The nuclear industry sold everyone in the UK 4 kWh/d for about 25 years, so the nuclear decommissioning authority's cost is 2.3 p/kWh. That's a hefty subsidy – though not, it must be said, as hefty as the subsidy currently given to offshore wind (7 p/kWh).

Safety

The safety of nuclear operations in Britain remains a concern. The THORP reprocessing facility at Sellafield, built in 1994 at a cost of £1.8 billion, had a growing leak from a broken pipe from August 2004 to April 2005. Over eight months, the leak let *85 000 litres* of uranium-rich fluid flow into a sump which was equipped with safety systems that were designed to detect immediately any leak of as little as *15 litres*. But the leak went undetected because the operators hadn't completed the checks that ensured the safety systems were working; and the operators were in the habit of ignoring safety alarms anyway.

The safety system came with belt and braces. Independent of the failed safety alarms, routine safety-measurements of fluids in the sump should have detected the abnormal presence of uranium within one month of the start of the leak; but the operators often didn't bother taking these routine measurements, because they felt too busy; and when they *did* take measurements that detected the abnormal presence of uranium in the sump (on 28 August 2004, 26 November 2004, and 24 February 2005), no action was taken.

By April 2005, *22 tons* of uranium had leaked, but still none of the leak-detection systems detected the leak. The leak was finally detected by *accountancy*, when the bean-counters noticed that they were getting 10% less uranium out than their clients claimed they'd put in! Thank goodness this private company had a profit motive, hey? The criticism from the

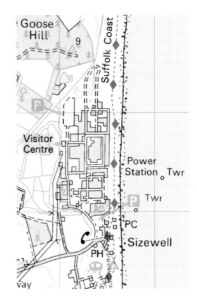

Figure 24.9. Sizewell's power stations. Sizewell A, in the foreground, had a capacity of 420 MW, and was shut down at the end of 2006. Sizewell B, behind, has a capacity of 1.2 GW. Photo by William Connolley.

Figure 24.10. Sizewell occupies less than 1 km². The blue grid's spacing is 1 km. © Crown copyright; Ordnance Survey.

Chief Inspector of Nuclear Installations was withering: "The Plant was operated in a culture that seemed to allow instruments to operate in alarm mode rather than questioning the alarm and rectifying the relevant fault."

If we let private companies build new reactors, how can we ensure that higher safety standards are adhered to? I don't know.

At the same time, we must not let ourselves be swept off our feet in horror at the danger of nuclear power. Nuclear power is not infinitely dangerous. It's just dangerous, much as coal mines, petrol repositories, fossil-fuel burning and wind turbines are dangerous. Even if we have no guarantee against nuclear accidents in the future, I think the right way to assess nuclear is to compare it objectively with other sources of power. Coal power stations, for example, expose the public to nuclear radiation, because coal ash typically contains uranium. Indeed, according to a paper published in the journal *Science*, people in America living near coal-fired power stations are exposed to higher radiation doses than those living near nuclear power plants.

When quantifying the public risks of different power sources, we need a new unit. I'll go with "deaths per GWy (gigawatt-year)." Let me try to convey what it would mean if a power source had a death rate of 1 death per GWy. One gigawatt-year is the energy produced by a 1 GW power station, if it operates flat-out for one year. Britain's electricity consumption is roughly 45 GW, or, if you like, 45 gigawatt-years per year. So if we got our electricity from sources with a death rate of 1 death per GWy, that would mean the British electricity supply system was killing 45 people per year. For comparison, 3000 people die per year on Britain's roads. So, if you are *not* campaigning for the abolition of roads, you may deduce that "1 death per GWy" is a death rate that, while sad, you might be content to live with. Obviously, 0.1 deaths per GWy would be preferable, but it takes only a moment's reflection to realize that, sadly, fossil-fuel energy production must have a cost greater than 0.1 deaths per GWy – just think of disasters on oil rigs; helicopters lost at sea; pipeline fires; refinery explosions; and coal mine accidents: there are tens of fossil-chain fatalities per year in Britain.

So, let's discuss the actual death rates of a range of electricity sources. The death rates vary a lot from country to country. In China, for example, the death rate in coal mines, per ton of coal delivered, is 50 times that of most nations. Figure 24.11 shows numbers from studies by the Paul Scherrer Institute and by a European Union project called ExternE, which made comprehensive estimates of all the impacts of energy production. According to the EU figures, coal, lignite, and oil have the highest death rates, followed by peat and biomass-power, with death rates above 1 per GWy. Nuclear and wind are the best, with death rates below 0.2 per GWy. Hydroelectricity is the best of all according to the EU study, but comes out worst in the Paul Scherrer Institute's study, because the latter surveyed a different set of countries.

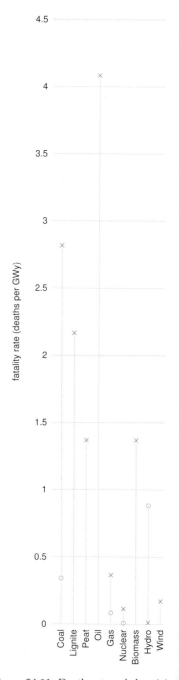

Figure 24.11. Death rates of electricity generation technologies. ×: European Union estimates by the ExternE project. ○: Paul Scherrer Institute.

Inherently safe nuclear power

Spurred on by worries about nuclear accidents, engineers have devised many new reactors with improved safety features. The GT-MHR power plant, for example, is claimed to be inherently safe; and, moreover it has a higher efficiency of conversion of heat to electricity than conventional nuclear plants [gt-mhr.ga.com].

Mythconceptions

Two widely-cited defects of nuclear power are construction costs, and waste. Let's examine some aspects of these issues.

Building a nuclear power station requires **huge** *amounts of concrete and steel, materials whose creation involves* **huge** *CO_2 pollution.*

The steel and concrete in a 1 GW nuclear power station have a carbon footprint of roughly 300 000 t CO_2.

Spreading this "huge" number over a 25-year reactor life we can express this contribution to the carbon intensity in the standard units (g CO_2 per kWh(e)),

$$\text{carbon intensity associated with construction} = \frac{300 \times 10^9 \, \text{g}}{10^6 \, \text{kW(e)} \times 220\,000 \, \text{h}}$$
$$= 1.4 \, \text{g/kWh(e)},$$

which is much smaller than the fossil-fuel benchmark of 400 g CO_2/kWh(e). The IPCC estimates that the *total* carbon intensity of nuclear power (including construction, fuel processing, and decommissioning) is less than 40 g CO_2/kWh(e) (Sims et al., 2007).

Please don't get me wrong: I'm not trying to be pro-nuclear. I'm just pro-arithmetic.

Isn't the waste from nuclear reactors a huge problem?

As we noted in the opening of this chapter, the volume of waste from nuclear reactors is relatively small. Whereas the ash from ten coal-fired power stations would have a mass of four million tons per year (having a volume of roughly 40 litres per person per year), the nuclear waste from Britain's ten nuclear power stations has a volume of just 0.84 litres per person per year – think of that as a bottle of wine per person per year (figure 24.13).

Most of this waste is low-level waste. 7% is intermediate-level waste, and just 3% of it – 25 ml per year – is high-level waste.

The high-level waste is the really nasty stuff. It's conventional to keep the high-level waste at the reactor for its first 40 years. It is stored in pools of water and cooled. After 40 years, the level of radioactivity has dropped 1000-fold. The level of radioactivity continues to fall; after 1000 years, the

Figure 24.12. Chernobyl power plant (top), and the abandoned town of Prypiat, which used to serve it (bottom). Photos by Nik Stanbridge.

radioactivity of the high-level waste is about the same as that of uranium ore. Thus waste storage engineers need to make a plan to secure high-level waste for about 1000 years.

Is this a difficult problem? 1000 years is certainly a long time compared with the lifetimes of governments and countries! But the volumes are so small, I feel nuclear waste is only a minor worry, compared with all the other forms of waste we are inflicting on future generations. At 25 ml per year, a lifetime's worth of high-level nuclear waste would amount to less than 2 litres. Even when we multiply by 60 million people, the lifetime volume of nuclear waste doesn't sound unmanageable: 105 000 cubic metres. That's the same volume as 35 olympic swimming pools. If this waste were put in a layer one metre deep, it would occupy just one tenth of a square kilometre.

There are already plenty of places that are off-limits to humans. I may not trespass in your garden. Nor should you in mine. We are neither of us welcome in Balmoral. "Keep out" signs are everywhere. Downing Street, Heathrow airport, military facilities, disused mines – they're all off limits. Is it impossible to imagine making another one-square-kilometre spot – perhaps deep underground – off limits for 1000 years?

Compare this 25 ml per year per person of high-level nuclear waste with the other traditional forms of waste we currently dump: municipal waste – 517 kg per year per person; hazardous waste – 83 kg per year per person.

People sometimes compare possible new nuclear waste with the nuclear waste we already have to deal with, thanks to our existing old reactors. Here are the numbers for the UK. The projected volume of "higher activity wastes" up to 2120, following decommissioning of existing nuclear facilities, is 478 000 m^3. Of this volume, 2% (about 10 000 m^3) will be the high level waste (1290 m^3) and spent fuel (8150 m^3) that together contain 92% of the activity. Building 10 new nuclear reactors (10 GW) would add another 31 900 m^3 of spent fuel to this total. That's the same volume as ten swimming pools.

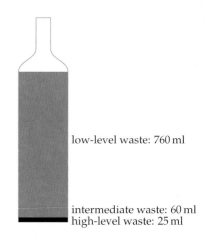

low-level waste: 760 ml

intermediate waste: 60 ml
high-level waste: 25 ml

Figure 24.13. British nuclear waste, per person, per year, has a volume just a little larger than one wine bottle.

If we got lots and lots of power from nuclear fission or fusion, wouldn't this contribute to global warming, because of all the extra energy being released into the environment?

That's a fun question. And because we've carefully expressed everything in this book in a single set of units, it's quite easy to answer. First, let's recap the key numbers about global energy balance from p20: the average solar power absorbed by atmosphere, land, and oceans is 238 W/m^2; doubling the atmospheric CO_2 concentration would effectively increase the net heating by 4 W/m^2. This 1.7% increase in heating is believed to be bad news for climate. Variations in solar power during the 11-year solar cycle have a range of 0.25 W/m^2. So now let's assume that in 100 years or so, the world population is 10 billion, and everyone is living at a European standard of living, using 125 kWh per day derived from fossil sources, from nuclear power, or from mined geothermal power. The area of the earth per person would be 51 000 m^2. Dividing the power per person by the area per person, we find that the extra power contributed by human energy use would be 0.1 W/m^2. That's one fortieth of the 4 W/m^2 that we're currently fretting about, and a little smaller than the 0.25 W/m^2 effect of solar variations. So yes, under these assumptions, human power production would *just* show up as a contributor to global climate change.

I heard that nuclear power can't be built at a sufficient rate to make a useful contribution.

The difficulty of building nuclear power fast has been exaggerated with the help of a misleading presentation technique I call "the magic playing field." In this technique, two things appear to be compared, but the basis of the comparison is switched halfway through. The Guardian's environment editor, summarizing a report from the *Oxford Research Group*, wrote "For nuclear power to make any significant contribution to a reduction in global carbon emissions in the next two generations, the industry would have to construct nearly 3000 new reactors – or about one a week for 60 years. A civil nuclear construction and supply programme on this scale is a pipe dream, and completely unfeasible. The highest historic rate is 3.4 new reactors a year." 3000 sounds much bigger than 3.4, doesn't it! In this application of the "magic playing field" technique, there is a switch not only of timescale but also of *region*. While the first figure (3000 new reactors over 60 years) is the number required *for the whole planet*, the second figure (3.4 new reactors per year) is the maximum rate of building by a *single country* (France)!

A more honest presentation would have kept the comparison on a per-planet basis. France has 59 of the world's 429 operating nuclear reactors, so it's plausible that the highest rate of reactor building for the whole planet was something like ten times France's, that is, 34 new reactors per year. And the required rate (3000 new reactors over 60 years) is 50 new reactors per year. So the assertion that "civil nuclear construction on this scale is a pipe dream, and completely unfeasible" is poppycock. Yes, it's a big construction rate, but it's in the same ballpark as historical construction rates.

How reasonable is my assertion that the world's maximum historical construction rate must have been about 34 new nuclear reactors per year? Let's look at the data. Figure 24.14 shows the power of the world's nuclear fleet as a function of time, showing only the power stations still operational in 2007. The rate of new build was biggest in 1984, and had a value of (drum-roll please...) about 30 GW per year – about 30 1-GW reactors. So there!

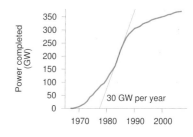

Figure 24.14. Graph of the total nuclear power in the world that was built since 1967 and that is still operational today. The world construction rate peaked at 30 GW of nuclear power per year in 1984.

What about nuclear fusion?

We say that we will put the sun into a box. The idea is pretty. The problem is, we don't know how to make the box.

Sébastien Balibar, Director of Research, CNRS

Fusion power is speculative and experimental. I think it is reckless to assume that the fusion problem *will* be cracked, but I'm happy to estimate how much power fusion could deliver, *if* the problem is cracked.

The two fusion reactions that are considered the most promising are:

the DT reaction, which fuses deuterium with tritium, making helium; and

the DD reaction, which fuses deuterium with deuterium.

Deuterium, a naturally occurring heavy isotope of hydrogen, can be obtained from seawater; tritium, a heavier isotope of hydrogen, isn't found in large quantities naturally (because it has a half-life of only 12 years) but it can be manufactured from lithium.

ITER is an international project to figure out how to make a steadily-working fusion reactor. The ITER prototype will use the DT reaction. DT is preferred over DD, because the DT reaction yields more energy and because it requires a temperature of "only" 100 million °C to get it going, whereas the DD reaction requires 300 million °C. (The maximum temperature in the sun is 15 million °C.)

Let's fantasize, and assume that the ITER project is successful. What sustainable power could fusion then deliver? Power stations using the DT reaction, fuelled by lithium, will run out of juice when the lithium runs out. Before that time, hopefully the second installment of the fantasy will have arrived: fusion reactors using deuterium alone.

I'll call these two fantasy energy sources "lithium fusion" and "deuterium fusion," naming them after the principal fuel we'd worry about in each case. Let's now estimate how much energy each of these sources could deliver.

Lithium fusion

World lithium reserves are estimated to be 9.5 million tons in ore deposits. If all these reserves were devoted to fusion over 1000 years, the power delivered would be 10 kWh/d per person.

There's another source for lithium: seawater, where lithium has a concentration of 0.17 ppm. To produce lithium at a rate of 100 million kg per year from seawater is estimated to have an energy requirement of 2.5 kWh(e) per gram of lithium. If the fusion reactors give back 2300 kWh(e) per gram of lithium, the power thus delivered would be 105 kWh/d per person (assuming 6 billion people). At this rate, the lithium in the oceans would last more than a million years.

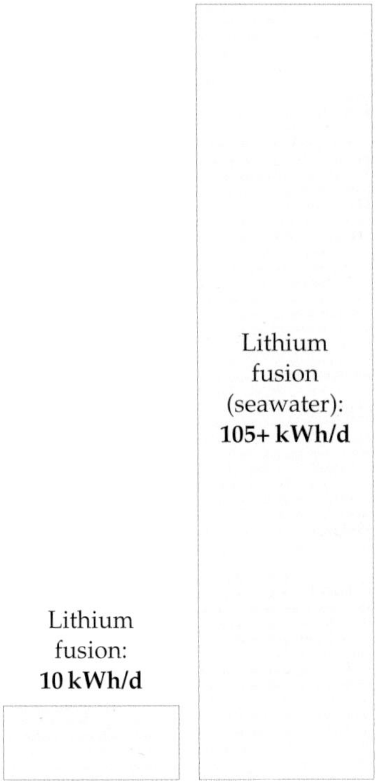

Figure 24.15. The inside of an experimental fusion reactor. Split image showing the JET vacuum vessel with a superimposed image of a JET plasma, taken with an ordinary TV camera. Photo: EFDA-JET.

Lithium
fusion
(seawater):
105+ kWh/d

Lithium
fusion:
10 kWh/d

Figure 24.16. Lithium-based fusion, if used fairly and "sustainably," could match our current levels of consumption. Mined lithium would deliver 10 kWh/d per person for 1000 years; lithium extracted from seawater could deliver 105 kWh/d per person for over a million years.

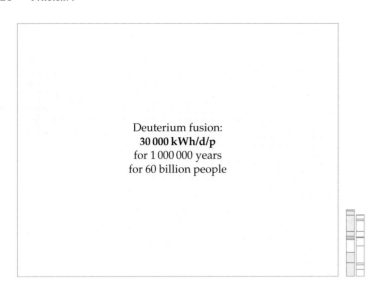

Deuterium fusion:
30 000 kWh/d/p
for 1 000 000 years
for 60 billion people

Figure 24.17. Deuterium-based fusion, if it is achievable, offers plentiful sustainable energy for millions of years. This diagram's scale is shrunk ten-fold in each dimension so as to fit fusion's potential contribution on the page. The red and green stacks from figure 18.1 are shown to the same scale, for comparison.

Deuterium fusion

If we imagine that scientists and engineers crack the problem of getting the DD reaction going, we have some very good news. There's 33 g of deuterium in every ton of water, and the energy that would be released from fusing just one gram of deuterium is a mind-boggling 100 000 kWh. Bearing in mind that the mass of the oceans is 230 million tons per person, we can deduce that there's enough deuterium to supply every person in a ten-fold increased world population with a power of 30 000 kWh per day (that's more than 100 times the average American consumption) for 1 million years (figure 24.17).

Notes and further reading

page no.

161 *Figure 24.1.* Source: World Nuclear Association [5qntkb]. The total capacity of operable nuclear reactors is 372 GW(e), using 65 000 tons of uranium per year. The USA has 99 GW, France 63.5 GW, Japan 47.6 GW, Russia 22 GW, Germany 20 GW, South Korea 17.5 GW, Ukraine 13 GW, Canada 12.6 GW, and UK 11 GW. In 2007 all the world's reactors generated 2608 TWh of electricity, which is an average of 300 GW, or 1.2 kWh per day per person.

162 *Fast breeder reactors obtain 60 times as much energy from the uranium.* Source: www.world-nuclear.org/info/inf98. html. Japan currently leads the development of fast breeder reactors.

– *A once-through one-gigawatt nuclear power station uses 162 tons per year of uranium.*
Source: www.world-nuclear.org/info/inf03.html. A 1 GW(e) station with a thermal efficiency of 33% running at a load factor of 83% has the following upstream footprint: mining – 16 600 tons of 1%-uranium ore; milling – 191 t of uranium oxide (containing 162 t of natural uranium); enrichment and fuel fabrication – 22.4 t of uranium oxide (containing 20 t of enriched uranium). The enrichment requires 115 000 SWU; see p102 for the energy cost of SWU (separative work units).

163 *it's been estimated that the low-grade uranium resource is more than 1000 times greater than the 22 million tons we just assumed.* Deffeyes and MacGregor (1980) estimate that the resource of uranium in concentrations of 30 ppm or more is 3×10^{10} tons. (The average ore grade processed in South Africa in 1985 and 1990 was 150 ppm. Phosphates typically average 100 ppm.)
Here's what the World Nuclear Association said on the topic of uranium reserves in June 2008:
"From time to time concerns are raised that the known resources might be insufficient when judged as a multiple of present rate of use. But this is the Limits to Growth fallacy, . . . which takes no account of the very limited nature of the knowledge we have at any time of what is actually in the Earth's crust. Our knowledge of geology is such that we can be confident that identified resources of metal minerals are a small fraction of what is there.
"Measured resources of uranium, the amount known to be economically recoverable from orebodies, are . . . dependent on the intensity of past exploration effort, and are basically a statement about what is known rather than what is there in the Earth's crust.
"The world's present measured resources of uranium (5.5 Mt) . . . are enough to last for over 80 years. This represents a higher level of assured resources than is normal for most minerals. Further exploration and higher prices will certainly, on the basis of present geological knowledge, yield further resources as present ones are used up."
"Economically rational players will only invest in finding these new reserves when they are most confident of gaining a return from them, which usually requires positive price messages caused by undersupply trends. If the economic system is working correctly and maximizing capital efficiency, there should never be more than a few decades of any resource commodity in reserves at any point in time."
[Exploration has a cost; exploring for uranium, for example, has had a cost of \$1–\$1.50 per kg of uranium (\$3.4/MJ), which is 2% of the spot price of \$78/kgU; in contrast, the finding costs of crude oil have averaged around \$6/barrel (\$1050/MJ) (12% of the spot price) over at least the past three decades.]
"Unlike the metals which have been in demand for centuries, society has barely begun to utilize uranium. There has been only one cycle of exploration-discovery-production, driven in large part by late 1970s price peaks.
"It is premature to speak about long-term uranium scarcity when the entire nuclear industry is so young that only one cycle of resource replenishment has been required." www.world-nuclear.org/info/inf75.html
Further reading: Herring (2004); Price and Blaise (2002); Cohen (1983).
The IPCC, citing the OECD, project that at the 2004 utilization levels, the uranium in conventional resources and phosphates would last 670 years in once-through reactors, 20 000 years in fast reactors with plutonium recycling, and 160 000 years in fast reactors recycling uranium and all actinides (Sims et al., 2007).

165 *Japanese researchers have found a technique for extracting uranium from seawater.* The price estimate of \$100 per kg is from Seko et al. (2003) and [y3wnzr]; the estimate of \$300 per kg is from OECD Nuclear Energy Agency (2006, p130). The uranium extraction technique involves dunking tissue in the ocean for a couple of months; the tissue is made of polymer fibres that are rendered sticky by irradiating them before they are dunked; the sticky fibres collect uranium to the tune of 2 g of uranium per kilogram of fibre.

–   *The expense of uranium extraction could be reduced by combining it with another use of seawater – for example, power-station cooling.* The idea of a nuclear-powered island producing hydrogen was floated by C. Marchetti. Breeder reactors would be cooled by seawater and would extract uranium from the cooling water at a rate of 600 t uranium per 500 000 Mt of seawater.

166 *Thorium reactors deliver 3.6×10^9 kWh of heat per ton of thorium.* Source: www.world-nuclear.org/info/inf62.html. There remains scope for advancement in thorium reactors, so this figure could be bumped up in the future.

–   *An alternative nuclear reactor for thorium, the "energy amplifier". . .* See Rubbia et al. (1995), web.ift.uib.no/~lillestol/Energy_Web/EA.html, [32t5zt], [2qr3yr], [ynk54y].

–   *World thorium resources in monazite.* source: US Geological Survey, Mineral Commodity Summaries, January 1999. [yl7tkm] Quoted in UIC Nuclear Issues Briefing Paper #67 November 2004.
"Other ore minerals with higher thorium contents, such as thorite, would be more likely sources if demand significantly increased."

[yju4a4] omits the figure for Turkey, which is found here: [yeyr7z].

167 *The nuclear industry sold everyone in the UK 4 kWh/d for about 25 years.* The total generated to 2006 was about 2200 TWh. Source: Stephen Salter's Energy Review for the Scottish National Party.

– *The nuclear decommissioning authority has an annual budget of £2 billion.* In fact, this clean-up budget seems to rise and rise. The latest figure for the total cost of decommissioning is £73 billion. news.bbc.co.uk/1/hi/uk/7215688.stm

168 *The criticism of the Chief Inspector of Nuclear Installations was withering...* (Weightman, 2007).

– *Nuclear power is not infinitely dangerous. It's just dangerous.* Further reading on risk: Kammen and Hassenzahl (1999).

– *People in America living near coal-fired power stations are exposed to higher radiation doses than those living near nuclear power plants.* Source: McBride et al. (1978). Uranium and thorium have concentrations of roughly 1 ppm and 2 ppm respectively in coal.
    Further reading: gabe.web.psi.ch/research/ra/ra_res.html,
    www.physics.ohio-state.edu/~wilkins/energy/Companion/E20.12.pdf.xpdf.

– *Nuclear power and wind power have the lowest death rates.* See also Jones (1984). These death rates are from studies that are predicting the future. We can also look in the past.
    In Britain, nuclear power has generated 200 GWy of electricity, and the nuclear industry has had 1 fatality, a worker who died at Chapelcross in 1978 [4f2ekz]. One death per 200 GWy is an impressively low death rate compared with the fossil fuel industry.
    Worldwide, the nuclear-power historical death rate is hard to estimate. The Three Mile Island meltdown killed no-one, and the associated leaks are estimated to have perhaps killed one person in the time since the accident. The accident at Chernobyl first killed 62 who died directly from exposure, and 15 local people who died later of thyroid cancer; it's estimated that nearby, another 4000 died of cancer, and that worldwide, about 5000 people (among 7 million who were exposed to fallout) died of cancer because of Chernobyl (Williams and Baverstock, 2006); but these deaths are impossible to detect because cancers, many of them caused by natural nuclear radiation, already cause 25% of deaths in Europe.
    One way to estimate a global death rate from nuclear power worldwide is to divide this estimate of Chernobyl's death-toll (9000 deaths) by the cumulative output of nuclear power from 1969 to 1996, which was 3685 GWy. This gives a death rate of 2.4 deaths per GWy.
    As for deaths attributed to wind, Caithness Windfarm Information Forum www.caithnesswindfarms.co.uk list 49 fatalities worldwide from 1970 to 2007 (35 wind industry workers and 14 members of the public). In 2007, Paul Gipe listed 34 deaths total worldwide [www.wind-works.org/articles/BreathLife.html]. In the mid-1990s the mortality rate associated with wind power was 3.5 deaths per GWy. According to Paul Gipe, the worldwide mortality rate of wind power dropped to 1.3 deaths per GWy by the end of 2000.
    So the historical death rates of both nuclear power and wind are higher than the predicted future death rates.

169 *The steel and concrete in a 1 GW nuclear power station have a carbon footprint of roughly 300 000 t CO_2.* A 1 GW nuclear power station contains 520 000 m^3 of concrete (1.2 million tons) and 67 000 tons of steel [2k8y7o]. Assuming 240 kg CO_2 per m^3 of concrete [3pvf4j], the concrete's footprint is around 100 000 t CO_2. From Blue Scope Steel [4r7zpg], the footprint of steel is about 2.5 tons of CO_2 per ton of steel. So the 67 000 tons of steel has a footprint of about 170 000 tons of CO_2.

170 *Nuclear waste discussion.* Sources: www.world-nuclear.org/info/inf04.html, [49hcnw], [3kduo7].
    New nuclear waste compared with old. Committee on Radioactive Waste Management (2006).

172 *World lithium reserves are estimated as 9.5 million tons.* The main lithium sources are found in Bolivia (56.6%), Chile (31.4%) and the USA (4.3%). www.dnpm.gov.br

– *There's another source for lithium: seawater...* Several extraction techniques have been investigated (Steinberg and Dang, 1975; Tsuruta, 2005; Chitrakar et al., 2001).

– *Fusion power from lithium reserves.*
  The energy density of natural lithium is about 7500 kWh per gram (Ongena and Van Oost, 2006). There's considerable variation among the estimates of how efficiently fusion reactors would turn this into electricity, ranging from 310 kWh(e)/g (Eckhartt, 1995) to 3400 kWh(e)/g of natural lithium (Steinberg and Dang, 1975). I've assumed 2300 kWh(e)/g, based on this widely quoted summary figure: "A 1 GW fusion plant will use about 100 kg of deuterium and 3 tons of natural lithium per year, generating about 7 billion kWh." [69vt8r], [6oby22], [63121p].

Further reading about fission: Hodgson (1999), Nuttall (2004), Rogner (2000), Williams (2000). Uranium Information Center
  – www.uic.com.au. www.world-nuclear.org, [wnchw].
  On costs: Zaleski (2005).
  On waste repositories: [shrln].
  On breeder reactors and thorium: www.energyfromthorium.com.

Further reading about fusion: www.fusion.org.uk, www.askmar.com/Fusion.html.

25 Living on other countries' renewables?

Whether the Mediterranean becomes an area of cooperation or confrontation in the 21st century will be of strategic importance to our common security.

Joschka Fischer, German Foreign Minister, February 2004

We've found that it's hard to get off fossil fuels by living on our own renewables. Nuclear has its problems too. So what else can we do? Well, how about living on someone else's renewables? (Not that we have any entitlement to someone else's renewables, of course, but perhaps they might be interested in selling them to us.)

Most of the resources for living sustainably are related to land area: if you want to use solar panels, you need land to put them on; if you want to grow crops, you need land again. Jared Diamond, in his book *Collapse*, observes that, while many factors contribute to the collapse of civilizations, a common feature of all collapses is that the human population density became too great.

Places like Britain and Europe are in a pickle because they have large population densities, and all the available renewables are diffuse – they have small power density (table 25.1). When looking for help, we should look to countries that have three things: *a)* low population density; *b)* large area; and *c)* a renewable power supply with high power density.

POWER PER UNIT LAND OR WATER AREA	
Wind	$2\,W/m^2$
Offshore wind	$3\,W/m^2$
Tidal pools	$3\,W/m^2$
Tidal stream	$6\,W/m^2$
Solar PV panels	$5–20\,W/m^2$
Plants	$0.5\,W/m^2$
Rain-water (highlands)	$0.24\,W/m^2$
Hydroelectric facility	$11\,W/m^2$
Solar chimney	$0.1\,W/m^2$
Concentrating solar power (desert)	**$15\,W/m^2$**

Table 25.1. Renewable facilities have to be country-sized because all renewables are so diffuse.

Region	Population	Area (km^2)	Density (persons per km^2)	Area per person (m^2)
Libya	5 760 000	1 750 000	3	305 000
Kazakhstan	15 100 000	2 710 000	6	178 000
Saudi Arabia	26 400 000	1 960 000	13	74 200
Algeria	32 500 000	2 380 000	14	73 200
Sudan	40 100 000	2 500 000	16	62 300
World	6 440 000 000	148 000 000	43	23 100
Scotland	5 050 000	78 700	64	15 500
European Union	496 000 000	4 330 000	115	8 720
Wales	2 910 000	20 700	140	7 110
United Kingdom	59 500 000	244 000	243	4 110
England	49 600 000	130 000	380	2 630

Table 25.2. Some regions, ordered from small to large population density. See p338 for more population densities.

Table 25.2 highlights some countries that fit the bill. Libya's population density, for example, is 70 times smaller than Britain's, and its area is 7 times bigger. Other large, area-rich, countries are Kazakhstan, Saudi Arabia, Algeria, and Sudan.

In all these countries, I think the most promising renewable is so-lar power, *concentrating solar power* in particular, which uses mirrors or lenses to focus sunlight. Concentrating solar power stations come in several flavours, arranging their moving mirrors in various geometries, and putting various power conversion technologies at the focus – Stirling engines, pressurized water, or molten salt, for example – but they all deliver fairly similar average powers per unit area, in the ballpark of $15\,\text{W/m}^2$.

A technology that adds up

"All the world's power could be provided by a square 100 km by 100 km in the Sahara." Is this true? Concentrating solar power in deserts delivers an average power per unit land area of roughly $15\,\text{W/m}^2$. So, allowing no space for anything else in such a square, the power delivered would be 150 GW. This is *not* the same as current world power consumption. It's not even near current world *electricity* consumption, which is 2000 GW. World power consumption today is 15 000 GW. So the correct statement about power from the Sahara is that today's consumption could be provided by a *1000 km by 1000 km* square in the desert, completely filled with concentrating solar power. That's four times the area of the UK. And if we are interested in living in an equitable world, we should presumably aim to supply more than *today's* consumption. To supply every person in the world with an average European's power consumption (125 kWh/d), the area required would be *two* 1000 km by 1000 km squares in the desert.

Fortunately, the Sahara is not the only desert, so maybe it's more relevant to chop the world into smaller regions, and ask what area is needed in each region's local desert. So, focusing on Europe, "what area is required in the North Sahara to supply *everyone in Europe and North Africa* with an average European's power consumption? Taking the population of Europe and North Africa to be 1 billion, the area required drops to $340\,000\,\text{km}^2$, which corresponds to a square **600 km by 600 km**. This area is equal to one Germany, to 1.4 United Kingdoms, or to **16 Waleses**.

The UK's share of this 16-Wales area would be one Wales: a 145 km by 145 km square in the Sahara would provide all the UK's current primary energy consumption. These squares are shown in figure 25.5. Notice that while the yellow square may look "little" compared with Africa, it does have the same area as Germany.

The DESERTEC plan

An organization called DESERTEC [www.desertec.org] is promoting a plan to use concentrating solar power in sunny Mediterranean countries, and high-voltage direct-current (HVDC) transmission lines (figure 25.7) to deliver the power to cloudier northern parts. HVDC technology has been in use since 1954 to transmit power both through overhead lines and through

Figure 25.3. Stirling dish engine. These beautiful concentrators deliver a power per unit land area of $14\,\text{W/m}^2$. Photo courtesy of Stirling Energy Systems. www.stirlingenergy.com

Figure 25.4. Andasol – a "100 MW" solar power station under construction in Spain. Excess thermal energy produced during the day will be stored in liquid salt tanks for up to seven hours, allowing a continuous and stable supply of electric power to the grid. The power station is predicted to produce 350 GWh per year (40 MW). The parabolic troughs occupy 400 hectares, so the power per unit land area will be $10\,\text{W/m}^2$. Upper photo: ABB. Lower photo: IEA SolarPACES.

Figure 25.5. The celebrated little square. This map shows a square of size 600 km by 600 km in Africa, and another in Saudi Arabia, Jordan, and Iraq. Concentrating solar power facilities completely filling one such square would provide enough power to give 1 billion people the average European's consumption of 125 kWh/d. The area of one square is the same as the area of Germany, and 16 times the area of Wales. Within each big square is a smaller 145 km by 145 km square showing the area required in the Sahara – one Wales – to supply all British power consumption.

submarine cables (such as the interconnector between France and England). It is already used to transmit electricity over 1000-km distances in South Africa, China, America, Canada, Brazil, and Congo. A typical 500 kV line can transmit a power of 2 GW. A pair of HVDC lines in Brazil transmits 6.3 GW.

HVDC is preferred over traditional high-voltage AC lines because less physical hardware is needed, less land area is needed, and the power losses of HVDC are smaller. The power losses on a 3500 km-long HVDC line, including conversion from AC to DC and back, would be about 15%. A further advantage of HVDC systems is that they help stabilize the electricity networks to which they are connected.

In the DESERTEC plans, the prime areas to exploit are coastal areas, because concentrating solar power stations that are near to the sea can deliver desalinated water as a by-product – valuable for human use, and for agriculture.

Table 25.6 shows DESERTEC's estimates of the potential power that

Country	Economic potential (TWh/y)	Coastal potential (TWh/y)
Algeria	169 000	60
Libya	140 000	500
Saudi Arabia	125 000	2 000
Egypt	74 000	500
Iraq	29 000	60
Morocco	20 000	300
Oman	19 000	500
Syria	10 000	0
Tunisia	9 200	350
Jordan	6 400	0
Yemen	5 100	390
Israel	3 100	1
UAE	2 000	540
Kuwait	1 500	130
Spain	1 300	70
Qatar	800	320
Portugal	140	7
Turkey	130	12
Total	620 000 (70 000 GW)	6 000 (650 GW)

Table 25.6. Solar power potential in countries around and near to Europe. The "economic potential" is the power that could be generated in suitable places where the direct normal irradiance is more than 2000 kWh/m^2/y.
The "coastal potential" is the power that could be generated within 20 m (vertical) of sea level; such power is especially promising because of the potential combination with desalination.
For comparison, the total power required to give 125 kWh per day to 1 billion people is 46 000 TWh/y (5 200 GW). 6000 TWh/y (650 GW) is 16 kWh per day per person for 1 billion people.

could be produced in countries in Europe and North Africa. The "economic potential" adds up to more than enough to supply 125 kWh per day to 1 billion people. The total "coastal potential" is enough to supply 16 kWh per day per person to 1 billion people.

Let's try to convey on a map what a realistic plan could look like. Imagine making solar facilities each having an area of 1500 km^2 – that's roughly the size of London. (Greater London has an area of 1580 km^2; the M25 orbital motorway around London encloses an area of 2300 km^2.) Let's call each facility a *blob*. Imagine that in each of these blobs, half the area is devoted to concentrating power stations with an average power density of 15 W/m^2, leaving space around for agriculture, buildings, railways, roads, pipelines, and cables. Allowing for 10% transmission loss between the blob and the consumer, each of these blobs generates an average power of 10 GW. Figure 25.8 shows some blobs to scale on a map. To give a sense of the scale of these blobs I've dropped a few in Britain too. *Four of these blobs would have an output roughly equal to Britain's total electricity consumption (16 kWh/d per person for 60 million people). Sixty-five blobs would provide all one billion people in Europe and North Africa with 16 kWh/d per person. Figure 25.8 shows 68 blobs in the desert.

Figure 25.7. Laying a high-voltage DC link between Finland and Estonia. A pair of these cables transmit a power of 350 MW. Photo: ABB.

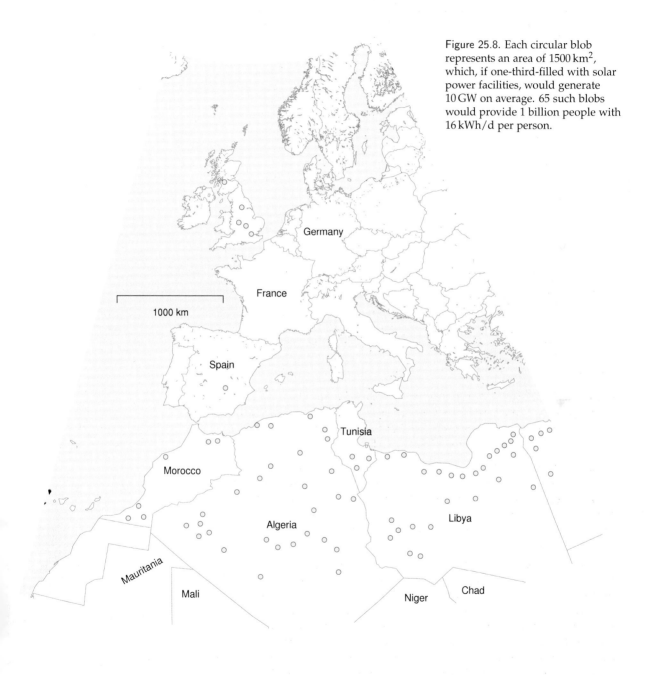

Figure 25.8. Each circular blob represents an area of 1500 km², which, if one-third-filled with solar power facilities, would generate 10 GW on average. 65 such blobs would provide 1 billion people with 16 kWh/d per person.

Concentrating photovoltaics

An alternative to concentrating thermal solar power in deserts is large-scale concentrating photovoltaic systems. To make these, we plop a high-quality electricity-producing solar cell at the focus of cheap lenses or mirrors. Faiman et al. (2007) say that "solar, in its concentrator photovoltaics variety, can be completely cost-competitive with fossil fuel [in desert states such as California, Arizona, New Mexico, and Texas] without the need for any kind of subsidy."

According to manufacturers Amonix, this form of concentrating solar power would have an average power per unit land area of $18 \, W/m^2$.

Another way to get a feel for required hardware is to personalize. One of the "25 kW" (peak) collectors shown in figure 25.9 generates on average about 138 kWh per day; the American lifestyle currently uses 250 kWh per day per person. So to get the USA off fossil fuels using solar power, we need roughly two of these 15 m × 15 m collectors per person.

Figure 25.9. A 25 kW (peak) concentrator photovoltaic collector produced by Californian company Amonix. Its 225 m^2 aperture contains 5760 Fresnel lenses with optical concentration ×260, each of which illuminates a 25%-efficient silicon cell. One such collector, in an appropriate desert location, generates 138 kWh per day – enough to cover the energy consumption of half an American. Note the human providing a scale. Photo by David Faiman.

Queries

I'm confused! In Chapter 6, you said that the best photovoltaic panels deliver 20 W/m^2 on average, in a place with British sunniness. Presumably in the desert the same panels would deliver 40 W/m^2. So how come the concentrating solar power stations deliver only 15–20 W/m^2? Surely concentrating power should be even better than plain flat panels?

Good question. The short answer is no. Concentrating solar power does not achieve a better power per unit land area than flat panels. The concentrating contraption has to track the sun, otherwise the sunlight won't be focused right; once you start packing land with sun-tracking contraptions, you have to leave gaps between them; lots of sunlight falls through the gaps and is lost. The reason that people nevertheless make concentrating solar power systems is that, today, flat photovoltaic panels are very expensive, and concentrating systems are cheaper. The concentrating people's goal is not to make systems with big power per unit land area. Land area is cheap (they assume). The goal is to deliver big power per dollar.

But if flat panels have bigger power density, why don't you describe covering the Sahara desert with them?

Because I am trying to discuss practical options for large-scale sustainable power production for Europe and North Africa by 2050. My guess is that by 2050, mirrors will still be cheaper than photovoltaic panels, so concentrating solar power is the technology on which we should focus.

What about solar chimneys?

A solar chimney or solar updraft tower uses solar power in a very simple way. A huge chimney is built at the centre of an area covered by a transparent roof made of glass or plastic; because hot air rises, hot air created in this greenhouse-like heat-collector whooshes up the chimney, drawing in cooler air from the perimeter of the heat-collector. Power is extracted from the air-flow by turbines at the base of the chimney. Solar chimneys are fairly simple to build, but they don't deliver a very impressive power per unit area. A pilot plant in Manzanares, Spain operated for seven years between 1982 and 1989. The chimney had a height of 195 m and a diameter of 10 m; the collector had a diameter of 240 m, and its roof had 6000 m² of glass and 40 000 m² of transparent plastic. It generated 44 MWh per year, which corresponds to a power per unit area of $0.1\,W/m^2$. Theoretically, the bigger the collector and the taller the chimney, the bigger the power density of a solar chimney becomes. The engineers behind Manzanares reckon that, at a site with a solar radiation of 2300 kWh/m² per year (262 W/m²), a 1000 m-high tower surrounded by a 7 km-diameter collector could generate 680 GWh per year, an average power of 78 MW. That's a power per unit area of about $1.6\,W/m^2$, which is similar to the power per unit area of windfarms in Britain, and one tenth of the power per unit area I said concentrating solar power stations would deliver. It's claimed that solar chimneys could generate electricity at a price similar to that of conventional power stations. I suggest that countries that have enough land and sunshine to spare should host a big bake-off contest between solar chimneys and concentrating solar power, to be funded by oil-producing and oil-consuming countries.

Figure 25.10. The Manzanares prototype solar chimney. Photos from `solarmillennium.de`.

What about getting power from Iceland, where geothermal power and hydroelectricity are so plentiful?

Indeed, Iceland already effectively exports energy by powering industries that make energy-intensive products. Iceland produces nearly one ton of aluminium per citizen per year, for example! So from Iceland's point of view, there are great profits to be made. But can Iceland save Europe? I would be surprised if Iceland's power production could be scaled up enough to make sizeable electricity exports even to Britain alone. As a benchmark, let's compare with the England–France Interconnector, which can deliver up to 2 GW across the English Channel. That maximum power is equivalent to 0.8 kWh per day per person in the UK, roughly 5% of British average electricity consumption. Iceland's average geothermal electricity generation is just 0.3 GW, which is less than 1% of Britain's average electricity consumption. Iceland's average electricity production is 1.1 GW. So to create a link sending power equal to the capacity of the French interconnector, Iceland would have to *triple* its electricity production. To provide us with 4 kWh per day per person (roughly what Britain gets from its own nuclear power stations), Iceland's electricity production would have to increase *ten-fold*. It is probably a good idea to build interconnectors to Iceland, but don't expect them to deliver more than a small contribution.

Figure 25.11. More geothermal power in Iceland. Photo by Rosie Ward.

Notes and further reading

page no.

178 *Concentrating solar power in deserts delivers an average power per unit area of roughly 15 W/m². My sources for this number are two companies making concentrating solar power for deserts.

www.stirlingenergy.com says one of its dishes with a 25 kW Stirling engine at its focus can generate 60 000 kWh/y in a favourable desert location. They could be packed at a concentration of one dish per 500 m². That's an average power of 14 W/m². They say that solar dish Stirling makes the best use of land area, in terms of energy delivered.

www.ausra.com uses flat mirrors to heat water to 285 °C and drive a steam turbine. The heated, pressurized water can be stored in deep metal-lined caverns to allow power generation at night. Describing a "240 MW(e)" plant proposed for Australia (Mills and Lièvre, 2004), the designers claim that 3.5 km² of mirrors would deliver 1.2 TWh(e); that's 38 W/m² of mirror. To find the power per unit land area, we need to allow for the gaps between the mirrors. Ausra say they need a 153 km by 153 km square in the desert to supply all US electric power (Mills and Morgan, 2008). Total US electricity is 3600 TWh/y, so they are claiming a power per unit land area of 18 W/m². This technology goes by the name *compact linear fresnel reflector* (Mills and Morrison, 2000; Mills et al., 2004; Mills and Morgan, 2008). Incidentally, rather than "concentrating solar power," the company Ausra prefers to use the term *solar thermal electricity* (STE); they emphasize the benefits of thermal storage, in contrast to concentrating photovoltaics, which don't come with a natural storage option.

Trieb and Knies (2004), who are strong proponents of concentrating solar power, project that the alternative concentrating solar power technologies would have powers per unit land area in the following ranges: parabolic troughs, 14–19 W/m²; linear fresnel collector, 19–28 W/m²; tower with heliostats, 9–14 W/m²; stirling dish, 9–14 W/m².

There are three European demonstration plants for concentrating solar power. Andasol – using parabolic troughs; Solúcar PS10, a tower near Seville; and Solartres, a tower using molten salt for heat storage. The Andasol parabolic-trough system shown in figure 25.4 is predicted to deliver 10 W/m². Solúcar's "11 MW" solar tower has 624 mirrors, each 121 m². The mirrors concentrate sunlight to a radiation density of up to 650 kW/m². The receiver receives a peak power of 55 MW. The power station can store 20 MWh of thermal energy, allowing it to keep going during 50 minutes of cloudiness. It was expected to generate 24.2 GWh of electricity per year, and it occupies 55 hectares. That's an average power per unit land area of 5 W/m². (Source: Abengoa Annual Report 2003.) Solartres will occupy 142 hectares and is expected to produce 96.4 GWh per year; that's a power density of 8 W/m². Andasol and Solartres will both use some natural gas in normal operation.

179 *HVDC is already used to transmit electricity over 1000-km distances in South Africa, China, America, Canada, Brazil, and Congo. Sources: Asplund (2004), Bahrman and Johnson (2007). Further reading on HVDC: Carlsson (2002).

Figure 25.12. Two engineers assembling an eSolar concentrating power station using heliostats (mirrors that rotate and tip to follow the sun). esolar.com make medium-scale power stations: a 33 MW (peak) power unit on a 64 hectare site. That's 51 W/m² peak, so I'd guess that in a typical desert location they would deliver about one quarter of that: 13 W/m².

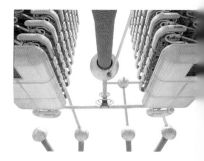

Figure 25.13. A high-voltage DC power system in China. Photo: ABB.

179   *Losses on a 3500 km-long HVDC line, including conversion from AC to DC and back, would be about 15%.* Sources: Trieb and Knies (2004); van Voorthuysen (2008).

182   *According to Amonix, concentrating photovoltaics would have an average power per unit land area of 18 W/m². * The assumptions of `www.amonix.com` are: the lens transmits 85% of the light; 32% cell efficiency; 25% collector efficiency; and 10% further loss due to shading. Aperture/land ratio of 1/3. Normal direct irradiance: 2222 kWh/m²/year. They expect each kW of peak capacity to deliver 2000 kWh/y (an average of 0.23 kW). A plant of 1 GW peak capacity would occupy 12 km² of land and deliver 2000 GWh per year. That's 18 W/m².

–   *Solar chimneys.* Sources: Schlaich J (2001); Schlaich et al. (2005); Dennis (2006), `www.enviromission.com.au`, `www.solarairpower.com`.

183   *Iceland's average geothermal electricity generation is just 0.3 GW. Iceland's average electricity production is 1.1 GW.* These are the statistics for 2006: 7.3 TWh of hydroelectricity and 2.6 TWh of geothermal electricity, with capacities of 1.16 GW and 0.42 GW, respectively. Source: Orkustofnun National Energy Authority [`www.os.is/page/energystatistics`].

Further reading: European Commission (2007), German Aerospace Center (DLR) Institute of Technical Thermodynamics Section Systems Analysis and Technology Assessment (2006), `www.solarmillennium.de`.

26 Fluctuations and storage

> *The wind, as a direct motive power, is wholly inapplicable to a system of machine labour, for during a calm season the whole business of the country would be thrown out of gear. Before the era of steam-engines, windmills were tried for draining mines; but though they were powerful machines, they were very irregular, so that in a long tract of calm weather the mines were drowned, and all the workmen thrown idle.*
>
> William Stanley Jevons, 1865

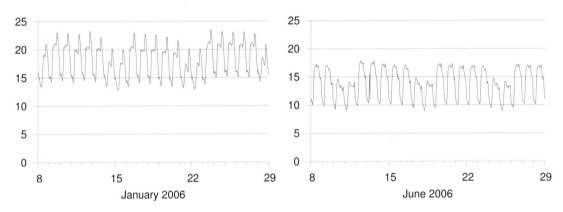

Figure 26.1. Electricity demand in Great Britain (in kWh/d per person) during two winter weeks and two summer weeks of 2006. The peaks in January are at 6pm each day. The five-day working week is evident in summer and winter. (If you'd like to obtain the national demand in GW, remember the top of the scale, 24 kWh/d per person, is the same as 60 GW per UK.)

If we kick fossil fuels and go all-out for renewables, *or* all-out for nuclear, *or* a mixture of the two, we may have a problem. Most of the big renewables are not turn-off-and-onable. When the wind blows and the sun comes out, power is there for the taking; but maybe two hours later, it's not available any more. Nuclear power stations are not usually designed to be turn-off-and-onable either. They are usually on all the time, and their delivered power can be turned down and up only on a timescale of hours. This is a problem because, on an electricity network, consumption and production must be exactly equal all the time. The electricity grid can't *store* energy. To have an energy plan that adds up every minute of every day, we therefore need *something easily turn-off-and-onable.* It's commonly assumed that the easily turn-off-and-onable something should be a *source* of power that gets turned off and on to compensate for the fluctuations of supply relative to demand (for example, a fossil fuel power station!). But another equally effective way to match supply and demand would be to have an easily turn-off-and-onable *demand* for power – a sink of power that can be turned off and on at the drop of a hat.

Either way, the easily turn-off-and-onable something needs to be a *big* something because electricity demand varies a lot (figure 26.1). The de-

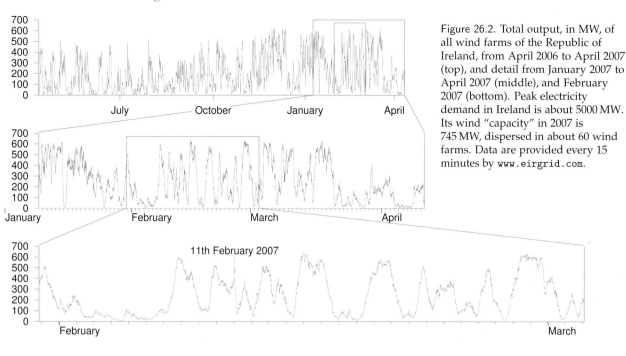

Figure 26.2. Total output, in MW, of all wind farms of the Republic of Ireland, from April 2006 to April 2007 (top), and detail from January 2007 to April 2007 (middle), and February 2007 (bottom). Peak electricity demand in Ireland is about 5000 MW. Its wind "capacity" in 2007 is 745 MW, dispersed in about 60 wind farms. Data are provided every 15 minutes by www.eirgrid.com.

mand sometimes changes significantly on a timescale of a few minutes. This chapter discusses how to cope with fluctuations in supply and demand, without using fossil fuels.

How much do renewables fluctuate?

However much we love renewables, we must not kid ourselves about the fact that wind does fluctuate.

Critics of wind power say: "Wind power is intermittent and unpredictable, so it can make no contribution to security of supply; if we create lots of wind power, we'll have to maintain lots of fossil-fuel power plant to replace the wind when it drops." Headlines such as "Loss of wind causes Texas power grid emergency" reinforce this view. Supporters of wind energy play down this problem: "Don't worry – *individual* wind farms may be intermittent, but taken together, the *sum* of all wind farms in different locations is much less intermittent."

Let's look at real data and try to figure out a balanced viewpoint. Figure 26.2 shows the summed output of the wind fleet of the Republic of Ireland from April 2006 to April 2007. Clearly wind *is* intermittent, even if we add up lots of turbines covering a whole country. The UK is a bit larger than Ireland, but the same problem holds there too. Between October 2006 and February 2007 there were 17 days when the output from Britain's 1632 windmills was less than 10% of their capacity. During that period there were five days when output was less than 5% and one day when it was

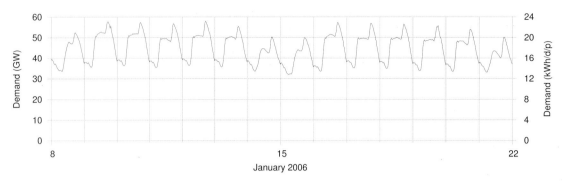

Figure 26.3. Electricity demand in
Great Britain during two winter
weeks of 2006. The left and right
scales show the demand in national
units (GW) and personal units
(kWh/d per person) respectively.
These are the same data as in
figure 26.1.

only 2%.

Let's quantify the fluctuations in country-wide wind power. The two
issues are short-term changes, and long-term lulls. Let's find the fastest
short-term change in a month of Irish wind data. On 11th February 2007,
the Irish wind power fell steadily from 415 MW at midnight to 79 MW at
4am. That's a slew rate of 84 MW per hour for a country-wide fleet of
capacity 745 MW. (By slew rate I mean the rate at which the delivered
power fell or rose – the slope of the graph on 11th February.) OK: if we
scale British wind power up to a capacity of 33 GW (so that it delivers
10 GW on average), we can expect to have occasional slew rates of

$$84\,\text{MW/h} \times \frac{33\,000\,\text{MW}}{745\,\text{MW}} = 3700\,\text{MW/h},$$

assuming Britain is like Ireland. So we need to be able to either power
up replacements for wind at a rate of 3.7 GW per hour – that's 4 nuclear
power stations going from no power to full power every hour, say – *or* we
need to be able to suddenly turn *down* our *demand* at a rate of 3.7 GW per
hour.

Could these windy demands be met? In answering this question we'll
need to talk more about "gigawatts." Gigawatts are big country-sized units
of power. They are to a country what a kilowatt-hour-per-day is to a person: a nice convenient unit. The UK's average electricity consumption is
about 40 GW. We can relate this national number to personal consumption: 1 kWh per day per person is equivalent to 2.5 GW nationally. So if
every person uses 16 kWh per day of electricity, then national consumption
is 40 GW.

Is a national slew-rate of 4 GW per hour completely outside human
experience? No. Every morning, as figure 26.3 shows, British demand
climbs by about 13 GW between 6.30am and 8.30am. That's a slew rate of
6.5 GW per hour. So our power engineers already cope, every day, with slew
rates bigger than 4 GW per hour on the national grid. An extra occasional
slew of 4 GW per hour induced by sudden wind variations is no reasonable
cause for ditching the idea of country-sized wind farms. It's a problem

just like problems that engineers have already solved. We simply need to figure out how to match ever-changing supply and demand in a grid with no fossil fuels. I'm not saying that the wind-slew problem is *already* solved – just that it is a problem of the same size as other problems that have been solved.

OK, before we start looking for solutions, we need to quantify wind's other problem: long-term lulls. At the start of February 2007, Ireland had a country-wide lull that lasted five days. This was not an unusual event, as you can see in figure 26.2. Lulls lasting two or three days happen several times a year.

There are two ways to get through lulls. Either we can store up energy somewhere before the lull, or we need to have a way of reducing demand during the entire lull. (Or a mix of the two.) If we have 33 GW of wind turbines delivering an average power of 10 GW then the amount of energy we must either store up in advance or do without during a five-day lull is

$$10\,\mathrm{GW} \times (5 \times 24\,\mathrm{h}) = 1200\,\mathrm{GWh}.$$

(The gigawatt-hour (GWh) is the cuddly energy unit for nations. Britain's electricity consumption is roughly 1000 GWh per day.)

To personalize this quantity, an energy store of 1200 GWh for the nation is equivalent to an energy store of 20 kWh per person. Such an energy store would allow the nation to go without 10 GW of electricity for 5 days; or equivalently, every individual to go without 4 kWh per day of electricity for 5 days.

Coping with lulls and slews

We need to solve two problems – lulls (long periods with small renewable production), and slews (short-term changes in either supply or demand). We've quantified these problems, assuming that Britain had roughly 33 GW of wind power. To cope with lulls, we must effectively store up roughly 1200 GWh of energy (20 kWh per person). The slew rate we must cope with is 6.5 GW per hour (or 0.1 kW per hour per person).

There are two solutions, both of which could scale up to solve these problems. The first solution is a centralized solution, and the second is decentralized. The first solution stores up energy, then copes with fluctuations by turning on and off a *source* powered from the energy store. The second solution works by turning on and off a piece of *demand*.

The first solution is *pumped storage*. The second uses the batteries of the *electric vehicles* that we discussed in Chapter 20. Before I describe these solutions, let's discuss a few other ideas for coping with slew.

Other supply-side ways of coping with slew

Some of the renewables are turn-off-and-onable. If we had a lot of renewable power that was easily turn-off-and-onable, all the problems of this chapter would go away. Countries like Norway and Sweden have large and deep hydroelectric supplies which they can turn on and off. What might the options be in Britain?

First, Britain could have lots of waste incinerators and biomass incinerators – power stations playing the role that is today played by fossil power stations. If these stations were designed to be turn-off-and-onable, there would be cost implications, just as there are costs when we have extra fossil power stations that are only working part-time: their generators would sometimes be idle and sometimes work twice as hard; and most generators aren't as efficient if you keep turning them up and down, compared with running them at a steady speed. OK, leaving cost to one side, the crucial question is how big a turn-off-and-onable resource we might have. If all municipal waste were incinerated, and an equal amount of agricultural waste were incinerated, then the average power from these sources would be about 3 GW. If we built capacity equal to *twice* this power, making incinerators capable of delivering 6 GW, and thus planning to have them operate only half the time, these would be able to deliver 6 GW throughout periods of high demand, then zero in the wee hours. These power stations could be designed to switch on or off within an hour, thus coping with slew rates of 6 GW per hour – but only for a maximum slew range of 6 GW! That's a helpful contribution, but not enough slew range in itself, if we are to cope with the fluctuations of 33 GW of wind.

What about hydroelectricity? Britain's hydroelectric stations have an average load factor of 20% so they certainly have the potential to be turned on and off. Furthermore, hydro has the wonderful feature that it can be turned on and off very quickly. Glendoe, a new hydro station with a capacity of 100 MW, will be able to switch from off to on in 30 seconds, for example. That's a slew rate of 12 GW per hour in just one power station! So a sufficiently large fleet of hydro power stations should be able to cope with the slew introduced by enormous wind farms. However, the capacity of the British hydro fleet is *not* currently big enough to make much contribution to our slew problem (assuming we want to cope with the rapid loss of say 10 or 33 GW of wind power). The total capacity of traditional hydroelectric stations in Britain is only about 1.5 GW.

So simply switching on and off other renewable power sources is not going to work in Britain. We need other solutions.

Pumped storage

Pumped storage systems use cheap electricity to shove water from a downhill lake to an uphill lake; then regenerate electricity when it's valuable,

station	power (GW)	head (m)	volume (million m^3)	energy stored (GWh)
Ffestiniog	0.36	320–295	1.7	1.3
Cruachan	0.40	365–334	11.3	10
Foyers	0.30	178–172	13.6	6.3
Dinorwig	1.80	542–494	6.7	9.1

Table 26.4. Pumped storage facilities in Britain. The maximum energy storable in today's pumped storage systems is about 30 GWh.

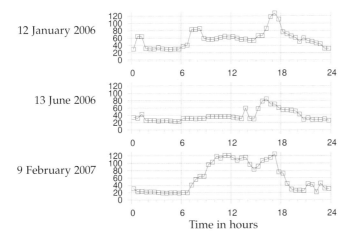

Figure 26.5. How pumped storage pays for itself. Electricity prices, in £ per MWh, on three days in 2006 and 2007.

using turbines just like the ones in hydroelectric power stations.

Britain has four pumped storage facilities, which can store 30 GWh between them (table 26.4, figure 26.6). They are typically used to store excess electricity at night, then return it during the day, especially at moments of peak demand – a profitable business, as figure 26.5 shows. The Dinorwig power station – an astonishing cathedral inside a mountain in Snowdonia – also plays an insurance role: it has enough oomph to restart the national grid in the event of a major failure. Dinorwig can switch on, from 0 to 1.3 GW power, in 12 seconds.

Dinorwig is the Queen of the four facilities. Let's review her vital statistics. The total energy that can be stored in Dinorwig is about 9 GWh. Its upper lake is about 500 m above the lower, and the working volume of 7 million m^3 flows at a maximum rate of 390 m^3/s, allowing power delivery at 1.7 GW for 5 hours. The efficiency of this storage system is 75%.

If all four pumped storage stations are switched on simultaneously, they can produce a power of 2.8 GW. They can switch on extremely fast, coping with any slew rate that demand-fluctuations or wind-fluctuations could come up with. However the capacity of 2.8 GW is not enough to replace 10 GW or 33 GW of wind power if it suddenly went missing. Nor is the total energy stored (30 GWh) anywhere near the 1200 GWh we are interested in storing in order to make it through a big lull. Could pumped

Llyn Stwlan

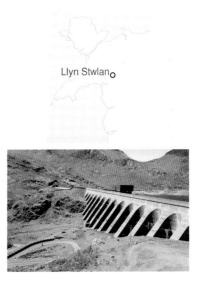

Figure 26.6. Llyn Stwlan, the upper reservoir of the Ffestiniog pumped storage scheme in north Wales. Energy stored: 1.3 GWh. Photo by Adrian Pingstone.

storage be ramped up? Can we imagine solving the entire lull problem using pumped storage alone?

Can we store 1200 GWh?

We are interested in making much bigger storage systems, storing a total of 1200 GWh (about 130 times what Dinorwig stores). And we'd like the capacity to be about 20 GW – about ten times bigger than Dinorwig's. So here is the pumped storage solution: we have to imagine creating roughly 12 new sites, each storing 100 GWh – roughly ten times the energy stored in Dinorwig. The pumping and generating hardware at each site would be the same as Dinorwig's.

Assuming the generators have an efficiency of 90%, table 26.7 shows a few ways of storing 100 GWh, for a range of height drops. (For the physics behind this table, see this chapter's endnotes.)

Ways to store 100 GWh			
drop from upper lake	working volume required (million m^3)	example size of lake area	depth
500 m	40	2 km^2×20 m	
500 m	40	4 km^2×10 m	
200 m	100	5 km^2×20 m	
200 m	100	10 km^2×10 m	
100 m	200	10 km^2×20 m	
100 m	200	20 km^2×10 m	

Table 26.7. Pumped storage. Ways to store 100 GWh. For comparison with column 2, the working volume of Dinorwig is 7 million m^3, and the volume of Lake Windermere is 300 million m^3. For comparison with column 3, Rutland water has an area of 12.6 km^2; Grafham water 7.4 km^2. Carron valley reservoir is 3.9 km^2. The largest lake in Great Britain is Loch Lomond, with an area of 71 km^2.

Is it plausible that twelve such sites could be found? Certainly, we could build several more sites like Dinorwig in Snowdonia alone. Table 26.8 shows two alternative sites near to Ffestiniog where two facilities equal to Dinorwig could have been built. These sites were considered alongside Dinorwig in the 1970s, and Dinorwig was chosen.

proposed location	power (GW)	head (m)	volume (million m^3)	energy stored (GWh)
Bowydd	2.40	250	17.7	12.0
Croesor	1.35	310	8.0	6.7

Table 26.8. Alternative sites for pumped storage facilities in Snowdonia. At both these sites the lower lake would have been a new artificial reservoir.

Pumped-storage facilities holding significantly more energy than Dinorwig could be built in Scotland by upgrading existing hydroelectric facilities. Scanning a map of Scotland, one candidate location would use Loch Sloy as its upper lake and Loch Lomond as its lower lake. There is already a small hydroelectric power station linking these lakes. Figure 26.9 shows these lakes and the Dinorwig lakes on the same scale. The height

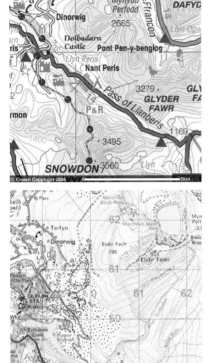

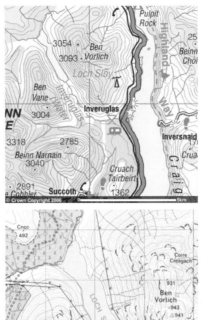

Figure 26.9. Dinorwig, in the Snowdonia National Park, compared with Loch Sloy and Loch Lomond. The upper maps show 10 km by 10 km areas. In the lower maps the blue grid is made of 1 km squares. Images produced from Ordnance Survey's Get-a-map service www.ordnancesurvey.co.uk/getamap. Images reproduced with permission of Ordnance Survey. © Crown Copyright 2006.

100 km

Dinorwig is the home of a 9 GWh storage system, using Marchlyn Mawr (615E, 620N) and Llyn Peris (590E, 598N) as its upper and lower reservoirs.

Loch Sloy illustrates the sort of location where a 40 GWh storage system could be created.

difference between Loch Sloy and Loch Lomond is about 270 m. Sloy's area is about 1.5 km², and it can already store an energy of 20 GWh. If Loch Sloy's dam were raised by another 40 m then the extra energy that could be stored would be about 40 GWh. The water level in Loch Lomond would change by at most 0.8 m during a cycle. This is less than the normal range of annual water level variations of Loch Lomond (2 m).

Figure 26.10 shows 13 locations in Scotland with potential for pumped storage. (Most of them already have a hydroelectric facility.) If ten of these had the same potential as I just estimated for Loch Sloy, then we could store 400 GWh – one third of the total of 1200 GWh that we were aiming for.

We could scour the map of Britain for other locations. The best locations would be near to big wind farms. One idea would be to make a new artificial lake in a hanging valley (across the mouth of which a dam would

Figure 26.10. Lochs in Scotland with potential for pumped storage.

be built) terminating above the sea, with the sea being used as the lower lake.

Thinking further outside the box, one could imagine getting away from lakes and reservoirs, putting half of the facility in an underground chamber. A pumped-storage chamber one kilometre below London has been mooted.

By building more pumped storage systems, it looks as if we could increase our maximum energy store from 30 GWh to 100 GWh or perhaps 400 GWh. Achieving the full 1200 GWh that we were hoping for looks tough, however. Fortunately there is another solution.

Figure 26.11. Okinawa pumped-storage power plant, whose lower reservoir is the ocean. Energy stored: 0.2 GWh. Photo by courtesy of J-Power. www.ieahydro.org.

Demand management using electric vehicles

To recap our requirements: we'd like to be able to store or do without about 1200 GWh, which is 20 kWh per person; and to cope with swings in supply of up to 33 GW – that's 0.5 kW per person. These numbers are delightfully similar in size to the energy and power requirements of electric cars. The electric cars we saw in Chapter 20 had energy stores of between 9 kWh and 53 kWh. A national fleet of 30 million electric cars would store an energy similar to 20 kWh per person! Typical battery chargers draw a power of 2 or 3 kW. So simultaneously switching on 30 million battery chargers would create a change in demand of about 60 GW! The average power required to power all the nation's transport, if it were all electric, is roughly 40 or 50 GW. There's therefore a close match between the adoption of electric cars proposed in Chapter 20 and the creation of roughly 33 GW of wind capacity, delivering 10 GW of power on average.

Here's one way this match could be exploited: electric cars could be plugged in to smart chargers, at home or at work. These smart chargers would be aware both of the value of electricity, and of the car user's requirements (for example, "my car must be fully charged by 7am on Monday morning"). The charger would sensibly satisfy the user's requirements by guzzling electricity whenever the wind blows, and switching off when the wind drops, or when other forms of demand increase. These smart chargers would provide a useful service in balancing to the grid, a service which could be rewarded financially.

We could have an especially robust solution if the cars' batteries were exchangeable. Imagine popping in to a filling station and slotting in a set of fresh batteries in exchange for your exhausted batteries. The filling station would be responsible for recharging the batteries; they could do this at the perfect times, turning up and down their chargers so that total supply and demand were always kept in balance. Using exchangeable batteries is an especially robust solution because there could be millions of spare batteries in the filling stations' storerooms. These spare batteries would provide an extra buffer to help us get through wind lulls. Some people say, "Horrors! How could I trust the filling station to look after my batteries for me? What if they gave me a duff one?" Well, you could equally well ask today "What if the filling station gave me petrol laced with water?" Myself, I'd much rather use a vehicle maintained by a professional than by a muppet like me!

Let's recap our options. We can balance fluctuating demand and fluctuating supply by switching on and off power *generators* (waste incinerators and hydroelectric stations, for example); by *storing* energy somewhere and regenerating it when it's needed; or by switching *demand* off and on.

The most promising of these options, in terms of scale, is switching on and off the power demand of electric-vehicle charging. 30 million cars, with 40 kWh of associated batteries each (some of which might be exchangeable batteries sitting in filling stations) adds up to 1200 GWh. If freight delivery were electrified too then the total storage capacity would be bigger still.

There is thus a beautiful match between wind power and electric vehicles. If we ramp up electric vehicles at the same time as ramping up wind power, roughly 3000 new vehicles for every 3 MW wind turbine, and if we ensure that the charging systems for the vehicles are smart, this synergy would go a long way to solving the problem of wind fluctuations. If my prediction about hydrogen vehicles is wrong, and hydrogen vehicles turn out to be the low-energy vehicles of the future, then the wind-with-electric-vehicles match-up that I've just described could of course be replaced by a wind-with-hydrogen match-up. The wind turbines would make electricity; and whenever electricity was plentiful, hydrogen would be produced and stored in tanks, for subsequent use in vehicles or in other applications, such as glass production.

Other demand-management and storage ideas

There are a few other demand-management and energy-storage options, which we'll survey now.

The idea of modifying the rate of production of stuff to match the power of a renewable source is not new. Many aluminium production plants are located close to hydroelectric power stations; the more it rains, the more aluminium is produced. Wherever power is used to create stuff that is storable, there's potential for switching that power-demand on and off in a smart way. For example, reverse-osmosis systems (which make pure water from sea-water – see p92) are major power consumers in many countries (though not Britain). Another storable product is heat. If, as suggested in Chapter 21, we electrify buildings' heating and cooling systems, especially water-heating and air-heating, then there's potential for lots of easily-turn-off-and-onable power demand to be attached to the grid. Well-insulated buildings hold their heat for many hours, so there's flexibility in the timing of their heating. Moreover, we could include large thermal reservoirs in buildings, and use heat-pumps to pump heat into or out of those reservoirs at times of electricity abundance; then use a second set of heat pumps to deliver heat or cold from the reservoirs to the places where heating or cooling are wanted.

Controlling electricity demand automatically would be easy. The simplest way to do this is to have devices such as fridges and freezers listen to the frequency of the mains. When there is a shortage of power on the grid, the frequency drops below its standard value of 50 Hz; when there is a power excess, the frequency rises above 50 Hz. (It's just like a dynamo on a bicycle: when you switch the lights on, you have to pedal harder to supply the extra power; if you don't then the bike goes a bit slower.) Fridges can be modified to nudge their internal thermostats up and down just a little in response to the mains frequency, in such a way that, without ever jeopardizing the temperature of your butter, they tend to take power at times that help the grid.

Can demand-management provide a significant chunk of virtual storage? How big a sink of power are the nation's fridges? On average, a typical fridge-freezer draws about 18 W; let's guess that the number of fridges is about 30 million. So the ability to switch off all the nation's fridges for a few minutes would be equivalent to 0.54 GW of automatic adjustable power. This is quite a lot of electrical power – more than 1% of the national total – and it is similar in size to the sudden increases in demand produced when the people, united in an act of religious observance (such as watching EastEnders), simultaneously switch on their kettles. Such "TV pick-ups" typically produce increases of demand of 0.6–0.8 GW. Automatically switching off every fridge would *nearly* cover these daily blips of concerted kettle boiling. These smart fridges could also help iron out short-time-scale fluctuations in wind power. The TV pick-ups associated with the holiest acts of observance (for example, watching England play footie against Sweden) can produce sudden increases in demand of over 2 GW. On such occasions, electricity demand and supply are kept in balance by unleashing the full might of Dinorwig.

To provide flexibility to the electricity-grid's managers, who perpetually turn power stations up and down to match supply to demand, many industrial users of electricity are on special contracts that allow the managers to switch off those users' demand at very short notice. In South Africa (where there are frequent electricity shortages), radio-controlled demand-management systems are being installed in hundreds of thousands of homes, to control air-conditioning systems and electric water heaters.

Denmark's solution

Here's how Denmark copes with the intermittency of its wind power. The Danes effectively pay to use other countries' hydroelectric facilities as storage facilities. Almost all of Denmark's wind power is exported to its European neighbours, some of whom have hydroelectric power, which they can turn down to balance things out. The saved hydroelectric power is then sold back to the Danes (at a higher price) during the next period of low wind and high demand. Overall, Danish wind is contributing useful energy, and the system as a whole has considerable security thanks to the capacity of the hydro system.

Could Britain adopt the Danish solution? We would need direct large-capacity connections to countries with lots of turn-off-and-on-able hydroelectric capacity; or a big connection to a Europe-wide electricity grid.

Norway has 27.5 GW of hydroelectric capacity. Sweden has roughly 16 GW. And Iceland has 1.8 GW. A 1.2 GW high-voltage DC interconnector to Norway was mooted in 2003, but not built. A connection to the Netherlands – the BritNed interconnector, with a capacity of 1 GW – will be built in 2010. Denmark's wind capacity is 3.1 GW, and it has a 1 GW connection to Norway, 0.6 GW to Sweden, and 1.2 GW to Germany, a total export capacity of 2.8 GW, very similar to its wind capacity. To be able to export all its excess wind power in the style of Denmark, Britain (assuming 33 GW of wind capacity) would need something like a 10 GW connection to Norway, 8 GW to Sweden, and 1 GW to Iceland.

A solution with two grids

A radical approach is to put wind power and other intermittent sources onto a separate *second* electricity grid, used to power systems that don't require reliable power, such as heating and electric vehicle battery-charging.

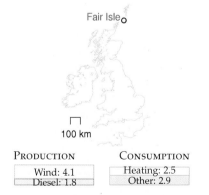

PRODUCTION	CONSUMPTION
Wind: 4.1	Heating: 2.5
Diesel: 1.8	Other: 2.9

Figure 26.12. Electrical production and consumption on Fair Isle, 1995–96. All numbers are in kWh/d per person. Production exceeds consumption because 0.6 kWh/d per person were dumped.

For over 25 years (since 1982), the Scottish island of Fair Isle (population 70, area 5.6 km²) has had *two* electricity networks that distribute power from two wind turbines and, if necessary, a diesel-powered electricity generator. Standard electricity service is provided on one network, and electric heating is delivered by a second set of cables. The electric heating is mainly served by excess electricity from the wind-turbines that would otherwise have had to be dumped. Remote frequency-sensitive programmable relays control individual water heaters and storage heaters in the individual buildings of the community. The mains frequency is used to inform heaters when they may switch on. In fact there are up to six frequency channels per household, so the system emulates seven grids. Fair Isle also successfully trialled a kinetic-energy storage system (a flywheel) to store energy during fluctuations of wind strength on a time-scale of 20 seconds.

Electrical vehicles as generators

If 30 million electric vehicles were willing, in times of national electricity shortage, to run their chargers in reverse and put power back into the grid, then, at 2 kW per vehicle, we'd have a potential power source of 60 GW – similar to the capacity of all the power stations in the country. Even if only one third of the vehicles were connected and available at one time, they'd still amount to a potential source of 20 GW of power. If each of those vehicles made an emergency donation of 2 kWh of energy – corresponding to perhaps 20% of its battery's energy-storage capacity – then the total energy provided by the fleet would be 20 GWh – twice as much as the energy in the Dinorwig pumped storage facility.

Other storage technologies

There are lots of ways to store energy, and lots of criteria by which storage solutions are judged. Figure 26.13 shows three of the most important criteria: energy density (how much energy is stored per kilogram of storage system); efficiency (how much energy you get back per unit energy put in); and lifetime (how many cycles of energy storage can be delivered before the system needs refurbishing). Other important criteria are: the maximum rate at which energy can be pumped into or out of the storage system, often expressed as a power per kg; the duration for which energy stays stored in the system; and of course the cost and safety of the system.

Flywheels

Figure 26.15 shows a monster flywheel used to supply brief bursts of power of up to 0.4 GW to power an experimental facility. It weighs 800 t. Spinning at 225 revolutions per minute, it can store 1000 kWh, and its energy density is about 1 Wh per kg.

Figure 26.15. One of the two flywheels at the fusion research facility in Culham, under construction. Photo: EFDA-JET. www.jet.efda.org.

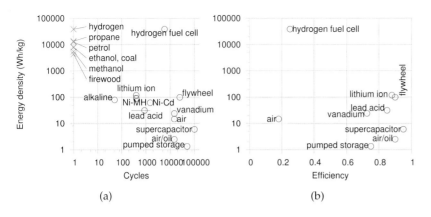

Figure 26.13. Some properties of storage systems and fuels. (a) Energy density (on a logarithmic scale) versus lifetime (number of cycles). (b) Energy density versus efficiency. The energy densities don't include the masses of the energy systems' containers, except in the case of "air" (compressed air storage). Taking into account the weight of a cryogenic tank for holding hydrogen, the energy density of hydrogen is reduced 39 000 Wh/kg to roughly 2400 Wh/kg.

fuel	calorific value	
	(kWh/kg)	(MJ/l)
propane	13.8	25.4
petrol	13.0	34.7
diesel oil (DERV)	12.7	37.9
kerosene	12.8	37
heating oil	12.8	37.3
ethanol	8.2	23.4
methanol	5.5	18.0
bioethanol		21.6
coal	8.0	
firewood	4.4	
hydrogen	39.0	
natural gas	14.85	0.04

(a)

Table 26.14. (a) Calorific values (energy densities, per kg and per litre) of some fuels (in kWh per kg and MJ per litre).
(b) Energy density of some batteries (in Wh per kg). 1 kWh = 1000 Wh.

battery type	energy density (Wh/kg)	lifetime (cycles)
nickel-cadmium	45–80	1500
NiMH	60–120	300–500
lead-acid	30–50	200–300
lithium-ion	110–160	300–500
lithium-ion-polymer	100–130	300–500
reusable alkaline	80	50

(b)

A flywheel system designed for energy storage in a racing car can store 400 kJ (0.1 kWh) of energy and weighs 24 kg (p126). That's an energy density of 4.6 Wh per kg.

High-speed flywheels made of composite materials have energy densities up to 100 Wh/kg.

Supercapacitors

Supercapacitors are used to store small amounts of electrical energy (up to 1 kWh) where many cycles of operation are required, and charging must be completed quickly. For example, supercapacitors are favoured over batteries for regenerative braking in vehicles that do many stops and starts. You can buy supercapacitors with an energy density of 6 Wh/kg.

A US company, EEStor, claims to be able to make much better supercapacitors, using barium titanate, with an energy density of 280 Wh/kg.

Vanadium flow batteries

VRB power systems have provided a 12 MWh energy storage system for the Sorne Hill wind farm in Ireland, whose current capacity is "32 MW," increasing to "39 MW." (VRB stands for vanadium redox battery.) This storage system is a big "flow battery," a redox regenerative fuel cell, with a couple of tanks full of vanadium in different chemical states. This storage system can smooth the output of its wind farm on a time-scale of minutes, but the longest time for which it could deliver one third of the capacity (during a lull in the wind) is one hour.

A 1.5 MWh vanadium system costing \$480 000 occupies 70 m^2 with a mass of 107 tons. The vanadium redox battery has a life of more than 10 000 cycles. It can be charged at the same rate that it is discharged (in contrast to lead-acid batteries which must be charged 5 times as slowly). Its efficiency is 70–75%, round-trip. The volume required is about 1 m^3 of 2-molar vanadium in sulphuric acid to store 20 kWh. (That's 20 Wh/kg.)

So to store 10 GWh would require 500 000 m^3 (170 swimming pools) – for example, tanks 2 m high covering a floor area of 500 m × 500 m.

Scaling up the vanadium technology to match a big pumped-storage system – 10 GWh – might have a noticeable effect on the world vanadium market, but there is no long-term shortage of vanadium. Current world-wide production of vanadium is 40 000 tons per year. A 10 GWh system would contain 36 000 tons of vanadium – about one year's worth of current production. Vanadium is currently produced as a by-product of other processes, and the total world vanadium resource is estimated to be 63 million tons.

"Economical" solutions

In the present world which doesn't put any cost on carbon pollution, the financial bar that a storage system must beat is an ugly alternative: storage can be emulated by simply putting up an extra gas-fired power station to meet extra demand, and shedding any excess electrical power by throwing it away in heaters.

Seasonal fluctuations

The fluctuations of supply and demand that have the longest timescale are seasonal. The most important fluctuation is that of building-heating, which goes up every winter. Current UK natural gas demand varies throughout the year, from a typical average of 36 kWh/d per person in July and August to an average of 72 kWh/d per person in December to February, with extremes of 30–80 kWh/d/p (figure 26.16).

Some renewables also have yearly fluctuations – solar power is stronger in summer and wind power is weaker.

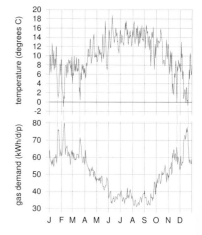

Figure 26.16. Gas demand (lower graph) and temperature (upper graph) in Britain during 2007.

How to ride through these very-long-timescale fluctuations? Electric vehicles and pumped storage are not going to help store the sort of quantities required. A useful technology will surely be long-term thermal storage. A big rock or a big vat of water can store a winter's worth of heat for a building – Chapter E discusses this idea in more detail. In the Netherlands, summer heat from roads is stored in aquifers until the winter; and delivered to buildings via heat pumps [2wmuw7].

Notes

page no.

187   *The total output of the wind fleet of the Republic of Ireland.* Data from eirgrid.com [2hxf6c].

—   *"Loss of wind causes Texas power grid emergency".* [2199ht] Actually, my reading of this news article is that this event, albeit unusual, was an example of *normal* power grid operation. The grid has industrial customers whose supply is interruptible, in the event of a mismatch between supply and demand. Wind output dropped by 1.4 GW at the same time that Texans' demand increased by 4.4 GW, causing exactly such a mismatch between supply and demand. The interruptible supplies were interrupted. Everything worked as intended.
   Here is another example, where better power-system planning would have helped: "Spain wind power hits record, cut ordered." [3x2kvv] Spain's average electricity consumption is 31 GW. On Tuesday 4th March 2008, its wind generators were delivering 10 GW. "Spain's power market has become particularly sensitive to fluctuations in wind."

—   *Supporters of wind energy play down this problem: "Don't worry – individual wind farms may be intermittent, but taken together, the sum of all wind farms is much less intermittent."* For an example, see the website yes2wind.com, which, on its page "debunking the myth that wind power isn't reliable" asserts that "the variation in output from wind farms distributed around the country is scarcely noticeable." www.yes2wind.com/intermittency_debunk.html

—   *... wind is intermittent, even if we add up lots of turbines covering a whole country. The UK is a bit larger than Ireland, but the same problem holds there too.* Source: Oswald et al. (2008).

191   *Dinorwig's pumped-storage efficiency is 75%.* Figure 26.17 shows data. Further information about Dinorwig and the alternate sites for pumped storage: Baines et al. (1983, 1986).

192   *Table 26.7.* The working volume required, V, is computed from the height drop h as follows. If ϵ is the efficiency of potential energy to electricity conversion,
$$V = 100\,\text{GWh}/(\rho g h \epsilon),$$
where ρ is the density of water and g is the acceleration of gravity. I assumed the generators have an efficiency of $\epsilon = 0.9$.

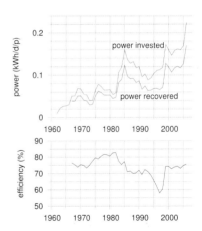

Figure 26.17. Efficiency of the four pumped storage systems of Britain.

192   *Table 26.8, Alternative sites for pumped storage facilities.* The proposed up-
      per reservoir for Bowydd was Llyn Newydd, grid reference SH 722 470; for
      Croesor: Llyn Cwm-y-Foel, SH 653 466.

193   *If ten Scottish pumped storage facilities had the same potential as Loch Sloy,
      then we could store 400 GWh.* This rough estimate is backed up by a study
      by Strathclyde University [5o2xgu] which lists 14 sites having an estimated
      storage capacity of 514 GWh.

196   *Fridges can be modified to nudge their internal thermostats up and down
      . . . in response to the mains frequency.* [2n3pmb] Further links: Dynamic De-
      mand www.dynamicdemand.co.uk; www.rltec.com; www.responsiveload.com.

197   *In South Africa . . . demand-management systems are being installed.*
      Source: [2k8h4o]

  –   *Almost all of Denmark's wind power is exported to its European neighbours.*
      Source: Sharman (2005).

198   *For over 25 years (since 1982), Fair Isle has had two electricity networks.*
      www.fairisle.org.uk/FIECo/
      Wind speeds are between 3 m/s and 16 m/s most of the time; 7 m/s is the
      most probable speed.

Figure 26.18. A possible site for
another 7 GWh pumped storage
facility. Croesor valley is in the
centre-left, between the sharp peak
(Cnicht) on the left and the broader
peaks (the Moelwyns) on the right.

199   *Figure 26.13. Storage efficiencies.* Lithium-ion batteries: 88% efficient.
      Source: www.national.com/appinfo/power/files/swcap_eet.pdf
      Lead-acid batteries: 85–95%.
      Source: www.windsun.com/Batteries/Battery_FAQ.htm
      Compressed air storage: 18% efficient. Source: Lemofouet-Gatsi and Rufer
      (2005); Lemofouet-Gatsi (2006). See also Denholm et al. (2005).

      Air/oil: hydraulic accumulators, as used for regenerative braking in trucks, are compressed-air storage devices that
      can be 90%-efficient round-trip and allow 70% of kinetic energy to be captured. Sources: Lemofouet-Gatsi (2006),
      [5cp27j].

  –   *Table 26.14.* Sources: Xtronics xtronics.com/reference/energy_density.htm; Battery University [2sxlyj]; flywheel
      information from Ruddell (2003).
      The latest batteries with highest energy density are lithium-sulphur and lithium-sulphide batteries, which have an
      energy density of 300 Wh/kg.
      Some disillusioned hydrogen-enthusiasts seem to be making their way up the periodic table and becoming boron-
      enthusiasts. Boron (assuming you will burn it to B_2O_3) has an energy density of 15 000 Wh per kg, which is nice and
      high. But I imagine that my main concern about hydrogen will apply to boron too: that the production of the fuel
      (here, boron from boron oxide) will be inefficient in energy terms, and so will the combustion process.

200   *Vanadium flow batteries.* Sources: www.vrbpower.com; *Ireland wind farm* [ktd7a]; *charging rate* [627ced]; *worldwide
      production* [5fasl7].

201   *. . . summer heat from roads is stored in aquifers. . .* [2wmuw7].

27 Five energy plans for Britain

If we are to get off our current fossil fuel addiction we need a plan for radical action. And the plan needs to add up. The plan also needs a political and financial roadmap. Politics and economics are not part of this book's brief, so here I will simply discuss what the technical side of a plan that adds up might look like.

There are many plans that add up. In this chapter I will describe five. Please don't take any of the plans I present as "the author's recommended solution." My sole recommendation is this:

Make sure your policies include a plan that adds up!

Each plan has a consumption side and a production side: we have to specify how much power our country will be consuming, and how that power is to be produced. To avoid the plans' taking many pages, I deal with a cartoon of our country, in which we consume power in just three forms: transport, heating, and electricity. This is a drastic simplification, omitting industry, farming, food, imports, and so forth. But I hope it's a helpful simplification, allowing us to compare and contrast alternative plans in one minute. Eventually we'll need more detailed plans, but today, we are so far from our destination that I think a simple cartoon is the best way to capture the issues.

I'll present a few plans that I believe are technically feasible for the UK by 2050. All will share the same consumption side. I emphasize again, this doesn't mean that I think this is the correct plan for consumption, or the only plan. I just want to avoid overwhelming you with a proliferation of plans. On the production side, I will describe a range of plans using different mixes of renewables, "clean coal," and nuclear power.

The current situation

The current situation in our cartoon country is as follows. Transport (of both humans and stuff) uses 40 kWh/d per person. Most of that energy is currently consumed as petrol, diesel, or kerosene. Heating of air and water uses 40 kWh/d per person. Much of that energy is currently provided by natural gas. Delivered electricity amounts to 18 kWh/d/p and uses fuel (mainly coal, gas, and nuclear) with an energy content of 45 kWh/d/p. The remaining 27 kWh/d/p goes up cooling towers (25 kWh/d/p) and is lost in the wires of the distribution network (2 kWh/d/p). The total energy input to this present-day cartoon country is 125 kWh/d per person.

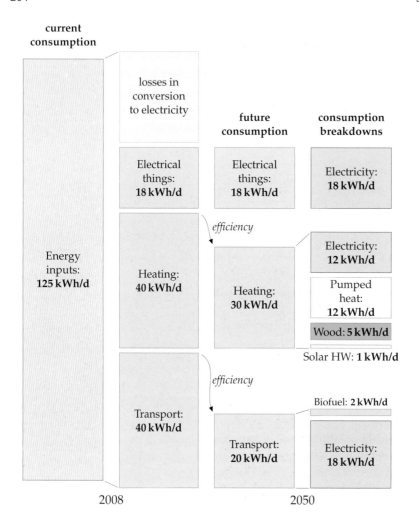

Figure 27.1. Current consumption per person in "cartoon Britain 2008" (left two columns), and a future consumption plan, along with a possible breakdown of fuels (right two columns). This plan requires that electricity supply be increased from 18 to 48 kWh/d per person of electricity.

Common features of all five plans

In my future cartoon country, the energy consumption is reduced by using more efficient technology for transport and heating.

In the five plans for the future, **transport** is largely electrified. Electric engines are more efficient than petrol engines, so the energy required for transport is reduced. Public transport (also largely electrified) is better integrated, better personalized, and better patronized. I've assumed that electrification makes transport about four times more efficient, and that economic growth cancels out some of these savings, so that the net effect is a halving of energy consumption for transport. There are a few essential vehicles that can't be easily electrified, and for those we make our own liquid fuels (for example biodiesel or biomethanol or cellulosic bioethanol). The energy for transport is 18 kWh/d/p of electricity and 2 kWh/d/p of liquid fuels. The electric vehicles' batteries serve as an energy storage facility, helping to cope with fluctuations of electricity supply and demand. The area required for the biofuel production is about 12% of the UK (500 m^2 per person), assuming that biofuel production comes from 1%-efficient plants and that conversion of plant to fuel is 33% efficient. Alternatively, the biofuels could be imported if we could persuade other countries to devote the required (Wales-sized) area of agricultural land to biofuels for us.

In all five plans, the energy consumption of **heating** is reduced by improving the insulation of all buildings, and improving the control of temperature (through thermostats, education, and the promotion of sweater-wearing by sexy personalities). New buildings (all those built from 2010 onwards) are really well insulated and require almost no space heating. Old buildings (which will still dominate in 2050) are mainly heated by air-source heat pumps and ground-source heat pumps. Some water heating is delivered by solar panels (2.5 square metres on every house), some by heat pumps, and some by electricity. Some buildings located near to managed forests and energy-crop plantations are heated by biomass. The power required for heating is thus reduced from 40 kWh/d/p to 12 kWh/d/p of electricity, 2 kWh/d/p of solar hot water, and 5 kWh/d/p of wood.

The wood for making heat (or possibly combined heat and power) comes from nearby forests and energy crops (perhaps miscanthus grass, willow, or poplar) covering a land area of 30 000 km^2, or 500 m^2 per person; this corresponds to 18% of the UK's agricultural land, which has an area of 2800 m^2 per person. The energy crops are grown mainly on the lower-grade land, leaving the higher-grade land for food-farming. Each 500 m^2 of energy crops yields 0.5 oven dry tons per year, which has an energy content of about 7 kWh/d; of this power, about 30% is lost in the process of heat production and delivery. The final heat delivered is 5 kWh/d per person.

In these plans, I assume the current demand for **electricity** for gadgets, light, and so forth is maintained. So we still require 18 kWh(e)/d/p of electricity. Yes, lighting efficiency is improved by a switch to light-emitting diodes for most lighting, and many other gadgets will get more efficient; but thanks to the blessings of economic growth, we'll have increased the number of gadgets in our lives – for example video-conferencing systems to help us travel less.

The total consumption of electricity under this plan goes *up* (because of the 18 kWh/d/p for electric transport and the 12 kWh/d/p for heat pumps) to 48 kWh/d/p (or 120 GW nationally). This is nearly a tripling of UK electricity consumption. Where's that energy to come from?

Let's describe some alternatives. Not all of these alternatives are "sustainable" as defined in this book; but they are all low-carbon plans.

Producing lots of electricity – the components

To make lots of electricity, each plan uses some amount of onshore and off-shore wind; some solar photovoltaics; possibly some solar power bought from countries with deserts; waste incineration (including refuse and agricultural waste); hydroelectricity (the same amount as we get today); perhaps wave power; tidal barrages, tidal lagoons, and tidal stream power; perhaps nuclear power; and perhaps some "clean fossil fuel," that is, coal burnt in power stations that do carbon capture and storage. Each plan aims for a total electricity production of 50 kWh/d/p on average – I got this figure by rounding up the 48 kWh/d/p of average demand, allowing for some loss in the distribution network.

Some of the plans that follow will import power from other countries. For comparison, it may be helpful to know how much of our current power is imported today. The answer is that, in 2006, the UK imported 28 kWh/d/p of fuel – 23% of its primary consumption. These imports are dominated by coal (18 kWh/d/p), crude oil (5 kWh/d/p), and natural gas (6 kWh/d/p). Nuclear fuel (uranium) is not usually counted as an import since it's easily stored.

In all five plans I will assume that we scale up municipal waste incineration so that almost all waste that can't usefully be recycled is incinerated rather than landfilled. Incinerating 1 kg per day per person of waste yields roughly 0.5 kWh/d per person of electricity. I'll assume that a similar amount of agricultural waste is also incinerated, yielding 0.6 kWh/d/p. Incinerating this waste requires roughly 3 GW of waste-to-energy capacity, a ten-fold increase over the incinerating power stations of 2008 (figure 27.2). London (7 million people) would have twelve 30-MW waste-to-energy plants like the SELCHP plant in South London (see p287). Birmingham (1 million people) would have two of them. Every town of 200 000 people would have a 10 MW waste-to-energy plant. Any fears that waste incineration at this scale would be difficult, dirty, or dangerous should be allayed by figure 27.3, which shows that many countries in Europe incinerate *far* more waste per person than the UK; these incineration-loving countries include Germany, Sweden, Denmark, the Netherlands, and Switzerland – not usually nations associated with hygiene problems! One good side-effect of this waste incineration plan is that it eliminates future methane emissions from landfill sites.

In all five plans, hydroelectricity contributes 0.2 kWh/d/p, the same as today.

Electric vehicles are used as a dynamically-adjustable load on the electricity network. The average power required to charge the electric vehicles is 45 GW (18 kWh/d/p). So fluctuations in renewables such as solar and wind can be balanced by turning up and down this load, as long as the fluctuations are not too big or lengthy. Daily swings in electricity demand are going to be bigger than they are today because of the replacement of

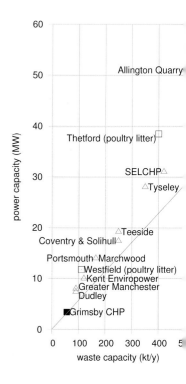

Figure 27.2. Waste-to-energy facilities in Britain. The line shows the average power production assuming 1 kg of waste → 0.5 kWh of electricity.

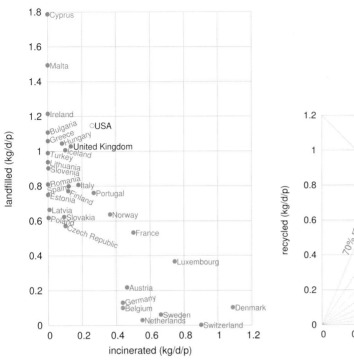

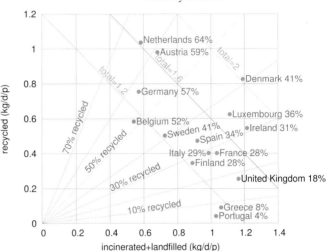

Figure 27.3. Left: Municipal solid waste put into landfill, versus amount incinerated, in kg per day per person, by country. Right: Amount of waste recycled versus amount landfilled or incinerated. Percentage of waste recycled is given beside each country's name.

gas for cooking and heating by electricity (see figure 26.16, p200). To ensure that surges in demand of 10 GW lasting up to 5 hours can be covered, all the plans would build five new pumped storage facilities like Dinorwig (or upgrade hydroelectric facilities to provide pumped storage). 50 GWh of storage is equal to five Dinorwigs, each with a capacity of 2 GW. Some of the plans that follow will require extra pumped storage beyond this. For additional insurance, all the plans would build an electricity interconnector to Norway, with a capacity of 2 GW.

Producing lots of electricity – plan D

Plan D ("D" stands for "domestic diversity") uses a lot of every possible domestic source of electricity, and depends relatively little on energy supply from other countries.

Here's where plan D gets its 50 kWh/d/p of electricity from. Wind: 8 kWh/d/p (20 GW average; 66 GW peak) (plus about 400 GWh of associated pumped storage facilities). Solar PV: 3 kWh/d/p. Waste incineration: 1.3 kWh/d/p. Hydroelectricity: 0.2 kWh/d/p. Wave: 2 kWh/d/p. Tide: 3.7 kWh/d/p. Nuclear: 16 kWh/d/p (40 GW). "Clean coal": 16 kWh/d/p (40 GW).

To get 8 kWh/d/p of wind requires a 30-fold increase in wind power over the installed power in 2008. Britain would have nearly 3 times as much wind hardware as Germany has now. Installing this much wind-power offshore over a period of 10 years would require roughly 50 jack-up barges.

Getting 3 kWh/d/p from solar photovoltaics requires 6 m^2 of 20%-efficient panels per person. Most south-facing roofs would have to be completely covered with panels; alternatively, it might be more economical, and cause less distress to the League for the Preservation of Old Buildings, to plant many of these panels in the countryside in the traditional Bavarian manner (figure 6.7, p41).

The waste incineration corresponds to 1 kg per day per person of domestic waste (yielding 0.5 kWh/d/p) and a similar amount of agricultural waste yielding 0.6 kWh/d/p; the hydroelectricity is 0.2 kWh/d/p, the same amount as we get from hydro today.

The wave power requires 16 000 Pelamis deep-sea wave devices occupying 830 km of Atlantic coastline (see the map on p73).

The tide power comes from 5 GW of tidal stream installations, a 2 GW Severn barrage, and 2.5 GW of tidal lagoons, which can serve as pumped storage systems too.

To get 16 kWh/d/p of nuclear power requires 40 GW of nukes, which is a roughly four-fold increase of the 2007 nuclear fleet. If we produced 16 kWh/d/p of nuclear power, we'd lie between Belgium, Finland, France and Sweden, in terms of per-capita production: Belgium and Finland each produce roughly 12 kWh/d/p; France and Sweden produce 19 kWh/d/p and 20 kWh/d/p respectively.

To get 16 kWh/d/p of "clean coal" (40 GW), we would have to take the current fleet of coal stations, which deliver about 30 GW, retrofit carbon-capture systems to them, which would reduce their output to 22 GW, then build another 18 GW of new clean-coal stations. This level of coal power requires an energy input of about 53 kWh/d/p of coal, which is a little bigger than the total rate at which we currently burn *all* fossil fuels at power stations, and well above the level we estimated as being "sustainable" in Chapter 23. This rate of consumption of coal is roughly three times the current rate of coal imports (18 kWh/d/p). If we didn't reopen UK coal mines, this plan would have 32% of UK electricity depending on imported coal. Reopened UK coal mines could deliver an energy input of about 8 kWh/d/p, so either way, the UK would not be self-sufficient for coal.

Do any features of this plan strike you as unreasonable or objectionable? If so, perhaps one of the next four plans is more to your liking.

Producing lots of electricity – plan N

Plan N is the "NIMBY" plan, for people who don't like industrializing the British countryside with renewable energy facilities, and who don't want

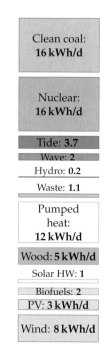

Figure 27.4. Plan D

new nuclear power stations either. Let's reveal the plan in stages.

First, we turn down all the renewable knobs from their very high settings in plan D to: wind: 2 kWh/d/p (5 GW average); solar PV: 0; wave: 0; tide: 1 kWh/d/p.

We've just lost ourselves 14 kWh/d/p (35 GW nationally) by turning down the renewables. (Don't misunderstand! Wind is still eight-fold increased over its 2008 levels.)

In the NIMBY plan, we reduce the contribution of nuclear power to 10 kWh/d/p (25 GW) – a reduction by 15 GW compared to plan D, but still a substantial increase over today's levels. 25 GW of nuclear power could, I think, be squeezed onto the existing nuclear sites, so as to avoid imposing on any new back yards. I left the clean-coal contribution unchanged at 16 kWh/d/p (40 GW). The electricity contributions of hydroelectricity and waste incineration remain the same as in plan D.

Where are we going to get an extra 50 GW from? The NIMBY says, "not in my back yard, but in someone else's." Thus the NIMBY plan pays other countries for imports of solar power from their deserts to the tune of 20 kWh/d/p (50 GW).

This plan requires the creation of five blobs each the size of London (44 km in diameter) in the transmediterranean desert, filled with solar power stations. It also requires power transmission systems to get 50 GW of power up to the UK. Today's high voltage electricity connection from France can deliver only 2 GW of power. So this plan requires a 25-fold increase in the capacity of the electricity connection from the continent. (Or an equivalent power-transport solution – perhaps ships filled with methanol or boron plying their way from desert shores.)

Having less wind power, plan N doesn't need to build in Britain the extra pumped-storage facilities mentioned in plan D, but given its dependence on sunshine, it still requires storage systems to be built somewhere to store energy from the fluctuating sun. Molten salt storage systems at the solar power stations are one option. Tapping into pumped storage systems in the Alps might also be possible. Converting the electricity to a storable fuel such as methanol is another option, though conversions entail losses and thus require more solar power stations.

This plan gets 32% + 40% = 72% of the UK's electricity from other countries.

Producing lots of electricity – plan L

Some people say "we don't want nuclear power!" How can we satisfy them? Perhaps it should be the job of this anti-nuclear bunch to persuade the NIMBY bunch that they do want renewable energy in our back yard after all.

We can create a nuclear-free plan by taking plan D, keeping all those renewables in our back yard, and doing a straight swap of nuclear for

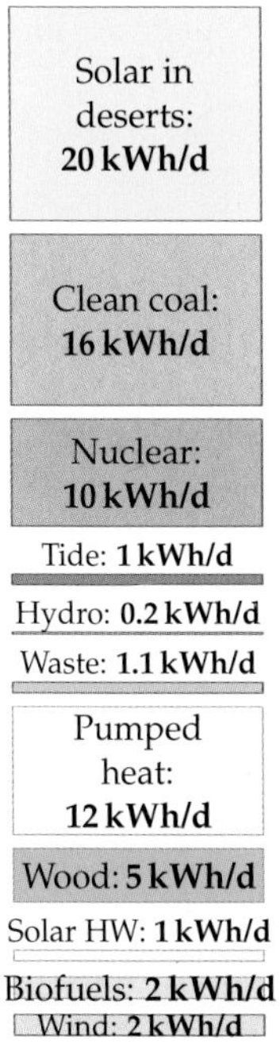

Figure 27.5. Plan N

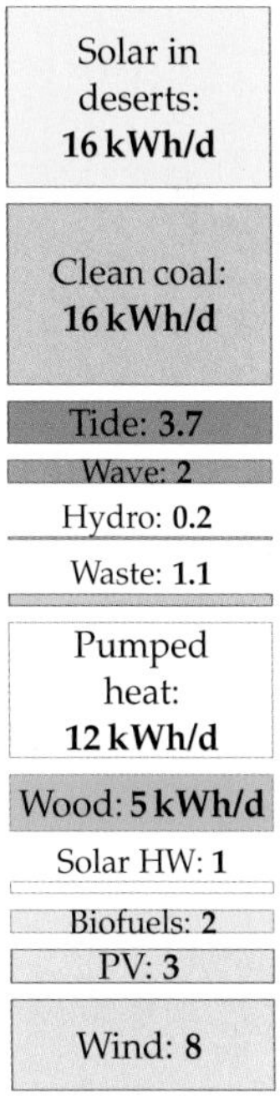

Figure 27.6. Plan L

desert power. As in plan N, the delivery of desert power requires a large increase in transmission systems between North Africa and Britain; the Europe–UK interconnectors would need to be increased from 2 GW to at least 40 GW.

Here's where plan L gets its 50 kWh/d/p of electricity from. Wind: 8 kWh/d/p (20 GW average) (plus about 400 GWh of associated pumped storage facilities). Solar PV: 3 kWh/d/p. Hydroelectricity and waste incineration: 1.3 kWh/d/p. Wave: 2 kWh/d/p. Tide: 3.7 kWh/d/p. "Clean coal": 16 kWh/d/p (40 GW). Solar power in deserts: 16 kWh/d/p (40 GW average power).

This plan imports 64% of UK electricity from other countries.

I call this "plan L" because it aligns fairly well with the policies of the Liberal Democrats – at least it did when I first wrote this chapter in mid-2007; recently, they've been talking about "real energy independence for the UK," and have announced a zero-carbon policy, under which Britain would be a net energy *exporter*; their policy does not detail how these targets would be met.

Producing lots of electricity – plan G

Some people say "we don't want nuclear power, *and* we don't want coal!" It sounds a desirable goal, but we need a plan to deliver it. I call this "plan G," because I guess the Green Party don't want nuclear or coal, though I think not all Greens would like the rest of the plan. Greenpeace, I know, *love* wind, so plan G is dedicated to them too, because it has *lots* of wind.

I make plan G by starting again from plan D, nudging up the wave contribution by 1 kWh/d/p (by pumping money into wave research and increasing the efficiency of the Pelamis converter) and bumping up wind power fourfold (relative to plan D) to 32 kWh/d/p, so that wind delivers 64% of all the electricity. This is a 120-fold increase of British wind power over today's levels. Under this plan, *world* wind power in 2008 is multiplied by 4, with all of the increase being placed on or around the British Isles.

The immense dependence of plan G on renewables, especially wind, creates difficulties for our main method of balancing supply and demand, namely adjusting the charging rate of millions of rechargeable batteries for transport. So in plan G we have to include substantial additional pumped-storage facilities, capable of balancing out the fluctuations in wind on a timescale of days. Pumped-storage facilities equal to 400 Dinorwigs can completely replace wind for a national lull lasting 2 days. Roughly 100 of Britain's major lakes and lochs would be required for the associated pumped-storage systems.

Plan G's electricity breaks down as follows. Wind: 32 kWh/d/p (80 GW average) (plus about 4000 GWh of associated pumped-storage facilities). Solar photovoltaics: 3 kWh/d/p. Hydroelectricity and waste incineration:

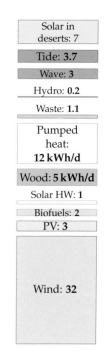

Figure 27.7. Plan G

1.3 kWh/d/p. Wave: 3 kWh/d/p. Tide: 3.7 kWh/d/p. Solar power in deserts: 7 kWh/d/p (17 GW).

This plan gets 14% of its electricity from other countries.

Producing lots of electricity – plan E

E stands for "economics." This fifth plan is a rough guess for what might happen in a liberated energy market with a strong carbon price. On a level economic playing field with a strong price signal preventing the emission of CO_2, we don't expect a diverse solution with a wide range of power-costs; rather, we expect an economically optimal solution that delivers the required power at the lowest cost. And when "clean coal" and nuclear go head to head on price, it's nuclear that wins. (Engineers at a UK electricity generator told me that the capital cost of regular *dirty* coal power stations is £1 billion per GW, about the same as nuclear; but the capital cost of "clean-coal" power, including carbon capture and storage, is roughly £2 billion per GW.) I've assumed that solar power in other people's deserts loses to nuclear power when we take into account the cost of the required 2000-km-long transmission lines (though van Voorthuysen (2008) reckons that with Nobel-prize-worthy developments in solar-powered production of chemical fuels, solar power in deserts would be the economic equal of nuclear power). Offshore wind also loses to nuclear, but I've assumed that onshore wind costs about the same as nuclear.

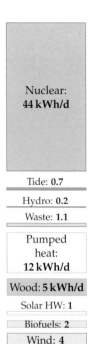

Figure 27.8. Plan E

Here's where plan E gets its 50 kWh/d/p of electricity from. Wind: 4 kWh/d/p (10 GW average). Solar PV: 0. Hydroelectricity and waste incineration: 1.3 kWh/d/p. Wave: 0. Tide: 0.7 kWh/d/p. And nuclear: 44 kWh/d/p (110 GW).

This plan has a ten-fold increase in our nuclear power over 2007 levels. Britain would have 110 GW, which is roughly double France's nuclear fleet. I included a little tidal power because I believe a well-designed tidal lagoon facility can compete with nuclear power.

In this plan, Britain has no energy imports (except for the uranium, which, as we said before, is not conventionally counted as an import).

Figure 27.9 shows all five plans.

How these plans relate to carbon-sucking and air travel

In a future world where carbon pollution is priced appropriately to prevent catastrophic climate change, we will be interested in any power scheme that can at low cost put extra carbon down a hole in the ground. Such carbon-neutralization schemes might permit us to continue flying at 2004 levels (while oil lasts). In 2004, average UK emissions of CO_2 from flying were about 0.5 t CO_2 per year per person. Accounting for the full greenhouse impact of flying, perhaps the effective emissions were about 1 t CO_2e per year per person. Now, in all five of these plans I assumed that one

1 t CO_2e means greenhouse-gas emissions equivalent to one ton of CO_2.

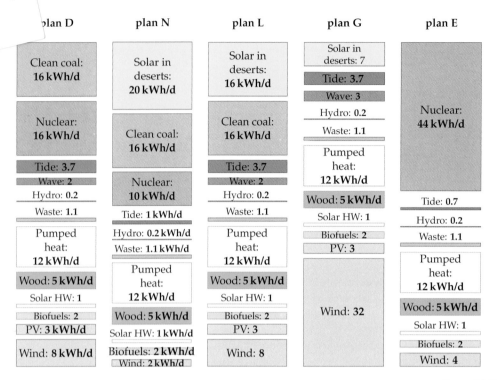

Figure 27.9. All five plans.

eighth of the UK was devoted to the production of energy crops which were then used for heating or for combined heat and power. If instead we directed all these crops to power stations with carbon capture and storage – the "clean-coal" plants that featured in three of the plans – then the amount of extra CO_2 captured would be about 1 t of CO_2 per year per person. If the municipal and agricultural waste incinerators were located at clean-coal plants too so that they could share the same chimney, perhaps the total captured could be increased to 2 t CO_2 per year per person. This arrangement would have additional costs: the biomass and waste might have to be transported further; the carbon-capture process would require a significant fraction of the energy from the crops; and the lost building-heating would have to be replaced by more air-source heat pumps. But, if carbon-neutrality is our aim, it would be worth planning ahead by seeking to locate new clean-coal plants with waste incinerators in regions close to potential biomass plantations.

"All these plans are absurd!"

If you don't like these plans, I'm not surprised. I agree that there is something unpalatable about every one of them. Feel free to make another plan that is more to your liking. But make sure it adds up!

Perhaps you will conclude that a viable plan has to involve less power consumption per capita. I might agree with that, but it's a difficult policy to sell – recall Tony Blair's response (p222) when someone suggested he should fly overseas for holidays less frequently!

Alternatively, you may conclude that we have too high a population density, and that a viable plan requires fewer people. Again, a difficult policy to sell.

Notes and further reading

page no.

206 *Incinerating 1 kg of waste yields roughly 0.5 kWh of electricity.*
The calorific value of municipal solid waste is about 2.6 kWh per kg; power stations burning waste produce electricity with an efficiency of about 20%. Source: SELCHP tour guide.

207 *Figure 27.3.* Data from Eurostat, www.epa.gov, and www.esrcsocietytoday.ac.uk/ESRCInfoCentre/.

210 *The policies of the Liberal Democrats.* See www.libdems.org.uk: [5os7dy], [yrw2oo].

28 Putting costs in perspective

A plan on a map

Let me try to make clear the scale of the previous chapter's plans by show-ing you a map of Britain bearing a sixth plan. This sixth plan lies roughly in the middle of the first five, so I call it plan M (figure 28.1).

The areas and rough costs of these facilities are shown in table 28.3. For simplicity, the financial costs are estimated using today's prices for comparable facilities, many of which are early prototypes. We can expect many of the prices to drop significantly. The rough costs given here are the building costs, and don't include running costs or decommissioning costs. The "per person" costs are found by dividing the total cost by 60 million. Please remember, this is not a book about economics – that would require another 400 pages! I'm providing these cost estimates only to give a *rough* indication of the price tag we should expect to see on a plan that adds up.

I'd like to emphasize that I am not advocating this particular plan – it includes several features that I, as dictator of Britain, would not select. I've deliberately included all available technologies, so that you can try out your own plans with other mixes.

For example, if you say "photovoltaics are going to be too expensive, I'd like a plan with more wave power instead," you can see how to do it: you need to increase the wave farms eight-fold. If you don't like the wind farms' locations, feel free to move them (but where to?). Bear in mind that putting more of them offshore will increase costs. If you'd like fewer wind farms, no problem – just specify which of the other technologies you'd like instead. You can replace five of the 100 km² wind farms by adding one more 1 GW nuclear power station, for example.

Perhaps you think that this plan (like each of the five plans in the previ-ous chapter) devotes unreasonably large areas to biofuels. Fine: you may therefore conclude that the demand for liquid fuels for transport must be reduced below the 2 kWh per day per person that this plan assumed; or that liquid fuels must be created in some other way.

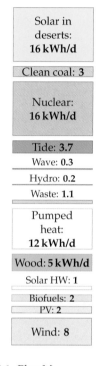

Figure 28.1. Plan M

Cost of switching from fossil fuels to renewables

Every wind farm costs a few million pounds to build and delivers a few megawatts. As a very rough ballpark figure in 2008, installing one watt of capacity costs one pound; one kilowatt costs 1000 pounds; a megawatt of wind costs a million; a gigawatt of nuclear costs a billion or perhaps two. Other renewables are more expensive. We (the UK) currently consume a total power of roughly 300 GW, most of which is fossil fuel. So we can anticipate that a major switching from fossil fuel to renewables and/or nu-clear is going to require roughly 300 GW of renewables and/or nuclear and

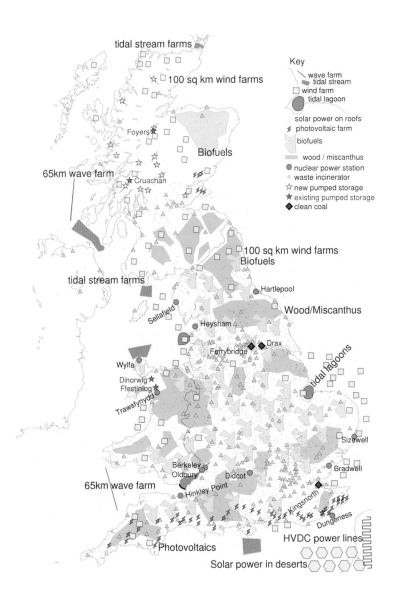

Figure 28.2. A plan that adds up, for Scotland, England, and Wales.
The grey-green squares are wind farms. Each is $100\,km^2$ in size and is shown to scale.
The red lines in the sea are wave farms, shown to scale.
Light-blue lightning-shaped polygons: solar photovoltaic farms – $20\,km^2$ each, shown to scale.
Blue sharp-cornered polygons in the sea: tide farms.
Blue blobs in the sea (Blackpool and the Wash): tidal lagoons.
Light-green land areas: woods and short-rotation coppices (to scale).
Yellow-green areas: biofuel (to scale).
Small blue triangles: waste incineration plants (not to scale).
Big brown diamonds: clean coal power stations, with cofiring of biomass, and carbon capture and storage (not to scale).
Purple dots: nuclear power stations (not to scale) – $3.3\,GW$ average production at each of 12 sites.
Yellow hexagons across the channel: concentrating solar power facilities in remote deserts (to scale, $335\,km^2$ each). The pink wiggly line in France represents new HVDC lines, $2000\,km$ long, conveying $40\,GW$ from remote deserts to the UK.
Yellow stars in Scotland: new pumped storage facilities.
Red stars: existing pumped storage facilities.
Blue dots: solar panels for hot water on all roofs.

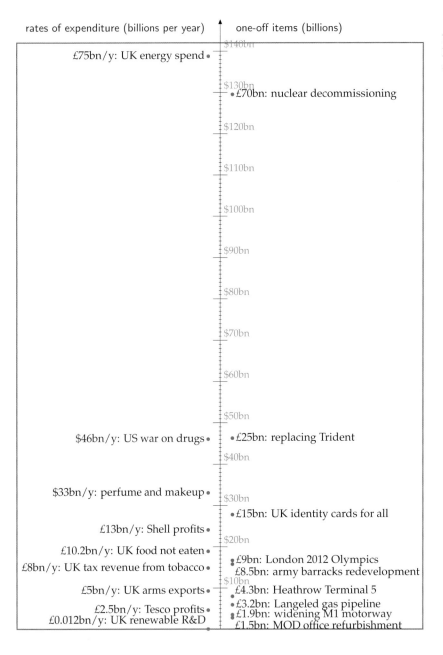

Figure 28.5. Things that run into billions. The scale down the centre has large ticks at $10 billion intervals and small ticks at $1 billion intervals.

rates of expenditure (billions per year)   one-off items (billions)

£75bn/y: UK energy spend •

$140bn

$130bn
• £70bn: nuclear decommissioning

$120bn

$110bn

$100bn

$90bn

$80bn

$70bn

$60bn

$50bn

$46bn/y: US war on drugs •   • £25bn: replacing Trident

$40bn

$33bn/y: perfume and makeup • $30bn
   • £15bn: UK identity cards for all

£13bn/y: Shell profits •
$20bn
£10.2bn/y: UK food not eaten •
£8bn/y: UK tax revenue from tobacco •   £9bn: London 2012 Olympics
   £8.5bn: army barracks redevelopment
$10bn
£5bn/y: UK arms exports •   £4.3bn: Heathrow Terminal 5
   • £3.2bn: Langeled gas pipeline
£2.5bn/y: Tesco profits •   £1.9bn: widening M1 motorway
£0.012bn/y: UK renewable R&D   £1.5bn: MOD office refurbishment

Special occasions

Cost of the London 2012 Olympics: £2.4 billion; no, I'm sorry, £5 billion [3x2cr4]; or perhaps £9 billion [2dd4mz]. (£150 per person in the UK.)

Business as usual

£2.5 billion/y: Tesco's profits (announced 2007). (£42 per year per person in the UK.)
£10.2 billion/y: spent by British people on food that they buy but do not eat. (£170 per year per person in the UK.)
£11 billion/y: BP's profits (2006).
£13 billion/y: Royal Dutch Shell's profits (2006).
$40 billion/y. Exxon's profits (2006).
$33 billion/y. World expenditure on perfumes and make-up.
$700 billion per year: USA's expenditure on foreign oil (2008). ($2300 per year per person in the USA.)

Government business as usual

£1.5 billion: the cost of refurbishment of Ministry of Defence offices. (Private Eye No. 1176, 19th January 2007, page 5.) (£25 per person in the UK.)
£15 billion: the cost of introducing UK identity card scheme [7v1xp]. (£250 per person in the UK.)

Planning for the future

£3.2 billion: the cost of the Langeled pipeline, which ships gas from Norwegian producers to Britain. The pipeline's capacity is 20 billion m^3 per year, corresponding to a power of 25 GW. [6x4nvu] [39g2wz] [3ac8sj]. (£53 per person in the UK.)

Tobacco taxes and related games

£8 billion/y: annual revenue from tobacco taxes in the UK [y7kg26]. (£130 per year per person in the UK.) The European Union spends almost €1 billion a year subsidising tobacco farming. www.ash.org.uk
$46 billion/y: Annual cost of the USA's "War on drugs." [r9fcf] ($150 per year per person in the USA.)

Space

$1.7 billion: the cost of one space shuttle. ($6 per person in the USA.)

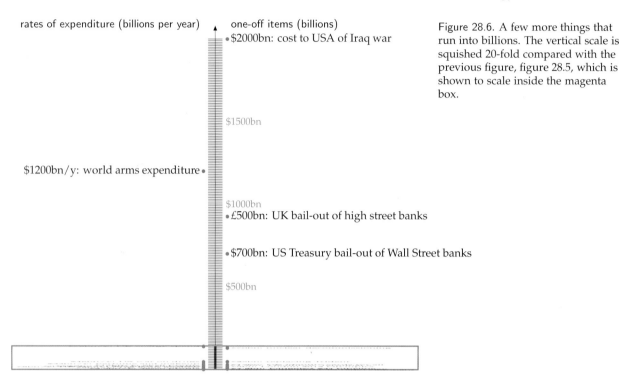

rates of expenditure (billions per year)   one-off items (billions)

• $2000bn: cost to USA of Iraq war

$1500bn

$1200bn/y: world arms expenditure •

$1000bn

• £500bn: UK bail-out of high street banks

• $700bn: US Treasury bail-out of Wall Street banks

$500bn

Figure 28.6. A few more things that run into billions. The vertical scale is squished 20-fold compared with the previous figure, figure 28.5, which is shown to scale inside the magenta box.

Banks

$700 billion: in October 2008, the US government committed $700 billion to bailing out Wall Street, and . . .

£500 billion: the UK government committed £500 billion to bailing out British banks.

Military

£5 billion per year: UK's arms exports (£83 per year per person in the UK), of which £2.5 billion go to the Middle East, and £1 billion go to Saudi Arabia. Source: Observer, 3 December 2006.

£8.5 billion: cost of redevelopment of army barracks in Aldershot and Salisbury Plain. (£140 per person in the UK.)

£3.8 billion: the cost of two new aircraft carriers (£63 per person in the UK). news.bbc.co.uk/1/low/scotland/6914788.stm

$4.5 billion per year: the cost of not making nuclear weapons – the US Department of Energy's budget allocates at least $4.5 billion per year to "stockpile stewardship" activities to maintain the nuclear stockpile *without* nuclear testing and *without* large-scale production of new weapons. ($15 per year per person in America.)

£10–25 billion: the cost of replacing Trident, the British nuclear weapon system. (£170–420 per person in the UK.) [ysncks].

$63 billion: American donation of "military aid" (i.e. weapons) to the Middle East over 10 years – roughly half to Israel, and half to Arab states. [2vq59t] ($210 per person in the USA.)

$1200 billion per year: world expenditure on arms [ym46a9]. ($200 per year per person in the world.)

$2000 billion or more: the cost, to the USA, of the [99bpt] Iraq war according to Nobel prize-winning economist Joseph Stiglitz. ($7000 per person in America.)

According to the Stern review, the global cost of averting dangerous climate change (if we act now) is $440 billion per year ($440 per year per person, if shared equally between the 1 billion richest people). In 2005, the US government alone spent $480 billion on wars and preparation for wars. The total military expenditure of the 15 biggest military-spending countries was $840 billion.

Expenditure that does **not** run into billions

£0.012 billion per year: the smallest item displayed in figure 28.5 is the UK government's annual investment in renewable-energy research and development. (£0.20 per person in the UK, per year.)

Notes and further reading

215   *Figure 28.2.* I've assumed that the solar photovoltaic farms have a power per unit area of $5 W/m^2$, the same as the Bavaria farm on p41, so each farm on the map delivers 100 MW on average. Their total average production would be 5 GW, which requires roughly 50 GW of peak capacity (that's 16 times Germany's PV capacity in 2006).
The yellow hexagons representing concentrating solar power have an average power of 5 GW each; it takes two of these hexagons to power one of the "blobs" of Chapter 25.

217   *A government report leaked by the Guardian...* The Guardian report, 13th August 2007, said [2bmuod] "Government officials have secretly briefed ministers that Britain has no hope of getting remotely near the new European Union renewable energy target that Tony Blair signed up to in the spring - and have suggested that they find ways of wriggling out of it."
The leaked document is at [3g8nn8].

219   *...perfume...* Source: Worldwatch Institute
www.worldwatch.org/press/news/2004/01/07/

221   *...wars and preparation for wars...* www.conscienceonline.org.uk

–   *Government investment in renewable-energy-related research and development.* In 2002–3, the UK Government's commitment to renewable-energy-related R&D was £12.2 million. Source: House of Lords Science and Technology Committee, 4th Report of Session 2003–04. [3jo7q2]
Comparably small is the government's allocation to the Low Carbon Buildings Programme, £0.018bn/y shared between wind, biomass, solar hot water/PV, ground-source heat pumps, micro-hydro and micro CHP.

29   *What to do now*

Unless we act now, not some time distant but now, these conse-
quences, disastrous as they are, will be irreversible. So there is nothing
more serious, more urgent or more demanding of leadership.

Tony Blair, 30 October 2006

a bit impractical actually...

Tony Blair, two months later,
responding to the suggestion that he should *show*
leadership by not flying to Barbados for holidays.

What we should do depends in part on our motivation. Recall that on page 5 we discussed three motivations for getting off fossil fuels: the end of cheap fossil fuels; security of supply; and climate change. Let's assume first that we have the climate-change motivation – that we want to reduce carbon emissions radically. (Anyone who doesn't believe in climate change can skip this section and rejoin the rest of us on page 223.)

What to do about carbon pollution

We are not on track to a zero-carbon future. Long-term investment is not happening. Carbon sequestration companies are not thriving, even though the advice from climate experts and economic experts alike is that sucking carbon dioxide from thin air will very probably be necessary to avoid dangerous climate change. Carbon is not even being captured at any coal power stations (except for one tiny prototype in Germany).

Why not?

The principal problem is that carbon pollution is not priced correctly. And there is no confidence that it's going to be priced correctly in the future. When I say "correctly," I mean that the price of emitting carbon dioxide should be big enough such that every running coal power station has carbon capture technology fitted to it.

Solving climate change is a complex topic, but in a single crude brush-stroke, here is the solution: the price of carbon dioxide must be such that people *stop burning coal without capture*. Most of the solution is captured in this one brush-stroke because, in the long term, coal is the big fossil fuel. (Trying to reduce emissions from oil and gas is of secondary importance because supplies of both oil and gas are expected to decline over the next 50 years.)

So what do politicians need to do? They need to ensure that all coal power stations have carbon capture fitted. The first step towards this goal is for government to finance a large-scale demonstration project to sort out the technology for carbon capture and storage; second, politicians need to change the long-term regulations for power stations so that the perfected technology is adopted everywhere. My simple-minded suggestion for this second step is to pass a law that says that – from some date – *all coal power stations must use carbon capture*. However, most democratic politicians seem to think that the way to close a stable door is to create a market in permits-to-leave-doors-open. So, if we conform to the dogma that climate change should be solved through markets, what's the market-based way to ensure we achieve our simple goal – all coal power stations to have carbon capture? Well, we can faff around with carbon trading – trading of permits to emit carbon and of certificates of carbon-capture, with one-tonne carbon-capture certificates being convertible into one-tonne carbon-emission permits. But coal station owners will invest in carbon capture and storage only if they are convinced that the price of carbon is going to be high enough for long enough that carbon-capturing facilities will pay for themselves. Experts say that a long-term guaranteed carbon price of something like $100 per ton of CO_2 will do the trick.

So politicians need to agree long-term reductions in CO_2 emissions that are sufficiently strong that investors have confidence that the price of carbon will rise permanently to at least $100 per ton of CO_2. Alternatively they could issue carbon pollution permits in an auction with a fixed minimum price. Another way would be for governments to underwrite investment in carbon capture by guaranteeing that they will redeem captured-carbon certificates for $100 per ton of CO_2, whatever happens to the market in carbon-emission permits.

I still wonder whether it would be wisest to close the stable door directly, rather than fiddling with an international market that is merely *intended* to encourage stable door-closing.

> *Britain's energy policy just doesn't stack up. It won't deliver security. It won't deliver on our commitments on climate change. It falls short of what the world's poorest countries need.*
>
> Lord Patten of Barnes, Chair of Oxford University task force on energy and climate change, 4 June 2007.

What to do about energy supply

Let's now expand our set of motivations, and assume that we want to get off fossil fuels in order to ensure security of energy supply.

What should we do to bring about the development of non-fossil energy supply, and of efficiency measures? One attitude is "Just let the market handle it. As fossil fuels become expensive, renewables and nuclear power will become relatively cheaper, and the rational consumer will prefer efficient technologies." I find it odd that people have such faith in markets, given how regularly markets give us things like booms and busts, credit crunches, and collapses of banks. Markets may be a good

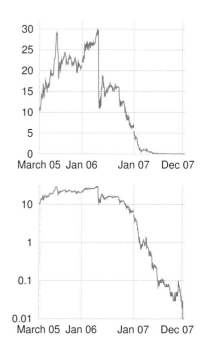

Figure 29.1. A fat lot of good that did! The price, in euro, of one ton of CO_2 under the first period of the European emissions trading scheme. Source: `www.eex.com`.

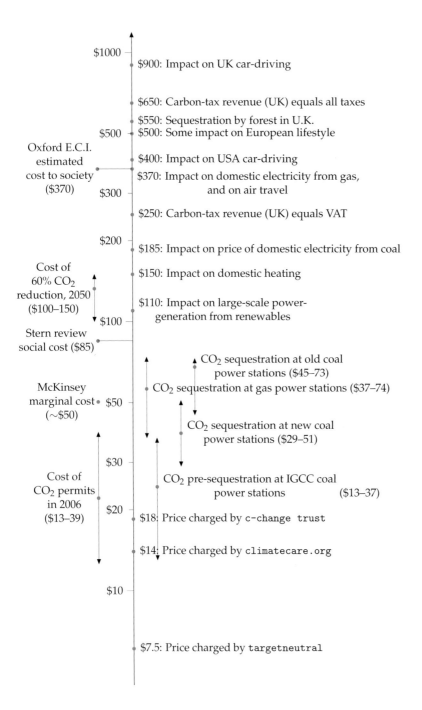

Figure 29.2. What price would CO_2 need to have in order to drive society to make significant changes in CO_2 pollution?

The diagram shows carbon dioxide costs (per tonne) at which particular investments will become economical, or particular behaviours will be significantly impacted, assuming that a major behavioural impact on activities like flying and driving results if the carbon cost doubles the cost of the activity.

As the cost rises through $20–70 per tonne, CO_2 would become sufficiently costly that it would be economical to add carbon sequestration to new and old power stations.

A price of $110 per tonne would transform large-scale renewable electricity-generation projects that currently cost 3p per kWh more than gas from pipedreams into financially viable ventures. For example, the proposed Severn barrage would produce tidal power with a cost of 6p per kWh, which is 3.3p above a typical selling price of 2.7p per kWh; if each 1000 kWh from the barrage avoided one ton of CO_2 pollution at a value of £60 per ton, the Severn barrage would more than pay for itself.

At $150 per tonne, domestic users of gas would notice the cost of carbon in their heating bills.

A price of $250 per tonne would increase the effective cost of a barrel of oil by $100.

At $370, carbon pollution would cost enough to significantly reduce people's inclination to fly.

At $500 per tonne, average Europeans who didn't change their lifestyle might spend 12% of income on the carbon costs of driving, flying, and heating their homes with gas.

And at $900 per tonne, the carbon cost of driving would be noticeable.

way of making some short-term decisions – about investments that will pay off within ten years or so – but can we expect markets to do a good job of making decisions about energy, decisions whose impacts last many decades or centuries?

If the free market is allowed to build houses, we end up with houses that are poorly insulated. Modern houses are only more energy-efficient thanks to legislation.

The free market isn't responsible for building roads, railways, dedicated bus lanes, car parks, or cycle paths. But road-building and the provision of car parks and cycle paths have a significant impact on people's transport choices. Similarly, planning laws, which determine *where* homes and workplaces may be created and *how densely* houses may be packed into land have an overwhelming influence on people's future travelling behaviour. If a new town is created that has no rail station, it is unlikely that the residents of that town will make long-distance journeys by rail. If housing and workplaces are more than a few miles apart, many people will feel that they have no choice but to drive to work.

One of the biggest energy-sinks is the manufacture of stuff; in a free market, many manufacturers supply us with stuff that has planned obsolescence, stuff that has to be thrown away and replaced, so as to make more business for the manufacturers.

So, while markets may play a role, it's silly to say "let the market handle it *all*." Surely we need to talk about legislation, regulations, and taxes.

Greening the tax system

> *We need to profoundly revise all of our taxes and charges. The aim is to tax pollution – notably fossil fuels – more, and tax work less.*

> Nicolas Sarkozy, President of France

At present it's much cheaper to buy a new microwave, DVD player, or vacuum cleaner than to get a malfunctioning one fixed. That's crazy.

This craziness is partly caused by our tax system, which taxes the labour of the microwave-repair man, and surrounds his business with time-consuming paperwork. He's doing a *good* thing, repairing my microwave! – yet the tax system makes it difficult for him to do business.

The idea of "greening the tax system" is to move taxes from "goods" like labour, to "bads" like environmental damage. Advocates of environmental tax reform suggest balancing tax cuts on "goods" by equivalent tax increases on "bads," so that the tax reforms are revenue-neutral.

Carbon tax

The most important tax to increase, if we want to promote fossil-fuel-free technologies, is a tax on carbon. The price of carbon needs to be high enough to promote investment in alternatives to fossil fuels, and investment in efficiency measures. Notice this is exactly the same policy as was suggested in the previous section. So, whether our motivation is fixing climate change, or ensuring security of supply, the policy outcome is the same: we need a carbon price that is stable and high. Figure 29.2 indicates very roughly the various carbon prices that are required to bring about various behaviour changes and investments; and the much lower prices charged by organizations that claim to "offset" greenhouse-gas emissions. How best to arrange a high carbon price? Is the European emissions trading scheme (figure 29.1) the way to go? This question lies in the domain of economists and international policy experts. The view of Cambridge economists Michael Grubb and David Newbery is that the European emissions trading scheme is not up to the job – "current instruments will not deliver an adequate investment response."

The Economist recommends a carbon tax as the primary mechanism for government support of clean energy sources. The Conservative Party's Quality of Life Policy Group also recommends increasing environmental taxes and reducing other taxes – "a shift from *pay as you earn* to *pay as you burn*." The Royal Commission on Environmental Pollution also says that the UK should introduce a carbon tax. "It should apply upstream and cover all sectors."

So, there's clear support for a big carbon tax, accompanied by reductions in employment taxes, corporation taxes, and value-added taxes. But taxes and markets alone are not going to bring about all the actions needed. The tax-and-market approach fails if consumers sometimes choose irrationally, if consumers value short-term cash more highly than long-term savings, or if the person choosing what to buy doesn't pay all the costs associated with their choice.

Indeed many brands are *"reassuringly expensive."* Consumer choice is not determined solely by price signals. Many consumers care more about image and perception, and some deliberately buy expensive.

Once an inefficient thing is bought, it's too late. It's essential that inefficient things should not be manufactured in the first place; or that the consumer, when buying, should feel influenced not to buy inefficient things.

Here are some further examples of failures of the free market.

The admission barrier

Imagine that carbon taxes are sufficiently high that a new super-duper low-carbon gizmo would cost 5% less than its long-standing high-carbon rival, the Dino-gizmo, *if* it were mass-produced in the same quantities. Thanks to clever technology, the Eco-gizmo's carbon emissions are a fantastic 90% lower than the Dino-gizmo's. It's clear that it would be good for society if everyone bought Eco-gizmos now. But at the moment, sales of the new Eco-gizmo are low, so the per-unit economic costs are higher than the

Dino-gizmo's. Only a few tree-huggers and lab coats will buy the Eco-Gizmo, and Eco-Gizmo Inc. will go out of business.

Perhaps government interventions are necessary to oil the transition and give innovation a chance. Support for research and development? Tax-incentives favouring the new product (like the tax-incentives that oiled the transition from leaded to unleaded petrol)?

The problem of small cost differences

Imagine that Eco-Gizmo Inc. makes it from tadpole to frog, and that carbon taxes are sufficiently high that an Eco-gizmo indeed costs 5% less than its long-standing high-carbon rival from Dino-appliances, Inc. Surely the carbon taxes will now do their job, and all consumers will buy the low-carbon gizmo? Ha! First, many consumers don't care too much about a 5% price difference. Image is everything. Second, if they feel at all threatened by the Eco-gizmo, Dino-appliances, Inc. will relaunch their Dino-gizmo, emphasizing that it's more patriotic, announcing that it's now available in green, and showing cool people sticking with the old faithful Dino-gizmo. "Real men buy Dino-gizmos." If this doesn't work, Dino will issue press-releases saying scientists haven't ruled out the possibility that long-term use of the Eco-gizmo might cause cancer, highlighting the case of an old lady who was tripped up by an Eco-gizmo, or suggesting that Eco-gizmos harm the lesser spotted fruit bat. Fear, Uncertainty, Doubt. As a back-up plan, Dino-appliances could always buy up the Eco-gizmo company. The winning product will have nothing to do with energy saving if the economic incentive to the consumer is only 5%.

How to fix this problem? Perhaps government should simply ban the sales of the Dino-gizmo (just as it banned sales of leaded-petrol cars)?

The problem of Larry and Tina

Imagine that Larry the landlord rents out a flat to Tina the tenant. Larry is responsible for maintaining the flat and providing the appliances in it, and Tina pays the monthly heating and electricity bills. Here's the problem: Larry feels no incentive to invest in modifications to the flat that would reduce Tina's bills. He could install more-efficient lightbulbs, and plug in a more economical fridge; these eco-friendly appliances would easily pay back their extra up-front cost over their long life; but it's Tina who would benefit, not Larry. Similarly, Larry feels little incentive to improve the flat's insulation or install double-glazing, especially when he takes into account the risk that Tina's boyfriend Wayne might smash one of the windows when drunk. In principle, in a perfect market, Larry and Tina would both make the "right" decisions: Larry would install all the energy-saving features, and would charge Tina a slightly higher monthly rent; Tina would recognize that the modern and well-appointed flat would be cheaper to live in and would thus be happy to pay the higher rent; Larry would demand an increased deposit in case of breakage of the expensive new windows; and Tina would respond rationally and banish Wayne. However, I don't think that Larry and Tina can ever deliver a perfect market. Tina is poor, so has difficulty paying large deposits. Larry strongly wishes to rent out the flat, so Tina mistrusts his assurances about the property's low energy bills, suspecting Larry of exaggeration.

So some sort of intervention is required, to get Larry and Tina to do the right thing – for example, government could legislate a huge tax on inefficient appliances; ban from sale all fridges that do not meet economy benchmarks; require all flats to meet high standards of insulation; or introduce a system of mandatory independent flat assessment, so that Tina could read about the flat's energy profile before renting.

Investment in research and development

We deplore the minimal amounts that the Government have committed to renewable-energy-related research and development (£12.2 million in 2002-03). ... If resources other than wind are to be exploited in the United Kingdom this has to change. We could not avoid the conclusion that the Government are not taking energy problems sufficiently seriously.

House of Lords Science and Technology Committee

The absence of scientific understanding often leads to superficial decision-making. The 2003 energy white paper was a good example of that. I would not like publicly to call it amateurish but it did not tackle the problem in a realistic way.

Sir David King, former Chief Scientist

Serving on the government's Renewables Advisory Board ... felt like watching several dozen episodes of Yes Minister *in slow motion. I do not think this government has ever been serious about renewables.*

Jeremy Leggett, founder of Solarcentury

I think the numbers speak for themselves. Just look at figure 28.5 (p218) and compare the billions spent on office refurbishments and military toys with the hundred-fold smaller commitment to renewable-energy-related research and development. It takes decades to develop renewable technologies such as tidal stream power, concentrating solar power, and photovoltaics. Nuclear fusion takes decades too. All these technologies need up-front support if they are going to succeed.

Individual action

People sometimes ask me "What should *I* do?" Table 29.3 indicates eight simple personal actions I'd recommend, and a *very* rough indication of the savings associated with each action. Terms and conditions apply. Your savings will depend on your starting point. The numbers in table 29.3 assume the starting point of an above-average consumer.

Simple action	possible saving
Put on a woolly jumper and turn down your heating's thermostat (to 15 or 17 °C, say). Put individual thermostats on all radiators. Make sure the heating's off when no-one's at home. Do the same at work.	20 kWh/d
Read all your meters (gas, electricity, water) every week, and identify easy changes to reduce consumption (e.g., switching things off). Compare competitively with a friend. Read the meters at your place of work too, creating a perpetual live energy audit.	4 kWh/d
Stop flying.	35 kWh/d
Drive less, drive more slowly, drive more gently, carpool, use an electric car, join a car club, cycle, walk, use trains and buses.	20 kWh/d
Keep using old gadgets (e.g. computers); don't replace them early.	4 kWh/d
Change lights to fluorescent or LED.	4 kWh/d
Don't buy clutter. Avoid packaging.	20 kWh/d
Eat vegetarian, six days out of seven.	10 kWh/d

Table 29.3. Eight simple personal actions.

Whereas the above actions are easy to implement, the ones in table 29.4 take a bit more planning, determination, and money.

Major action	possible saving
Eliminate draughts.	5 kWh/d
Double glazing.	10 kWh/d
Improve wall, roof, and floor insulation.	10 kWh/d
Solar hot water panels.	8 kWh/d
Photovoltaic panels.	5 kWh/d
Knock down old building and replace by new.	35 kWh/d
Replace fossil-fuel heating by ground-source or air-source heat pumps.	10 kWh/d

Table 29.4. Seven harder actions.

Finally, table 29.5 shows a few runners-up: some simple actions with small savings.

Action	possible saving
Wash laundry in cold water.	0.5 kWh/d
Stop using a tumble-dryer; use a clothes-line or airing cupboard.	0.5 kWh/d

Table 29.5. A few more simple actions with small savings.

Notes and further reading

page no.

222 *"a bit impractical actually"* The full transcript of the interview with Tony Blair (9 January 2007) is here [2ykfgw]. Here are some more quotes from it:

Interviewer: Have you thought of perhaps not flying to Barbados for a holiday and not using all those air miles?

Tony Blair: I would, frankly, be reluctant to give up my holidays abroad.

Interviewer: It would send out a clear message though wouldn't it, if we didn't see that great big air journey off to the sunshine? . . . – a holiday closer to home?

Tony Blair: Yeah – but I personally think these things are a bit impractical actually to expect people to do that. I think that what we need to do is to look at how you make air travel more energy efficient, how you develop the new fuels that will allow us to burn less energy and emit less. How – for example – in the new frames for the aircraft, they are far more energy efficient.

I know everyone always – people probably think the Prime Minister shouldn't go on holiday at all, but I think if what we do in this area is set people unrealistic targets, you know if we say to people we're going to cancel all the cheap air travel . . . You know, I'm still waiting for the first politician who's actually running for office who's going to come out and say it – and they're not.

The other quote: "Unless we act now, not some time distant but now, these consequences, disastrous as they are, will be irreversible. So there is nothing more serious, more urgent or more demanding of leadership." is Tony Blair speaking at the launch of the Stern review, 30 October 2006 [2nsvx2]. See also [yxq5xk] for further comment.

225 *Environmental tax reform.* See the Green Fiscal Commission, www.greenfiscalcommission.org.uk.

226 The Economist *recommends a carbon tax.* "Nuclear power's new age," *The Economist*, September 8th 2007.

– *The Conservative Party's Quality of Life Policy Group* – Gummer et al. (2007).

30 Energy plans for Europe, America, and the World

Figure 30.1 shows the power consumptions of lots of countries or regions, versus their gross domestic products (GDPs). It is a widely held assumption that human development and growth are good things, so when sketching world plans for sustainable energy I am going to assume that all the countries with low GDP per capita are going to progress rightwards in figure 30.1. And as their GDPs increase, it's inevitable that their power consumptions will increase too. It's not clear what consumption we should plan for, but I think that the average European level (125 kWh per day per person) seems a reasonable assumption; alternatively, we could assume that efficiency measures, like those envisaged in Cartoon Britain in Chapters 19–28, allow all countries to attain a European standard of living with a lower power consumption. In the consumption plan on p204, Cartoon Britain's consumption fell to about 68 kWh/d/p. Bearing in mind that Cartoon Britain doesn't have much industrial activity, perhaps it would be sensible to assume a slightly higher target, such as Hong Kong's 80 kWh/d/p.

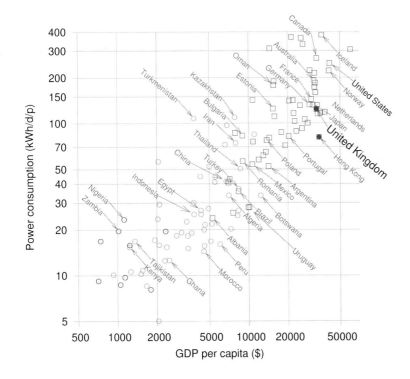

Figure 30.1. Power consumption per capita versus GDP per capita, in purchasing-power-parity US dollars. Data from UNDP Human Development Report, 2007. Squares show countries having "high human development;" circles, "medium" or "low." Both variables are on logarithmic scales. Figure 18.4 shows the same data on normal scales.

Redoing the calculations for Europe

Can Europe live on renewables?

Europe's average population density is roughly half of Britain's, so there is more land area in which to put enormous renewable facilities. The area of the European Union is roughly $9000\,m^2$ per person. However, many of the renewables have lower power density in Europe than in Britain: most of Europe has less wind, less wave, and less tide. Some parts do have more hydro (in Scandanavia and Central Europe); and some have more solar. Let's work out some rough numbers.

Wind

The heart of continental Europe has lower typical windspeeds than the British Isles – in much of Italy, for example, windspeeds are below $4\,m/s$. Let's guess that one fifth of Europe has big enough wind-speeds for economical wind-farms, having a power density of $2\,W/m^2$, and then assume that we give those regions the same treatment we gave Britain in Chapter 4, filling 10% of them with wind farms. The area of the European Union is roughly $9000\,m^2$ per person. So wind gives

$$\frac{1}{5} \times 10\% \times 9000\,m^2 \times 2\,W/m^2 = 360\,W$$

which is $9\,kWh/d$ per person.

Hydroelectricity

Hydroelectric production in Europe totals $590\,TWh/y$, or $67\,GW$; shared between 500 million, that's $3.2\,kWh/d$ per person. This production is dominated by Norway, France, Sweden, Italy, Austria, and Switzerland. If every country doubled its hydroelectric facilities – which I think would be difficult – then hydro would give $6.4\,kWh/d$ per person.

Wave

Taking the whole Atlantic coastline (about $4000\,km$) and multiplying by an assumed average production rate of $10\,kW/m$, we get $2\,kWh/d$ per person. The Baltic and Mediterranean coastlines have no wave resource worth talking of.

Tide

Doubling the estimated total resource around the British Isles ($11\,kWh/d$ per person, from Chapter 14) to allow for French, Irish and Norwegian tidal resources, then sharing between a population of 500 million, we get

2.6 kWh/d per person. The Baltic and Mediterranean coastlines have no tidal resource worth talking of.

Solar photovoltaics and thermal panels on roofs

Most places are sunnier than the UK, so solar panels would deliver more power in continental Europe. 10 m^2 of roof-mounted photovoltaic panels would deliver about 7 kWh/d in all places south of the UK. Similarly, 2 m^2 of water-heating panels could deliver on average 3.6 kWh/d of low-grade thermal heat. (I don't see much point in suggesting having more than 2 m^2 per person of water-heating panels, since this capacity would already be enough to saturate typical demand for hot water.)

What else?

The total so far is $9 + 6.4 + 2 + 2.6 + 7 + 3.6 = 30.6$ kWh/d per person. The only resources not mentioned so far are geothermal power, and large-scale solar farming (with mirrors, panels, or biomass).

Geothermal power might work, but it's still in the research stages. I suggest treating it like fusion power: a good investment, but not to be relied on.

So what about solar farming? We could imagine using 5% of Europe (450 m^2 per person) for solar photovoltaic farms like the Bavarian one in figure 6.7 (which has a power density of 5 W/m^2). This would deliver an average power of

$$5 \text{ W/m}^2 \times 450 \text{ m}^2 = 54 \text{ kWh/d per person.}$$

Solar PV farming would, therefore, add up to something substantial. The main problem with photovoltaic panels is their cost. Getting power during the winter is also a concern!

Energy crops? Plants capture only 0.5 W/m^2 (figure 6.11). Given that Europe needs to feed itself, the non-food energy contribution from plants in Europe can never be enormous. Yes, there will be some oil-seed rape here and some forestry there, but I don't imagine that the total non-food contribution of plants could be more than 12 kWh/d per person.

The bottom line

Let's be realistic. Just like Britain, *Europe can't live on its own renewables*. So if the aim is to get off fossil fuels, Europe needs nuclear power, or solar power in other people's deserts (as discussed on p179), or both.

Figure 30.2. A solar water heater providing hot water for a family in Michigan. The system's pump is powered by the small photovoltaic panel on the left.

Redoing the calculations for North America

The average American uses 250 kWh/d per day. Can we hit that target with
renewables? What if we imagine imposing shocking efficiency measures
(such as efficient cars and high-speed electric trains) such that Americans
were reduced to the misery of living on the mere 125 kWh/d of an average
European or Japanese citizen?

Wind

A study by Elliott et al. (1991) assessed the wind energy potential of the
USA. The windiest spots are in North Dakota, Wyoming, and Montana.
They reckoned that, over the whole country, 435 000 km² of windy land
could be exploited without raising too many hackles, and that the electricity generated would be 4600 TWh per year, which is 42 kWh per day
per person if shared between 300 million people. Their calculations assumed an average power density of 1.2 W/m², incidentally – smaller than
the 2 W/m² we assumed in Chapter 4. The area of these wind farms,
435 000 km², is roughly the same as the area of California. The amount
of wind hardware required (assuming a load factor of 20%) would be a
capacity of about 2600 GW, which would be a 200-fold increase in wind
hardware in the USA.

Offshore wind

If we assume that shallow offshore waters with an area equal to the sum
of Delaware and Connecticut (20 000 km², a substantial chunk of all shallow waters on the east coast of the USA) are filled with offshore wind
farms having a power density of 3 W/m², we obtain an average power of
60 GW. That's 4.8 kWh/d per person if shared between 300 million people.
The wind hardware required would be 15 times the total wind hardware
currently in the USA.

Geothermal

I mentioned the MIT geothermal energy study (Massachusetts Institute of
Technology, 2006) in Chapter 16. The authors are upbeat about the potential of geothermal energy in North America, especially in the western
states where there is more hotter rock. "With a reasonable investment
in R&D, enhanced geothermal systems could provide 100 GW(e) or more
of cost-competitive generating capacity in the next 50 years. Further, enhanced geothermal systems provide a secure source of power for the long
term." Let's assume they are right. 100 GW of electricity is 8 kWh/d per
person when shared between 300 million.

Hydro

The hydroelectric facilities of Canada, the USA, and Mexico generate about 660 TWh per year. Shared between 500 million people, that amounts to 3.6 kWh/d per person. Could the hydroelectric output of North America be doubled? If so, hydro would provide 7.2 kWh/d per person.

What else?

The total so far is $42 + 4.8 + 8 + 7.2 = 62$ kWh/d per person. Not enough for even a European existence! I could discuss various other options such as the sustainable burning of Canadian forests in power stations. But rather than prolong the agony, let's go immediately for a technology that adds up: concentrating solar power.

Figure 30.3 shows the area within North America that would provide everyone there (500 million people) with an average power of 250 kWh/d.

The bottom line

North America's *non-solar* renewables aren't enough for North America to live on. But when we include a massive expansion of solar power, there's enough. So North America needs solar in its own deserts, or nuclear power, or both.

Redoing the calculations for the world

How can 6 billion people obtain the power for a European standard of living – 80 kWh per day per person, say?

Wind

The exceptional spots in the world with strong steady winds are the central states of the USA (Kansas, Oklahoma); Saskatchewan, Canada; the southern extremities of Argentina and Chile; northeast Australia; northeast and northwest China; northwest Sudan; southwest South Africa; Somalia; Iran; and Afghanistan. And everywhere offshore except for a tropical strip 60 degrees wide centred on the equator.

For our global estimate, let's go with the numbers from Greenpeace and the European Wind Energy Association: "the total available wind resources worldwide are estimated at 53 000 TWh per year." That's 24 kWh/d per person.

Hydro

Worldwide, hydroelectricity currently contributes about 1.4 kWh/d per person.

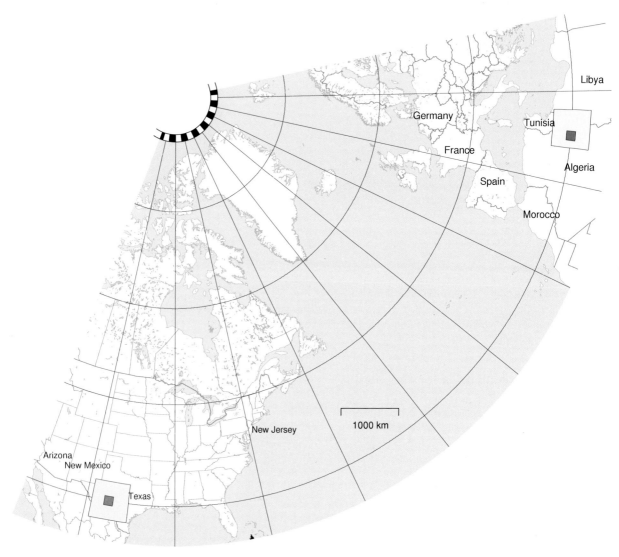

Figure 30.3. The little square strikes again. The 600 km by 600 km square in North America, completely filled with concentrating solar power, would provide enough power to give 500 million people the average American's consumption of 250 kWh/d.

This map also shows the square of size 600 km by 600 km in Africa, which we met earlier. I've assumed a power density of 15 W/m², as before.

The area of one yellow square is a little bigger than the area of Arizona, and 16 times the area of New Jersey. Within each big square is a smaller 145 km by 145 km square showing the area required in the desert – one New Jersey – to supply 30 million people with 250 kWh per day per person.

From the website www.ieahydro.org, "The International Hydropower Association and the International Energy Agency estimate the world's total technical feasible hydro potential at 14 000 TWh/year [6.4 kWh/d per person on the globe], of which about 8000 TWh/year [3.6 kWh/d per person] is currently considered economically feasible for development. Most of the potential for development is in Africa, Asia and Latin America."

Tide

There are several places in the world with tidal resources on the same scale as the Severn estuary (figure 14.8). In Argentina there are two sites: San José and Golfo Nuevo; Australia has the Walcott Inlet; the USA & Canada share the Bay of Fundy; Canada has Cobequid; India has the Gulf of Khambat; the USA has Turnagain Arm and Knik Arm; and Russia has Tugur.

And then there is the world's tidal whopper, a place called Penzhinsk in Russia with a resource of 22 GW – ten times as big as the Severn!

Kowalik (2004) estimates that worldwide, 40–80 GW of tidal power could be generated. Shared between 6 billion people, that comes to 0.16–0.32 kWh/d per person.

Wave

We can estimate the total extractable power from waves by multiplying the length of exposed coastlines (roughly 300 000 km) by the typical power per unit length of coastline (10 kW per metre): the raw power is thus about 3000 GW.

Assuming 10% of this raw power is intercepted by systems that are 50%-efficient at converting power to electricity, wave power could deliver 0.5 kWh/d per person.

Geothermal

According to D. H. Freeston of the Auckland Geothermal Institute, geothermal power amounted on average to about 4 GW, worldwide, in 1995 – which is 0.01 kWh/d per person.

If we assume that the MIT authors on p234 were right, and if we assume that the whole world is like America, then geothermal power offers 8 kWh/d per person.

Solar for energy crops

People get all excited about energy crops like jatropha, which, it's claimed, wouldn't need to compete with food for land, because it can be grown on wastelands. People need to look at the numbers before they get excited.

31   The last thing we should talk about

Capturing carbon dioxide from thin air is the last thing we should talk about.

When I say this, I am deliberately expressing a double meaning. First, the energy requirements for carbon capture from thin air are so enormous, it seems almost absurd to talk about it (and there's the worry that raising the possibility of fixing climate change by this sort of geoengineering might promote inaction today). But second, I do think we should talk about it, contemplate how best to do it, and fund research into how to do it better, because capturing carbon from thin air may turn out to be our last line of defense, if climate change is as bad as the climate scientists say, and if humanity fails to take the cheaper and more sensible options that may still be available today.

Before we discuss capturing carbon from thin air, we need to understand the global carbon picture better.

Understanding CO$_2$

When I first planned this book, my intention was to ignore climate change altogether. In some circles, "Is climate change happening?" was a controversial question. As were "Is it caused by humans?" and "Does it matter?" And, dangling at the end of a chain of controversies, "What should we do about it?" I felt that sustainable energy was a compelling issue by itself, and it was best to avoid controversy. My argument was to be: "Never mind when fossil fuels are going to run out; never mind whether climate change is happening; *burning fossil fuels is not sustainable anyway*; let's imagine living sustainably, and figure out how much sustainable energy is available."

However, climate change has risen into public consciousness, and it raises all sorts of interesting back-of-envelope questions. So I decided to discuss it a little in the preface and in this closing chapter. Not a complete discussion, just a few interesting numbers.

Units

Carbon pollution charges are usually measured in dollars or euros per ton of CO$_2$, so I'll use the *ton of CO$_2$* as the main unit when talking about per-capita carbon pollution, and the *ton of CO$_2$ per year* to measure rates of pollution. (The average European's greenhouse emissions are equivalent to 11 tons per year of CO$_2$; or 30 kg per day of CO$_2$.) But when talking about carbon in fossil fuels, vegetation, soil, and water, I'll talk about tons of carbon. One ton of CO$_2$ contains 12/44 tons of carbon, a bit more than a quarter of a ton. On a planetary scale, I'll talk about gigatons of carbon (Gt C). A gigaton of carbon is a billion tons. Gigatons are hard to imagine, but if you want to bring it down to a human scale, imagine burning one

Figure 31.1. The weights of an atom of carbon and a molecule of CO$_2$ are in the ratio 12 to 44, because the carbon atom weighs 12 units and the two oxygen atoms weigh 16 each. $12 + 16 + 16 = 44$.

ton of coal (which is what you might use to heat a house over a year). Now imagine everyone on the planet burning one ton of coal per year: that's 6 Gt C per year, because the planet has 6 billion people.

Where is the carbon?

Where is all the carbon? We need to know how much is in the oceans, in the ground, and in vegetation, compared to the atmosphere, if we want to understand the consequences of CO_2 emissions.

Figure 31.2 shows where the carbon is. Most of it – 40 000 Gt – is in the ocean (in the form of dissolved CO_2 gas, carbonates, living plant and

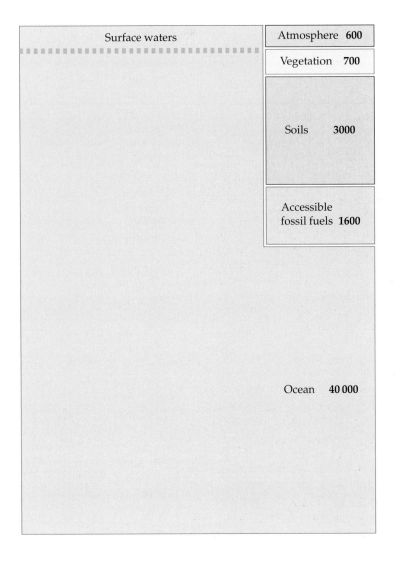

Figure 31.2. Estimated amounts of carbon, in gigatons, in accessible places on the earth. (There's a load more carbon in rocks too; this carbon moves round on a timescale of millions of years, with a long-term balance between carbon in sediment being subducted at tectonic plate boundaries, and carbon popping out of volcanoes from time to time. For simplicity I ignore this geological carbon.)

blah
blah
blah
blah

animal life, and decaying materials). Soils and vegetation together contain about 3700 Gt. Accessible fossil fuels – mainly coal – contain about 1600 Gt. Finally, the atmosphere contains about 600 Gt of carbon.

Until recently, all these pools of carbon were roughly in balance: all flows of carbon out of a pool (say, soils, vegetation, or atmosphere) were balanced by equal flows into that pool. The flows into and out of the fossil fuel pool were both negligible. Then humans started burning fossil fuels. This added two extra *unbalanced* flows, as shown in figure 31.3.

The rate of fossil fuel burning was roughly 1 GtC/y in 1920, 2 GtC/y in 1955, and 8.4 GtC in 2006. (These figures include a small contribution from cement production, which releases CO_2 from limestone.)

How has this significant extra flow of carbon modified the picture shown in figure 31.2? Well, it's not exactly known. Figure 31.3 shows the key things that *are* known. Much of the extra 8.4 GtC per year that we're putting into the atmosphere stays in the atmosphere, raising the atmospheric concentration of carbon-dioxide. The atmosphere equilibrates fairly rapidly with the surface waters of the oceans (this equilibration takes only five or ten years), and there is a net flow of CO_2 from the atmosphere into the surface waters of the oceans, amounting to 2 GtC per year. (Recent research indicates this rate of carbon-uptake by the oceans may be reducing, however.) This unbalanced flow into the surface waters causes ocean acidification, which is bad news for coral. Some extra carbon is moving into vegetation and soil too, perhaps about 1.5 GtC per year, but these flows are less well measured. Because roughly half of the carbon emissions are staying in the atmosphere, continued carbon pollution at a rate of 8.4 GtC per year will continue to increase CO_2 levels in the atmosphere, and in the surface waters.

What is the long-term destination of the extra CO_2? Well, since the amount in fossil fuels is so much smaller than the total in the oceans, "in the long term" the extra carbon will make its way into the ocean, and the amounts of carbon in the atmosphere, vegetation, and soil will return to normal. However, "the long term" means thousands of years. Equilibration between atmosphere and the *surface* waters is rapid, as I said, but figures 31.2 and 31.3 show a dashed line separating the surface waters of the ocean from the rest of the ocean. On a time-scale of 50 years, this boundary is virtually a solid wall. Radioactive carbon dispersed across the globe by the atomic bomb tests of the 1960s and 70s has penetrated the oceans to a depth of only about 400 m. In contrast the average depth of the oceans is about 4000 m.

The oceans circulate slowly: a chunk of deep-ocean water takes about 1000 years to roll up to the surface and down again. The circulation of the deep waters is driven by a combination of temperature gradients and salinity gradients, so it's called the thermohaline circulation (in contrast to the circulations of the surface waters, which are wind-driven).

This slow turn-over of the oceans has a crucial consequence: we have

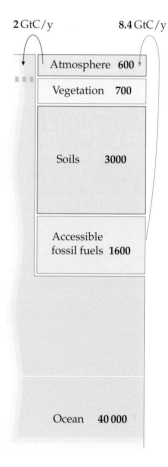

Figure 31.3. The arrows show two extra carbon flows produced by burning fossil fuels. There is an imbalance between the 8.4 GtC/y emissions into the atmosphere from burning fossil fuels and the 2 GtC/y take-up of CO_2 by the oceans. This cartoon omits the less-well quantified flows between atmosphere, soil, vegetation, and so forth.

enough fossil fuels to seriously influence the climate over the next 1000 years.

Where is the carbon going

Figure 31.3 is a gross simplification. For example, humans are causing additional flows not shown on this diagram: the burning of peat and forests in Borneo in 1997 alone released about 0.7 GtC. Accidentally-started fires in coal seams release about 0.25 GtC per year.

Nevertheless, this cartoon helps us understand roughly what will happen in the short term and the medium term under various policies. First, if carbon pollution follows a "business as usual" trajectory, burning another 500 Gt of carbon over the next 50 years, we can expect the carbon to continue to trickle gradually into the surface waters of the ocean at a rate of 2 GtC per year. By 2055, at least 100 Gt of the 500 would have gone into the surface waters, and CO_2 concentrations in the atmosphere would be roughly double their pre-industrial levels.

If fossil-fuel burning were reduced to zero in the 2050s, the 2 Gt flow from atmosphere to ocean would also reduce significantly. (I used to imagine that this flow into the ocean would persist for decades, but that would be true only if the surface waters were out of equilibrium with the atmosphere; but, as I mentioned earlier, the surface waters and the atmosphere reach equilibrium within just a few years.) Much of the 500 Gt we put into the atmosphere would only gradually drift into the oceans over the next few thousand years, as the surface waters roll down and are replaced by new water from the deep.

Thus our perturbation of the carbon concentration might eventually be righted, but only after thousands of years. And that's assuming that this large perturbation of the atmosphere doesn't drastically alter the ecosystem. It's conceivable, for example, that the acidification of the surface waters of the ocean might cause a sufficient extinction of ocean plant-life that a new vicious cycle kicks in: acidification means extinguished plant-life, means plant-life absorbs less CO_2 from the ocean, means oceans become even more acidic. Such vicious cycles (which scientists call "positive feedbacks" or "runaway feedbacks") have happened on earth before: it's believed, for example, that ice ages ended relatively rapidly because of positive feedback cycles in which rising temperatures caused surface snow and ice to melt, which reduced the ground's reflection of sunlight, which meant the ground absorbed more heat, which led to increased temperatures. (Melted snow – water – is much darker than frozen snow.) Another positive feedback possibility to worry about involves methane hydrates, which are frozen in gigaton quantities in places like Arctic Siberia, and in 100-gigaton quantities on continental shelves. Global warming greater than 1 °C would possibly melt methane hydrates, which release methane into the atmosphere, and methane increases global warming more strongly

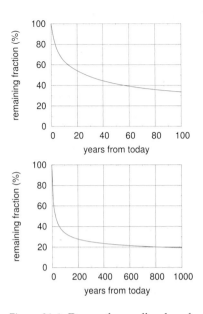

Figure 31.4. Decay of a small pulse of CO_2 added to today's atmosphere, according to the Bern model of the carbon cycle. Source: Hansen et al. (2007).

than CO_2 does.

This isn't the place to discuss the uncertainties of climate change in any more detail. I highly recommend the books *Avoiding Dangerous Climate Change* (Schellnhuber et al., 2006) and *Global Climate Change* (Dessler and Parson, 2006). Also the papers by Hansen et al. (2007) and Charney et al. (1979).

The purpose of this chapter is to discuss the idea of fixing climate change by sucking carbon dioxide from thin air; we discuss the energy cost of this sucking next.

The cost of sucking

Today, pumping carbon out of the ground is big bucks. In the future, perhaps pumping carbon *into* the ground is going to be big bucks. Assuming that inadequate action is taken now to halt global carbon pollution, perhaps a coalition of the willing will in a few decades pay to create a giant vacuum cleaner, and clean up everyone's mess.

Before we go into details of how to capture carbon from thin air, let's discuss the unavoidable energy cost of carbon capture. Whatever technologies we use, they have to respect the laws of physics, and unfortunately grabbing CO_2 from thin air and concentrating it requires energy. The laws of physics say that the energy required must be at least 0.2 kWh per kg of CO_2 (table 31.5). Given that real processes are typically 35% efficient at best, I'd be amazed if the energy cost of carbon capture is ever reduced below 0.55 kWh per kg.

Now, let's assume that we wish to neutralize a typical European's CO_2 output of 11 tons per year, which is 30 kg per day per person. The energy required, assuming a cost of 0.55 kWh per kg of CO_2, is 16.5 kWh per day per person. This is exactly the same as British electricity consumption. So powering the giant vacuum cleaner may require us to *double* our electricity production – or at least, to somehow obtain extra power equal to our current electricity production.

If the cost of running giant vacuum cleaners can be brought down, brilliant, let's make them. But no amount of research and development can get round the laws of physics, which say that grabbing CO_2 from thin air and concentrating it into liquid CO_2 requires at least 0.2 kWh per kg of CO_2.

Now, what's the best way to suck CO_2 from thin air? I'll discuss four technologies for building the giant vacuum cleaner:

   A. chemical pumps;

   B. trees;

   C. accelerated weathering of rocks;

   D. ocean nourishment.

A. Chemical technologies for carbon capture

The chemical technologies typically deal with carbon dioxide in two steps.

$$0.03\% \; CO_2 \quad \xrightarrow{\text{concentrate}} \quad \text{Pure } CO_2 \quad \xrightarrow{\text{compress}} \quad \text{Liquid } CO_2$$

First, they *concentrate* CO_2 from its low concentration in the atmosphere; then they *compress* it into a small volume ready for shoving somewhere (either down a hole in the ground or deep in the ocean). Each of these steps has an energy cost. The costs required by the laws of physics are shown in table 31.5.

In 2005, the best published methods for CO_2 capture from thin air were quite inefficient: the energy cost was about 3.3 kWh per kg, with a financial cost of about \$140 per ton of CO_2. At this energy cost, capturing a European's 30 kg per day would cost 100 kWh per day – almost the same as the European's energy consumption of 125 kWh per day. Can better vacuum cleaners be designed?

Recently, Wallace Broecker, climate scientist, "perhaps the world's foremost interpreter of the Earth's operation as a biological, chemical, and physical system," has been promoting an as yet unpublished technology developed by physicist Klaus Lackner for capturing CO_2 from thin air. Broecker imagines that the world could carry on burning fossil fuels at much the same rate as it does now, and 60 million CO_2-scrubbers (each the size of an up-ended shipping container) will vacuum up the CO_2. What energy does Lackner's process require? In June 2007 Lackner told me that his lab was achieving 1.3 kWh per kg, but since then they have developed a new process based on a resin that absorbs CO_2 when dry and releases CO_2 when moist. Lackner told me in June 2008 that, in a dry climate, the concentration cost has been reduced to about 0.18–0.37 kWh of low-grade heat per kg CO_2. The compression cost is 0.11 kWh per kg. Thus Lackner's total cost is 0.48 kWh or less per kg. For a European's emissions of 30 kg CO_2 per day, we are still talking about a cost of 14 kWh per day, of which 3.3 kWh per day would be electricity, and the rest heat.

Hurray for technical progress! But please don't think that this is a *small* cost. We would require roughly a 20% increase in world energy production, just to run the vacuum cleaners.

	cost (kWh/kg)
concentrate	0.13
compress	0.07
total	0.20

Table 31.5. The inescapable energy-cost of concentrating and compressing CO_2 from thin air.

B. What about trees?

Trees are carbon-capturing systems; they suck CO_2 out of thin air, and they don't violate any laws of physics. They are two-in-one machines: they are carbon-capture facilities powered by built-in solar power stations. They capture carbon using energy obtained from sunlight. The fossil fuels that we burn were originally created by this process. So, the suggestion is, how about trying to do the opposite of fossil fuel burning? How about creating

wood and burying it in a hole in the ground, while, next door, humanity continues digging up fossil wood and setting fire to it? It's daft to imagine creating buried wood at the same time as digging up buried wood. Even so, let's work out the land area required to solve the climate problem with trees.

The best plants in Europe capture carbon at a rate of roughly 10 tons of dry wood per hectare per year – equivalent to about 15 tons of CO_2 per hectare per year – so to fix a European's output of 11 tons of CO_2 per year we need 7500 square metres of forest per person. This required area of 7500 square metres per person is *twice the area of Britain* per person. And then you'd have to find somewhere to permanently store 7.5 tons of wood per person per year! At a density of 500 kg per m^3, each person's wood would occupy 15 m^3 per year. A lifetime's wood – which, remember, must be safely stored away and never burned – would occupy 1000 m^3. That's five times the entire volume of a typical house. If anyone proposes using trees to undo climate change, they need to realise that country-sized facilities are required. I don't see how it could ever work.

1 hectare = $10\,000\,m^2$

C. Enhanced weathering of rocks

Is there a sneaky way to avoid the significant energy cost of the chemical approach to carbon-sucking? Here is an interesting idea: pulverize rocks that are capable of absorbing CO_2, and leave them in the open air. This idea can be pitched as the acceleration of a natural geological process. Let me explain.

Two flows of carbon that I omitted from figure 31.3 are the flow of carbon from rocks into oceans, associated with the natural weathering of rocks, and the natural precipitation of carbon into marine sediments, which eventually turn back into rocks. These flows are relatively small, involving about 0.2 GtC per year (0.7 Gt CO_2 per year). So they are dwarfed by current human carbon emissions, which are about 40 times bigger. But the suggestion of enhanced-weathering advocates is that we could fix climate change by speeding up the rate at which rocks are broken down and absorb CO_2. The appropriate rocks to break down include olivines or magnesium silicate minerals, which are widespread. The idea would be to find mines in places surrounded by many square kilometres of land on which crushed rocks could be spread, or perhaps to spread the crushed rocks directly on the oceans. Either way, the rocks would absorb CO_2 and turn into carbonates and the resulting carbonates would end up being washed into the oceans. To pulverized the rocks into appropriately small grains for the reaction with CO_2 to take place requires only 0.04 kWh per kg of sucked CO_2. Hang on, isn't that smaller than the 0.20 kWh per kg required by the laws of physics? Yes, but nothing is wrong: the rocks themselves are the sources of the missing energy. Silicates have higher energy than carbonates, so the rocks pay the energy cost of sucking the CO_2 from thin air.

I like the small energy cost of this scheme but the difficult question is, who would like to volunteer to cover their country with pulverized rock?

D. Ocean nourishment

One problem with chemical methods, tree-growing methods, and rock-pulverizing methods for sucking CO_2 from thin air is that all would require a lot of work, and no-one has any incentive to do it – unless an international agreement pays for the cost of carbon capture. At the moment, carbon prices are too low.

A final idea for carbon sucking might sidestep this difficulty. The idea is to persuade the ocean to capture carbon a little faster than normal as a by-product of fish farming.

Some regions of the world have food shortages. There are fish shortages in many areas, because of over-fishing during the last 50 years. The idea of *ocean nourishment* is to fertilize the oceans, supporting the base of the food chain, enabling the oceans to support more plant life and more fish, and incidentally to fix more carbon. Led by Australian scientist Ian Jones, the ocean nourishment engineers would like to pump a nitrogen-containing fertilizer such as urea into appropriate fish-poor parts of the ocean. They claim that $900\,km^2$ of ocean can be nourished to take up about $5\,Mt\,CO_2/y$. Jones and his colleagues reckon that the ocean nourishment process is suitable for any areas of the ocean deficient in nitrogen. That includes most of the North Atlantic. Let's put this idea on a map. UK carbon emissions are about $600\,Mt\,CO_2/y$. So complete neutralization of UK carbon emissions would require 120 such areas in the ocean. The map

Figure 31.6. 120 areas in the Atlantic Ocean, each $900\,km^2$ in size. These make up the estimated area required in order to fix Britain's carbon emissions by ocean nourishment.

in figure 31.6 shows these areas to scale alongside the British Isles. As usual, a plan that actually adds up requires country-sized facilities! And we haven't touched on how we would make all the required urea.

While it's an untested idea, and currently illegal, I do find ocean nourishment interesting because, in contrast to geological carbon storage, it's a technology that might be implemented even if the international community doesn't agree on a high value for cleaning up carbon pollution; fishermen might nourish the oceans purely in order to catch more fish.

> *Commentators can be predicted to oppose manipulations of the ocean, focusing on the uncertainties rather than on the potential benefits. They will be playing to the public's fear of the unknown. People are ready to passively accept an escalation of an established practice (e.g., dumping CO_2 in the atmosphere) while being wary of innovations that might improve their future well being. They have an uneven aversion to risk.*
>
> Ian Jones

> *We, humanity, cannot release to the atmosphere all, or even most, fossil fuel CO_2. To do so would guarantee dramatic climate change, yielding a different planet...*
>
> J. Hansen et al (2007)

> *"Avoiding dangerous climate change" is impossible – dangerous climate change is already here. The question is, can we avoid* **catastrophic** *climate change?*
>
> David King, UK Chief Scientist, 2007

Notes

page no.

240 *climate change ... was a controversial question.* Indeed there still is a "yawning gap between mainstream opinion on climate change among the educated elites of Europe and America" [voxbz].

241 *Where is the carbon?* Sources: Schellnhuber et al. (2006), Davidson and Janssens (2006).

242 *The rate of fossil fuel burning...* Source: Marland et al. (2007).

– *Recent research indicates carbon-uptake by the oceans may be reducing.* www.timesonline.co.uk/tol/news/uk/scienc article1805870.ece, www.sciencemag.org/cgi/content/abstract/1136188, [yofchc], Le Quéré et al. (2007).

– *roughly half of the carbon emissions are staying in the atmosphere.* It takes 2.1 billion tons of carbon in the atmosphere (7.5 GtCO₂) to raise the atmospheric CO_2 concentration by one part per million (1 ppm). If all the CO_2 we pumped into the atmosphere stayed there, the concentration would be rising by more than 3 ppm per year – but it is actually rising at only 1.5 ppm per year.

– *Radioactive carbon ...has penetrated to a depth of only about 400 m.* The mean value of the penetration depth of bomb ^{14}C for all observational sites during the late 1970s is 390±39 m (Broecker et al., 1995). From [3e28ed].

244 *Global warming greater than 1°C would possibly melt methane hydrates.* Source: Hansen et al. (2007, p1942).

245 *Table 31.5. Inescapable cost of concentrating and compressing CO_2 from thin air.* The unavoidable energy requirement to concentrate CO_2 from 0.03% to 100% at atmospheric pressure is $kT \ln 100/0.03$ per molecule, which is 0.13 kWh per kg. The ideal energy cost of compression of CO_2 to 110 bar (a pressure mentioned for geological storage) is 0.067 kWh/kg. So the total ideal cost of CO_2 capture and compression is 0.2 kWh/kg. According to the IPCC special report on carbon capture and storage, the practical cost of the second step, compression of CO_2 to 110 bar, is 0.11 kWh per kg. (0.4 GJ per t CO_2; 18 kJ per mole CO_2; 7 kT per molecule.)

245 *Shoving the CO_2 down a hole in the ground or deep in the ocean.* See Williams (2000) for discussion. "For a large fraction of injected CO_2 to remain in the ocean, injection must be at great depths. A consensus is developing that the best near-term strategy would be to discharge CO_2 at depths of 1000–1500 metres, which can be done with existing technology."
See also the Special Report by the IPCC: www.ipcc.ch/ipccreports/srccs.htm.

– *In 2005, the best methods for carbon capture were quite inefficient: the energy cost was about 3.3 kWh per kg, with a financial cost of about $140 per ton of CO_2.* Sources: Keith et al. (2005), Lackner et al. (2001), Herzog (2003), Herzog (2001), David and Herzog (2000).

– *Wallace Broecker, climate scientist...* www.af-info.or.jp/eng/honor/hot/enrbro.html. His book promoting artificial trees: Broecker and Kunzig (2008).

246 *The best plants in Europe capture carbon at a rate of roughly 10 tons of dry wood per hectare per year.* Source: Select Committee on Science and Technology.

– *Enhanced weathering of rocks.* See Schuiling and Krijgsman (2006).

247 *Ocean nourishment.* See Judd et al. (2008). See also Chisholm et al. (2001). The risks of ocean nourishment are discussed in Jones (2008).

32 Saying yes

Because Britain currently gets 90% of its energy from fossil fuels, it's no surprise that getting off fossil fuels requires big, big changes – a total change in the transport fleet; a complete change of most building heating systems; and a 10- or 20-fold increase in green power.

Given the general tendency of the public to say "no" to wind farms, "no" to nuclear power, "no" to tidal barrages – "no" to anything other than fossil fuel power systems – I am worried that we won't actually get off fossil fuels when we need to. Instead, we'll settle for half-measures: slightly-more-efficient fossil-fuel power stations, cars, and home heating systems; a fig-leaf of a carbon trading system; a sprinkling of wind turbines; an inadequate number of nuclear power stations.

We need to choose a plan that adds up. It *is* possible to make a plan that adds up, but it's not going to be easy.

We need to stop saying no and start saying yes. We need to stop the Punch and Judy show and get building.

If you would like an honest, realistic energy policy that adds up, please tell all your political representatives and prospective political candidates.

Acknowledgments

For leading me into environmentalism, I thank Robert MacKay, Gale Ryba, and Mary Archer.

For decades of intense conversation on every detail, thank you to Matthew Bramley, Mike Cates, and Tim Jervis.

For good ideas, for inspiration, for suggesting good turns of phrase, for helpful criticism, for encouragement, I thank the following people, all of whom have shaped this book. John Hopfield, Sanjoy Mahajan, Iain Murray, Ian Fells, Tony Benn, Chris Bishop, Peter Dayan, Zoubin Ghahramani, Kimber Gross, Peter Hodgson, Jeremy Lefroy, Robert MacKay, William Nuttall, Mike Sheppard, Ed Snelson, Quentin Stafford-Fraser, Prashant Vaze, Mark Warner, Seb Wills, Phil Cowans, Bart Ullstein, Helen de Mattos, Daniel Corbett, Greg McMullen, Alan Blackwell, Richard Hills, Philip Sargent, Denis Mollison, Volker Heine, Olivia Morris, Marcus Frean, Erik Winfree, Caryl Walter, Martin Hellman, Per Sillrén, Trevor Whittaker, Daniel Nocera, Jon Gibbins, Nick Butler, Sally Daultrey, Richard Friend, Guido Bombi, Alessandro Pastore, John Peacock, Carl Rasmussen, Phil C. Stuart, Adrian Wrigley, Jonathan Kimmitt, Henry Jabbour, Ian Bryden, Andrew Green, Montu Saxena, Chris Pickard, Kele Baker, Davin Yap, Martijn van Veen, Sylvia Frean, Janet Lefroy, John Hinch, James Jackson, Stephen Salter, Derek Bendall, Deep Throat, Thomas Hsu, Geoffrey Hinton, Radford Neal, Sam Roweis, John Winn, Simon Cran-McGreehin, Jackie Ford, Lord Wilson of Tillyorn, Dan Kammen, Harry Bhadeshia, Colin Humphreys, Adam Kalinowski, Anahita New, Jonathan Zwart, John Edwards, Danny Harvey, David Howarth, Andrew Read, Jenny Smithers, William Connolley, Ariane Kossack, Sylvie Marchand, Phil Hobbs, David Stern, Ryan Woodard, Noel Thompson, Matthew Turner, Frank Stajano, Stephen Stretton, Terry Barker, Jonathan Köhler, Peter Pope, Aleks Jakulin, Charles Lee, Dave Andrews, Dick Glick, Paul Robertson, Jürg Matter, Alan and Ruth Foster, David Archer, Philip Sterne, Oliver Stegle, Markus Kuhn, Keith Vertanen, Anthony Rood, Pilgrim Beart, Ellen Nisbet, Bob Flint, David Ward, Pietro Perona, Andrew Urquhart, Michael McIntyre, Andrew Blake, Anson Cheung, Daniel Wolpert, Rachel Warren, Peter Tallack, Philipp Hennig, Christian Steinrücken, Tamara Broderick, Demosthenis Pafitis, David Newbery, Annee Blott, Henry Leveson-Gower, John Colbert, Philip Dawid, Mary Waltham, Philip Slater, Christopher Hobbs, Margaret Hobbs, Paul Chambers, Michael Schlup, Fiona Harvey, Jeremy Nicholson, Ian Gardner, Sir John Sulston, Michael Fairbank, Menna Clatworthy, Gabor Csanyi, Stephen Bull, Jonathan Yates, Michael Sutherland, Michael Payne, Simon Learmount, John Riley, Lord John Browne, Cameron Freer, Parker Jones, Andrew Stobart, Peter Ravine, Anna Jones, Peter Brindle, Eoin Pierce, Willy Brown, Graham Treloar, Robin Smale, Dieter Helm, Gordon Taylor, Saul Griffith, David Cebonne, Simon Mercer, Alan Storkey, Giles Hodgson, Amos Storkey, Chris Williams, Tristan Collins, Darran Messem, Simon Singh, Gos Micklem, Peter Guthrie, Shin-Ichi Maeda, Candida Whitmill, Beatrix Schlarb-Ridley, Fabien Petitcolas, Sandy Polak, Dino Seppi, Tadashi Tokieda, Lisa Willis, Paul Weall, Hugh Hunt, Jon Fairbairn, Miloš T. Kojašević, Andrew Howe, Ian Leslie, Andrew Rice, Miles Hember, Hugo Willson, Win Rampen, Nigel Goddard, Richard Dietrich, Gareth Gretton, David Sterratt, Jamie Turner, Alistair Morfey, Rob Jones, Paul McKeigue, Rick Jefferys, Robin S Berlingo, Frank Kelly, Michael Kelly, Scott Kelly, Anne Miller, Malcolm Mackley, Tony Juniper, Peter Milloy, Cathy Kunkel, Tony Dye, Rob Jones, Garry Whatford, Francis Meyer, Wha-Jin Han, Brendan McNamara, Michael Laughton, Dermot McDonnell, John McCone, Andreas Kay, John McIntyre, Denis Bonnelle, Ned Ekins-Daukes, John Daglish, Jawed Karim, Tom Yates, Lucas Kruijswijk, Sheldon Greenwell, Charles Copeland, Georg Heidenreich, Colin Dunn, Steve Foale, Leo Smith, Mark McAndrew, Bengt Gustafsson, Roger Pharo, David Calderwood, Graham Pendlebury, Brian Collins, Paul Hasley, Martin Dowling, Martin Whiteland, Andrew Janca, Keith Henson, Graeme Mitchison, Valerie MacKay, Dewi Williams, Nick Barnes, Niall Mansfield, Graham Smith, Wade Amos, Sven Weier, Richard McMahon, Andrew Wallace, Corinne Meakins, Eoin O'Carroll, Iain McClatchie, Alexander Ac, Mark Suthers, Gustav Grob, Ibrahim Dincer, Ian Jones, Adnan Midilli, Chul Park, David Gelder, Damon Hart-

Davis, George Wallis, Philipp Spöth, James Wimberley, Richard Madeley, Jeremy Leggett, Michael Meacher, Dan Kelley, Tony Ward-Holmes, Charles Barton, James Wimberley, Jay Mucha, Johan Simu, Stuart Lawrence, Nathaniel Taylor, Dickon Pinner, Michael Davey, Michael Riedel, William Stoett, Jon Hilton, Mike Armstrong, Tony Hamilton, Joe Burlington, David Howey, Jim Brough, Mark Lynas, Hezlin Ashraf-Ball, Jim Oswald, John Lightfoot, Carol Atkinson, Nicola Terry, George Stowell, Damian Smith, Peter Campbell, Ian Percival, David Dunand, Nick Cook, Leon di Marco, Dave Fisher, John Cox, Jonathan Lee, Richard Procter, Matt Taylor, Carl Scheffler, Chris Burgoyne, Francisco Monteiro, Ian McChesney, and Liz Moyer. Thank you all.

For help with finding climate data, I thank Emily Shuckburgh. I'm very grateful to Kele Baker for gathering the electric car data in figure 20.21. I also thank David Sterratt for research contributions, and Niall Mansfield, Jonathan Zwart, and Anna Jones for excellent editorial advice.

The errors that remain are of course my own.

I am especially indebted to Seb Wills, Phil Cowans, Oliver Stegle, Patrick Welche, and Carl Scheffler for keeping my computers working.

I thank the African Institute for Mathematical Sciences, Cape Town, and the Isaac Newton Institute for Mathematical Sciences, Cambridge, for hospitality.

Many thanks to the Digital Technology Group, Computer Laboratory, Cambridge and Heriot–Watt University Physics Department for providing weather data online. I am grateful to Jersey Water and Guernsey Electricity for tours of their facilities.

Thank you to Gilby Productions for providing the TinyURL service. TinyURL is a trademark of Gilby Productions. Thank you to Eric Johnston and Satellite Signals Limited for providing a nice interface for maps [www.satsig.net].

Thank you to David Stern for the portrait, to Becky Smith for iconic artwork, and to Claire Jervis for the photos on pages ix, 31, 90, 95, 153, 245, 289, and 325. For other photos, thanks to Robert MacKay, Eric LeVin, Marcus Frean, Rosie Ward, Harry Bhadeshia, Catherine Huang, Yaan de Carlan, Pippa Swannell, Corinne Le Quéré, David Faiman, Kele Baker, Tim Jervis, and anonymous contributors to Wikipedia. I am grateful to the office of the Mayor of London for providing copies of advertisements.

The artwork on page 240 is "Maid in London," and on page 288, "Sunflowers," by Banksy www.banksy.co.uk. Thank you, Banksy!

Offsetting services were provided by cheatneutral.com.

This book is written in LaTeX on the Ubuntu GNU/Linux operating system using free software. The figures were drawn with gnuplot and metapost. Many of the maps were created with Paul Wessel and Walter Smith's gmt software. Thank you also to Martin Weinelt and OMC. Thank you to Donald Knuth, Leslie Lamport, Richard Stallman, Linus Torvalds, and all those who contribute to free software.

Finally I owe the biggest debt of gratitude to the Gatsby Charitable Foundation, who supported me and my research group before, during, and after the writing of this book.

Part III

Technical chapters

A   Cars II

We estimated that a car driven 100 km uses about 80 kWh of energy.

Where does this energy go? How does it depend on properties of the car? Could we make cars that are 100 times more efficient? Let's make a simple cartoon of car-driving, to describe where the energy goes. The energy in a typical fossil-fuel car goes to four main destinations, all of which we will explore:

1. speeding up then slowing down using the brakes;

2. air resistance;

3. rolling resistance;

4. heat – 75% of the energy is thrown away as heat, because the energy-conversion chain is inefficient.

Initially our cartoon will ignore rolling resistance; we'll add in this effect later in the chapter.

Assume the driver accelerates rapidly up to a cruising speed v, and maintains that speed for a distance d, which is the distance between traffic lights, stop signs, or congestion events. At this point, he slams on the brakes and turns all his kinetic energy into heat in the brakes. (This vehicle doesn't have fancy regenerative braking.) Once he's able to move again, he accelerates back up to his cruising speed, v. This acceleration gives the car kinetic energy; braking throws that kinetic energy away.

Energy goes not only into the brakes: while the car is moving, it makes air swirl around. A car leaves behind it a tube of swirling air, moving at a speed similar to v. Which of these two forms of energy is the bigger: kinetic energy of the swirling air, or heat in the brakes? Let's work it out.

- The car speeds up and slows down once in each duration d/v. The rate at which energy pours into the brakes is:

$$\frac{\text{kinetic energy}}{\text{time between braking events}} = \frac{\frac{1}{2}m_{\mathrm{c}}v^2}{d/v} = \frac{\frac{1}{2}m_{\mathrm{c}}v^3}{d}, \qquad (A.1)$$

where m_{c} is the mass of the car.

Figure A.1. A Peugot 206 has a drag coefficient of 0.33. Photo by Christopher Batt.

The key formula for most of the calculations in this book is:

$$\text{kinetic energy} = \frac{1}{2}mv^2.$$

For example, a car of mass $m = 1000$ kg moving at 100 km per hour or $v = 28$ m/s has an energy of

$$\frac{1}{2}mv^2 \simeq 390\,000\,\mathrm{J} \simeq 0.1\,\mathrm{kWh}.$$

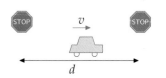

Figure A.2. Our cartoon: a car moves at speed v between stops separated by a distance d.

Figure A.3. A car moving at speed v creates behind it a tube of swirling air; the cross-sectional area of the tube is similar to the frontal area of the car, and the speed at which air in the tube swirls is roughly v.

- The tube of air created in a time t has a volume Avt, where A is the cross-sectional area of the tube, which is similar to the area of the front view of the car. (For a streamlined car, A is usually a little smaller than the frontal area A_{car}, and the ratio of the tube's effective cross-sectional area to the car area is called the drag coefficient c_d. Throughout the following equations, A means the effective area of the car, $c_d A_{car}$.) The tube has mass $m_{air} = \rho Avt$ (where ρ is the density of air) and swirls at speed v, so its kinetic energy is:

$$\frac{1}{2}m_{air}v^2 = \frac{1}{2}\rho Avt\, v^2,$$

and the rate of generation of kinetic energy in swirling air is:

$$\frac{\frac{1}{2}\rho Avtv^2}{t} = \frac{1}{2}\rho Av^3.$$

So the total rate of energy production by the car is:

$$\text{power going into brakes} + \text{power going into swirling air} = \tfrac{1}{2}m_c v^3/d + \tfrac{1}{2}\rho Av^3. \quad (A.2)$$

Both forms of energy dissipation scale as v^3. So this cartoon predicts that a driver who halves his speed v makes his power consumption 8 times smaller. If he ends up driving the same total distance, his journey will take twice as long, but the total energy consumed by his journey will be four times smaller.

Which of the two forms of energy dissipation – brakes or air-swirling – is the bigger? It depends on the ratio of

$$(m_c/d)/(\rho A).$$

If this ratio is much bigger than 1, then more power is going into brakes; if it is smaller, more power is going into swirling air. Rearranging this ratio, it is bigger than 1 if

$$m_c > \rho Ad.$$

Now, Ad is the volume of the tube of air swept out from one stop sign to the next. And ρAd is the mass of that tube of air. So we have a very simple situation: energy dissipation is dominated by kinetic-energy-being-dumped-into-the-brakes if the mass of the car is *bigger* than the mass of the tube of air from one stop sign to the next; and energy dissipation is dominated by making-air-swirl if the mass of the car is *smaller* (figure A.4).

Let's work out the special distance d^* between stop signs, below which the dissipation is braking-dominated and above which it is air-swirling dominated (also known as drag-dominated). If the frontal area of the car is:

$$A_{car} = 2\,\text{m wide} \times 1.5\,\text{m high} = 3\,\text{m}^2$$

I'm using this formula:

$$\text{mass} = \text{density} \times \text{volume}$$

The symbol ρ (Greek letter 'rho') denotes the density.

Figure A.4. To know whether energy consumption is braking-dominated or air-swirling-dominated, we compare the mass of the car with the mass of the tube of air between stop-signs.

Figure A.5. Power consumed by a car is proportional to its cross-sectional area, during motorway driving, and to its mass, during town driving. Guess which gets better mileage – the VW on the left, or the spaceship?

and the drag coefficient is $c_d = 1/3$ and the mass is $m_c = 1000\,\text{kg}$ then the special distance is:

$$d^* = \frac{m_c}{\rho c_d A_{car}} = \frac{1000\,\text{kg}}{1.3\,\text{kg/m}^3 \times \frac{1}{3} \times 3\,\text{m}^2} = 750\,\text{m}.$$

So "city-driving" is dominated by kinetic energy and braking if the distance between stops is less than 750 m. Under these conditions, it's a good idea, if you want to save energy:

1. to reduce the mass of your car;

2. to get a car with regenerative brakes (which roughly halve the energy lost in braking – see Chapter 20); and

3. to drive more slowly.

When the stops are significantly more than 750 m apart, energy dissipation is drag-dominated. Under these conditions, it doesn't much matter what your car weighs. Energy dissipation will be much the same whether the car contains one person or six. Energy dissipation can be reduced:

1. by reducing the car's drag coefficient;

2. by reducing its cross-sectional area; or

3. by driving more slowly.

The actual energy consumption of the car will be the energy dissipation in equation (A.2), cranked up by a factor related to the inefficiency of the engine and the transmission. Typical petrol engines are about 25% efficient, so of the chemical energy that a car guzzles, three quarters is wasted in making the car's engine and radiator hot, and just one quarter goes into "useful" energy:

$$\text{total power of car} \simeq 4 \left[\frac{1}{2} m_c v^3 / d + \frac{1}{2} \rho A v^3 \right].$$

Let's check this theory of cars by plugging in plausible numbers for motorway driving. Let $v = 70\,\text{miles per hour} = 110\,\text{km/h} = 31\,\text{m/s}$ and $A = c_d A_{car} = 1\,\text{m}^2$. The power consumed by the engine is estimated to be roughly

$$4 \times \frac{1}{2} \rho A v^3 = 2 \times 1.3\,\text{kg/m}^3 \times 1\,\text{m}^2 \times (31\,\text{m/s})^3 = 80\,\text{kW}.$$

If you drive the car at this speed for one hour every day, then you travel 110 km and use 80 kWh of energy per day. If you drove at half this speed for two hours per day instead, you would travel the same distance and use up 20 kWh of energy. This simple theory seems consistent with the

ENERGY-PER-DISTANCE	
Car at 110 km/h	$\leftrightarrow$ 80 kWh/(100 km)
Bicycle at 21 km/h	$\leftrightarrow$ 2.4 kWh/(100 km)

PLANES AT 900 KM/H	
A380	27 kWh/100 seat-km

Table A.6. Facts worth remembering: car energy consumption.

mileage figures for cars quoted in Chapter 3. Moreover, the theory gives insight into how the energy consumed by your car could be reduced. The theory has a couple of flaws which we'll explore in a moment.

Could we make a new car that consumes 100 times less energy and still goes at 70 mph? **No.** Not if the car has the same shape. On the motorway at 70 mph, the energy is going mainly into making air swirl. Changing the materials the car is made from makes no difference to that. A miraculous improvement to the fossil-fuel engine could perhaps boost its efficiency from 25% to 50%, bringing the energy consumption of a fossil-fuelled car down to roughly 40 kWh per 100 km.

Electric vehicles have some wins: while the weight of the energy store, per useful kWh stored, is about 25 times bigger than that of petrol, the weight of an electric engine can be about 8 times smaller. And the energy-chain in an electric car is much more efficient: electric motors can be 90% efficient.

We'll come back to electric cars in more detail towards the end of this chapter.

Bicycles and the scaling trick

Here's a fun question: what's the energy consumption of a bicycle, in kWh per 100 km? Pushing yourself along on a bicycle requires energy for the same reason as a car: you're making air swirl around. Now, we could do all the calculations from scratch, replacing car-numbers by bike-numbers. But there's a simple trick we can use to get the answer for the bike from the answer for the car. The energy consumed by a car, per distance travelled, is the power-consumption associated with air-swirling,

$$4 \times \frac{1}{2}\rho A v^3,$$

divided by the speed, v; that is,

$$\text{energy per distance} = 4 \times \frac{1}{2}\rho A v^2.$$

The "4" came from engine inefficiency; ρ is the density of air; the area $A = c_d A_{car}$ is the effective frontal area of a car; and v is its speed.

Now, we can compare a bicycle with a car by dividing $4 \times \frac{1}{2}\rho A v^2$ for the bicycle by $4 \times \frac{1}{2}\rho A v^2$ for the car. All the fractions and ρs cancel, if the efficiency of the carbon-powered bicyclist's engine is similar to the efficiency of the carbon-powered car engine (which it is). The ratio is:

$$\frac{\text{energy per distance of bike}}{\text{energy per distance of car}} = \frac{c_d^{bike} A_{bike} v_{bike}^2}{c_d^{car} A_{car} v_{car}^2}.$$

The trick we are using is called "scaling." If we know how energy consumption scales with speed and area, then we can predict energy con-

DRAG COEFFICIENTS	
CARS	
Honda Insight	0.25
Prius	0.26
Renault 25	0.28
Honda Civic (2006)	0.31
VW Polo GTi	0.32
Peugeot 206	0.33
Ford Sierra	0.34
Audi TT	0.35
Honda Civic (2001)	0.36
Citroën 2CV	0.51
Cyclist	0.9
Long-distance coach	0.425
PLANES	
Cessna	0.027
Learjet	0.022
Boeing 747	0.031

DRAG-AREAS (m^2)	
Land Rover Discovery	1.6
Volvo 740	0.81
Typical car	**0.8**
Honda Civic	0.68
VW Polo GTi	0.65
Honda Insight	0.47

Table A.7. Drag coefficients and drag areas.

sumption of objects with completely different speeds and areas. Specifically, let's assume that the area ratio is

$$\frac{A_{bike}}{A_{car}} = \frac{1}{4}.$$

(Four cyclists can sit shoulder to shoulder in the width of one car.) Let's
assume the bike is not very well streamlined:

$$\frac{c_d^{bike}}{c_d^{car}} = \frac{1}{1/3}.$$

And let's assume the speed of the bike is 21 km/h (13 miles per hour), so

$$\frac{v_{bike}}{v_{car}} = \frac{1}{5}.$$

Then

$$\frac{\text{energy-per-distance of bike}}{\text{energy-per-distance of car}} = \left(\frac{c_d^{bike}}{c_d^{car}} \frac{A_{bike}}{A_{car}}\right)\left(\frac{v_{bike}}{v_{car}}\right)^2$$

$$= \left(\frac{3}{4}\right) \times \left(\frac{1}{5}\right)^2$$

$$= \frac{3}{100}$$

So a cyclist at 21 km/h consumes about 3% of the energy per kilometre of
a lone car-driver on the motorway – about **2.4 kWh per 100 km**.
    If you would like a vehicle whose fuel efficiency is 30 times better than
a car's, it's simple: ride a bike.

What about rolling resistance?

Some things we've completely ignored so far are the energy consumed in
the tyres and bearings of the car, the energy that goes into the noise of
wheels against asphalt, the energy that goes into grinding rubber off the
tyres, and the energy that vehicles put into shaking the ground. Collec-
tively, these forms of energy consumption are called *rolling resistance*. The
standard model of rolling resistance asserts that the force of rolling resistance is simply proportional to the weight of the vehicle, independent of

wheel	C_{rr}
train (steel on steel)	0.002
bicycle tyre	0.005
truck rubber tyres	0.007
car rubber tyres	0.010

Table A.8. The rolling resistance is equal to the weight multiplied by the
coefficient of rolling resistance, C_{rr}. The rolling resistance includes the force
due to wheel flex, friction losses in the wheel bearings, shaking and vibration
of both the roadbed and the vehicle (including energy absorbed by the
vehicle's shock absorbers), and sliding of the wheels on the road or rail. The
coefficient varies with the quality of the road, with the material the wheel is
made from, and with temperature. The numbers given here assume smooth
roads. [2bhu35]

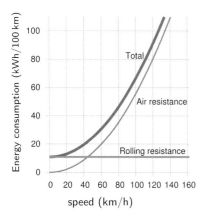

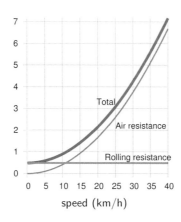

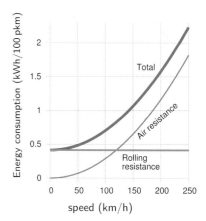

Figure A.9. Simple theory of car fuel consumption (energy per distance) when driving at steady speed. Assumptions: the car's engine uses energy with an efficiency of 0.25, whatever the speed; $c_d A_{car} = 1\,\text{m}^2$; $m_{car} = 1000\,\text{kg}$; and $C_{rr} = 0.01$.

Figure A.10. Simple theory of bike fuel consumption (energy per distance). Vertical axis is energy consumption in kWh per 100 km. Assumptions: the bike's engine (that's you!) uses energy with an efficiency of 0.25,; the drag-area of the cyclist is $0.75\,\text{m}^2$; the cyclist+bike's mass is 90 kg; and $C_{rr} = 0.005$.

Figure A.11. Simple theory of train energy consumption, *per passenger*, for an eight-carriage train carrying 584 passengers. Vertical axis is energy consumption in kWh per 100 p-km. Assumptions: the train's engine uses energy with an efficiency of 0.90; $c_d A_{train} = 11\,\text{m}^2$; $m_{train} = 400\,000\,\text{kg}$; and $C_{rr} = 0.002$.

the speed. The constant of proportionality is called the coefficient of rolling resistance, C_{rr}. Table A.8 gives some typical values.

The coefficient of rolling resistance for a car is about 0.01. The effect of rolling resistance is just like perpetually driving up a hill with a slope of one in a hundred. So rolling friction is about 100 newtons per ton, independent of speed. You can confirm this by pushing a typical one-ton car along a flat road. Once you've got it moving, you'll find you can keep it moving with one hand. (100 newtons is the weight of 100 apples.) So at a speed of 31 m/s (70 mph), the power required to overcome rolling resistance, for a one-ton vehicle, is

$$\text{force} \times \text{velocity} = (100\ \text{newtons}) \times (31\,\text{m/s}) = 3100\,\text{W};$$

which, allowing for an engine efficiency of 25%, requires 12 kW of power to go into the engine; whereas the power required to overcome drag was estimated on p256 to be 80 kW. So, at high speed, about 15% of the power is required for rolling resistance.

Figure A.9 shows the theory of fuel consumption (energy per unit distance) as a function of steady speed, when we add together the air resistance and rolling resistance.

The speed at which a car's rolling resistance is equal to air resistance is

given by

$$C_{rr}m_c g = \frac{1}{2}\rho c_d A v^2,$$

that is,

$$v = \sqrt{2\frac{C_{rr}m_c g}{\rho c_d A}} = 7\,\text{m/s} = 16\,\text{miles per hour}.$$

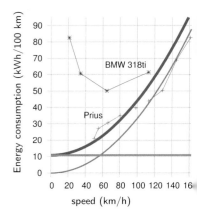

Figure A.12. Current cars' fuel consumptions do not vary as speed squared. Prius data from B.Z. Wilson; BMW data from Phil C. Stuart. The smooth curve shows what a speed-squared curve would look like, assuming a drag-area of 0.6 m².

Bicycles

For a bicycle ($m = 90$ kg, $A = 0.75$ m²), the transition from rolling-resistance-dominated cycling to air-resistance-dominated cycling takes place at a speed of about 12 km/h. At a steady speed of 20 km/h, cycling costs about 2.2 kWh per 100 km. By adopting an aerodynamic posture, you can reduce your drag area and cut the energy consumption down to about 1.6 kWh per 100 km.

Trains

For an eight-carriage train as depicted in figure 20.4 ($m = 400\,000$ kg, $A = 11$ m²), the speed above which air resistance is greater than rolling resistance is

$$v = 33\,\text{m/s} = 74\,\text{miles per hour}.$$

For a single-carriage train ($m = 50\,000$ kg, $A = 11$ m²), the speed above which air resistance is greater than rolling resistance is

$$v = 12\,\text{m/s} = 26\,\text{miles per hour}.$$

Dependence of power on speed

When I say that halving your driving speed should reduce fuel consumption (in miles per gallon) to *one quarter* of current levels, some people feel sceptical. They have a point: most cars' engines have an optimum revolution rate, and the choice of gears of the car determines a range of speeds at which the optimum engine efficiency can be delivered. If my suggested experiment of halving the car's speed takes the car out of this designed range of speeds, the consumption might not fall by as much as four-fold. My tacit assumption that the engine's efficiency is the same at all speeds and all loads led to the conclusion that it's always good (in terms of miles per gallon) to travel slower; but if the engine's efficiency drops off at low speeds, then the most fuel-efficient speed might be at an intermediate speed that makes a compromise between going slow and keeping the engine efficient. For the BMW 318ti in figure A.12, for example, the optimum speed is about 60 km/h. But if society were to decide that car speeds should be reduced, there is nothing to stop engines and gears being redesigned so that the peak engine efficiency was found at the right speed. As further evidence

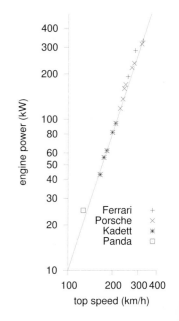

Figure A.13. Powers of cars (kW) versus their top speeds (km/h). Both scales are logarithmic. The power increases as the third power of the speed. To go twice as fast requires eight times as much engine power. From Tennekes (1997).

that the power a car requires really does increase as the cube of speed, figure A.13 shows the engine power versus the top speeds of a range of cars. The line shows the relationship "power proportional to v^3."

Electric cars: is range a problem?

People often say that the range of electric cars is not big enough. Electric car advocates say "no problem, we can just put in more batteries" – and that's true, but we need to work out what effect the extra batteries have on the energy consumption. The answer depends sensitively on what energy density we assume the batteries deliver: for an energy density of 40 Wh/kg (typical of lead-acid batteries), we'll see that it's hard to push the range beyond 200 or 300 km; but for an energy density of 120 Wh/kg (typical of various lithium-based batteries), a range of 500 km is easily achievable.

Let's assume that the mass of the car and occupants is 740 kg, *without* any batteries. In due course we'll add 100 kg, 200 kg, 500 kg, or perhaps 1000 kg of batteries. Let's assume a typical speed of 50 km/h (30 mph); a drag-area of 0.8 m², a rolling resistance of 0.01; a distance between stops of 500 m; an engine efficiency of 85%; and that during stops and starts, regenerative braking recovers half of the kinetic energy of the car. Charging up the car from the mains is assumed to be 85% efficient. Figure A.14 shows the transport cost of the car versus its range, as we vary the amount of battery on board. The upper curve shows the result for a battery whose energy density is 40 Wh/kg (old-style lead-acid batteries). The range is limited by a wall at about 500 km. To get close to this maximum range, we have to take along comically large batteries: for a range of 400 km, for example, 2000 kg of batteries are required, and the transport cost is above 25 kWh per 100 km. If we are content with a range of 180 km, however, we can get by with 500 kg of batteries. Things get much better when we switch to lighter lithium-ion batteries. At an energy density of 120 Wh/kg, electric cars with 500 kg of batteries can easily deliver a range of 500 km. The transport cost is predicted to be about 13 kWh per 100 km.

It thus seems to me that the range problem has been solved by the advent of modern batteries. It would be nice to have even better batteries, but an energy density of 120 Wh per kg is already good enough, as long as we're happy for the batteries in a car to weigh up to 500 kg. In practice I imagine most people would be content to have a range of 300 km, which can be delivered by 250 kg of batteries. If these batteries were divided into ten 25 kg chunks, separately unpluggable, then a car user could keep just four of the ten chunks on board when he's doing regular commuting (100 kg gives a range of 140 km); and collect an extra six chunks from a battery-recharging station when he wants to make longer-range trips. During long-range trips, he would exchange his batteries for a fresh set at a battery-exchange station every 300 km or so.

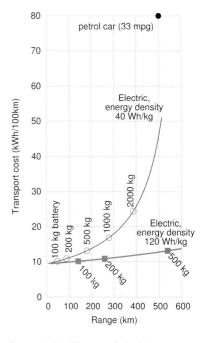

Figure A.14. Theory of electric car range (horizontal axis) and transport cost (vertical axis) as a function of battery mass, for two battery technologies. A car with 500 kg of old batteries, with an energy density of 40 Wh per kg, has a range of 180 km. With the same weight of modern batteries, delivering 120 Wh per kg, an electric car can have a range of more than 500 km. Both cars would have an energy cost of about 13 kWh per 100 km. These numbers allow for a battery charging efficiency of 85%.

Notes and further reading

page no.

256   *Typical petrol engines are about 25% efficient.*   Encarta [6by8x] says "The efficiencies of good modern Otto-cycle engines range between 20 and 25%." The petrol engine of a Toyota Prius, famously one of the most efficient car engines, uses the Atkinson cycle instead of the Otto cycle; it has a peak power output of 52 kW and has an efficiency of 34% when delivering 10 kW [348whs]. The most efficient diesel engine in the world is 52%-efficient, but it's not suitable for cars as it weighs 2300 tons: the Wartsila–Sulzer RTA96-C turbocharged diesel engine (figure A.15) is intended for container ships and has a power output of 80 MW.

–   *Regenerative brakes roughly halve the energy lost in braking.* Source: E4tech (2007).

257   *Electric engines can be about 8 times lighter than petrol engines.*
A 4-stroke petrol engine has a power-to-mass ratio of roughly 0.75 kW/kg. The best electric motors have an efficiency of 90% and a power-to-mass ratio of 6 kW/kg. So replacing a 75 kW petrol engine with a 75 kW electric motor saves 85 kg in weight. Sadly, the power to weight ratio of batteries is about 1 kW per kg, so what the electric vehicle gained on the motor, it loses on the batteries.

259   *The bike's engine uses energy with an efficiency of 0.25.* This and the other assumptions about cycling are confirmed by di Prampero et al. (1979). The drag-area of a cyclist in racing posture is $c_d A = 0.3 \, \text{m}^2$. The rolling resistance of a cyclist on a high-quality racing cycle (total weight 73 kg) is 3.2 N.

260   *Figure A.12.*
Prius data from B. Z. Wilson [home.hiwaay.net/~bzwilson/prius/]. BMW data from Phil C. Stuart [www.randomuseless.info/318ti/economy.html].

Further reading: Gabrielli and von Kármán (1950).

Figure A.15. The Wartsila-Sulzer RTA96-C 14-cylinder two-stroke diesel engine. 27 m long and 13.5 m high. www.wartsila.com

B   Wind II

The physics of wind power

To estimate the energy in wind, let's imagine holding up a hoop with area A, facing the wind whose speed is v. Consider the mass of air that passes through that hoop in one second. Here's a picture of that mass of air just before it passes through the hoop:

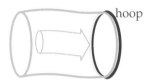

And here's a picture of the same mass of air one second later:

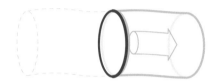

The mass of this piece of air is the product of its density ρ, its area A, and its length, which is v times t, where t is one second.

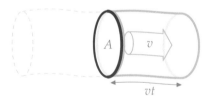

The kinetic energy of this piece of air is

$$\frac{1}{2}mv^2 = \frac{1}{2}\rho Avt\, v^2 = \frac{1}{2}\rho Atv^3. \tag{B.1}$$

So the power of the wind, for an area A – that is, the kinetic energy passing across that area per unit time – is

$$\frac{\frac{1}{2}mv^2}{t} = \frac{1}{2}\rho Av^3. \tag{B.2}$$

This formula may look familiar – we derived an identical expression on p255 when we were discussing the power requirement of a moving car.

What's a typical wind speed? On a windy day, a cyclist really notices the wind direction; if the wind is behind you, you can go much faster than

I'm using this formula again:

$$\text{mass} = \text{density} \times \text{volume}$$

miles/hour	km/h	m/s	Beaufort scale
2.2	3.6	1	force 1
7	11	3	force 2
11	18	5	force 3
13	21	6	force 4
16	25	7	
22	36	10	force 5
29	47	13	force 6
36	31	16	force 7
42	68	19	force 8
49	79	22	force 9
60	97	27	force 10
69	112	31	force 11
78	126	35	force 12

Figure B.1. Speeds.

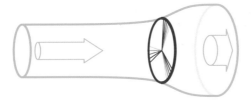

Figure B.2. Flow of air past a windmill. The air is slowed down and splayed out by the windmill.

normal; the speed of such a wind is therefore comparable to the typical speed of the cyclist, which is, let's say, 21 km per hour (13 miles per hour, or 6 metres per second). In Cambridge, the wind is only occasionally this big. Nevertheless, let's use this as a typical British figure (and bear in mind that we may need to revise our estimates).

The density of air is about 1.3 kg per m^3. (I usually round this to 1 kg per m^3, which is easier to remember, although I haven't done so here.) Then the typical power of the wind per square metre of hoop is

$$\frac{1}{2}\rho v^3 = \frac{1}{2}1.3\,\text{kg/m}^3 \times (6\,\text{m/s})^3 = 140\,\text{W/m}^2. \qquad (B.3)$$

Not all of this energy can be extracted by a windmill. The windmill slows the air down quite a lot, but it has to leave the air with *some* kinetic energy, otherwise that slowed-down air would get in the way. Figure B.2 is a cartoon of the actual flow past a windmill. The maximum fraction of the incoming energy that can be extracted by a disc-like windmill was worked out by a German physicist called Albert Betz in 1919. If the departing wind speed is one third of the arriving wind speed, the power extracted is 16/27 of the total power in the wind. 16/27 is 0.59. In practice let's guess that a windmill might be 50% efficient. In fact, real windmills are designed with particular wind speeds in mind; if the wind speed is significantly greater than the turbine's ideal speed, it has to be switched off.

As an example, let's assume a diameter of $d = 25$ m, and a hub height of 32 m, which is roughly the size of the lone windmill above the city of Wellington, New Zealand (figure B.3). The power of a single windmill is

efficiency factor × power per unit area × area

$$= 50\% \times \frac{1}{2}\rho v^3 \times \frac{\pi}{4}d^2 \qquad (B.4)$$

$$= 50\% \times 140\,\text{W/m}^2 \times \frac{\pi}{4}(25\,\text{m})^2 \qquad (B.5)$$

$$= 34\,\text{kW}. \qquad (B.6)$$

Indeed, when I visited this windmill on a very breezy day, its meter showed it was generating 60 kW.

To estimate how much power we can get from wind, we need to decide how big our windmills are going to be, and how close together we can pack them.

Figure B.3. The Brooklyn windmill above Wellington, New Zealand, with people providing a scale at the base. On a breezy day, this windmill was producing 60 kW, (1400 kWh per day). Photo by Philip Banks.

How densely could such windmills be packed? Too close and the up-wind ones will cast wind-shadows on the downwind ones. Experts say that windmills can't be spaced closer than 5 times their diameter without losing significant power. At this spacing, the power that windmills can generate per unit land area is

$$\frac{\text{power per windmill (B.4)}}{\text{land area per windmill}} = \frac{\frac{1}{2}\rho v^3 \frac{\pi}{8}d^2}{(5d)^2} \tag{B.7}$$

$$= \frac{\pi}{200}\frac{1}{2}\rho v^3 \tag{B.8}$$

$$= 0.016 \times 140\,\text{W/m}^2 \tag{B.9}$$

$$= 2.2\,\text{W/m}^2. \tag{B.10}$$

This number is worth remembering: a wind farm with a wind speed of 6 m/s produces a power of 2 W per m^2 of land area. Notice that our answer does not depend on the diameter of the windmill. The ds cancelled because bigger windmills have to be spaced further apart. Bigger windmills might be a good idea in order to catch bigger windspeeds that exist higher up (the taller a windmill is, the bigger the wind speed it encounters), or because of economies of scale, but those are the only reasons for preferring big windmills.

This calculation depended sensitively on our estimate of the windspeed. Is 6 m/s plausible as a long-term typical windspeed in windy parts of Britain? Figures 4.1 and 4.2 showed windspeeds in Cambridge and Cairngorm. Figure B.6 shows the mean winter and summer windspeeds in eight more locations around Britain. I fear 6 m/s was an overestimate of the typical speed in most of Britain! If we replace 6 m/s by Bedford's

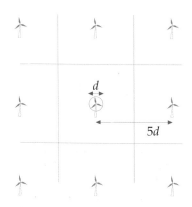

Figure B.4. Wind farm layout.

POWER PER UNIT AREA	
wind farm (speed 6 m/s)	2 W/m^2

Table B.5. Facts worth remembering: wind farms.

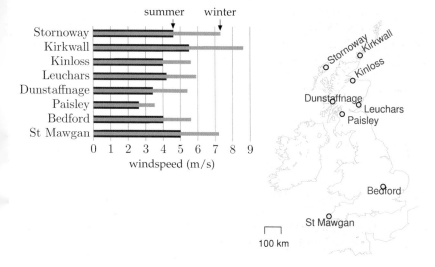

Figure B.6. Average summer windspeed (dark bar) and average winter windspeed (light bar) in eight locations around Britain. Speeds were measured at the standard weatherman's height of 10 metres. Averages are over the period 1971–2000.

4 m/s as our estimated windspeed, we must scale our estimate down, multiplying it by $(4/6)^3 \simeq 0.3$. (Remember, wind power scales as wind-speed cubed.)

On the other hand, to estimate the typical power, we shouldn't take the mean wind speed and cube it; rather, we should find the mean cube of the windspeed. The average of the cube is bigger than the cube of the average. But if we start getting into these details, things get even more complicated, because real wind turbines don't actually deliver a power proportional to wind-speed cubed. Rather, they typically have just a range of wind-speeds within which they deliver the ideal power; at higher or lower speeds real wind turbines deliver less than the ideal power.

Variation of wind speed with height

Taller windmills see higher wind speeds. The way that wind speed increases with height is complicated and depends on the roughness of the surrounding terrain and on the time of day. As a ballpark figure, doubling the height typically increases wind-speed by 10% and thus increases the power of the wind by 30%.

Some standard formulae for speed v as a function of height z are:

1. According to the wind shear formula from NREL [ydt7uk], the speed varies as a power of the height:

$$v(z) = v_{10} \left(\frac{z}{10\,\text{m}}\right)^{\alpha},$$

where v_{10} is the speed at 10 m, and a typical value of the exponent α is 0.143 or 1/7. The one-seventh law ($v(z)$ is proportional to $z^{1/7}$) is used by Elliott et al. (1991), for example.

2. The wind shear formula from the Danish Wind Industry Association [yaoonz] is

$$v(z) = v_{\text{ref}}\frac{\log(z/z_0)}{\log(z_{\text{ref}}/z_0)},$$

where z_0 is a parameter called the roughness length, and v_{ref} is the speed at a reference height z_{ref} such as 10 m. The roughness length for typical countryside (agricultural land with some houses and sheltering hedgerows with some 500-m intervals – "roughness class 2") is $z_0 = 0.1$ m.

In practice, these two wind shear formulae give similar numerical answers. That's not to say that they are accurate at all times however. Van den Berg (2004) suggests that different wind profiles often hold at night.

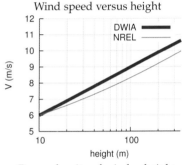

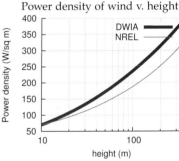

Figure B.7. Top: Two models of wind speed and wind power as a function of height. DWIA = Danish Wind Industry Association; NREL = National Renewable Energy Laboratory. For each model the speed at 10 m has been fixed to 6 m/s. For the Danish Wind model, the roughness length is set to $z_0 = 0.1$ m. Bottom: The power density (the power per unit of upright area) according to each of these models.

Figure B.8. The qr5 from quietrevolution.co.uk. Not a typical windmill.

Standard windmill properties

The typical windmill of today has a rotor diameter of around 54 metres centred at a height of 80 metres; such a machine has a "capacity" of 1 MW. The "capacity" or "peak power" is the *maximum* power the windmill can generate in optimal conditions. Usually, wind turbines are designed to start running at wind speeds somewhere around 3 to 5 m/s and to stop if the wind speed reaches gale speeds of 25 m/s. The actual average power delivered is the "capacity" multiplied by a factor that describes the fraction of the time that wind conditions are near optimal. This factor, sometimes called the "load factor" or "capacity factor," depends on the site; a typical load factor for a *good* site in the UK is 30%. In the Netherlands, the typical load factor is 22%; in Germany, it is 19%.

Other people's estimates of wind farm power per unit area

In the government's study [www.world-nuclear.org/policy/DTI-PIU.pdf] the UK onshore wind resource is estimated using an assumed wind farm power per unit area of at most 9 W/m^2 (capacity, not average production). If the capacity factor is 33% then the average power production would be 3 W/m^2.

The London Array is an offshore wind farm planned for the outer Thames Estuary. With its 1 GW capacity, it is expected to become the world's largest offshore wind farm. The completed wind farm will consist of 271 wind turbines in 245 km^2 [6o86ec] and will deliver an average power of 3100 GWh per year (350 MW). (Cost £1.5 bn.) That's a power per unit area of 350 MW/245 km^2 = 1.4 W/m^2. This is lower than other offshore farms because, I guess, the site includes a big channel (Knock Deep) that's too deep (about 20 m) for economical planting of turbines.

> *I'm more worried about what these plans [for the proposed London Array wind farm] will do to this landscape and our way of life than I ever was about a Nazi invasion on the beach.*
>
> Bill Boggia of Graveney, where the undersea cables of the wind farm will come ashore.

Queries

What about micro-generation? If you plop one of those mini-turbines on your roof, what energy can you expect it to deliver?

Assuming a windspeed of 6 m/s, which, as I said before, is *above* the average for most parts of Britain; and assuming a diameter of 1 m, the power delivered would be 50 W. That's 1.3 kWh per day – not very much. And in reality, in a typical urban location in England, a microturbine delivers just 0.2 kWh per day – see p66.

Perhaps the worst windmills in the world are a set in Tsukuba City, Japan, which actually consume more power than they generate. Their installers were so embarrassed by the stationary turbines that they imported power to make them spin so that they looked like they were working! [6bkvbn]

Figure B.9. An Ampair "600 W" micro-turbine. The average power generated by this micro-turbine in Leamington Spa is 0.037 kWh per day (1.5 W).

Notes and further reading

page no.

264 *The maximum fraction of the incoming energy that can be extracted by a disc-like windmill...* There is a nice explanation of this on the Danish Wind Industry Association's website. [yekdaa].

267 *Usually, wind turbines are designed to start running at wind speeds around 3 to 5 m/s.* [ymfbsn].

–   *a typical load factor for a good site is 30%.* In 2005, the average load factor of all major UK wind farms was 28% [ypvbvd]. The load factor varied during the year, with a low of 17% in June and July. The load factor for the best region in the country – Caithness, Orkney and the Shetlands – was 33%. The load factors of the two offshore wind farms operating in 2005 were 36% for North Hoyle (off North Wales) and 29% for Scroby Sands (off Great Yarmouth). Average load factors in 2006 for ten regions were: Cornwall 25%; Mid-Wales 27%; Cambridgeshire and Norfolk 25%; Cumbria 25%; Durham 16%; Southern Scotland 28%; Orkney and Shetlands 35%; Northeast Scotland 26%; Northern Ireland 31%; offshore 29%. [wbd8o]

Watson et al. (2002) say a minimum annual mean wind speed of 7.0 m/s is currently thought to be necessary for commercial viability of wind power. About 33% of UK land area has such speeds.

Figure B.10. A 5.5-m diameter Iskra 5 kW turbine [www.iskrawind.com] having its annual check-up. This turbine, located in Hertfordshire (not the windiest of locations in Britain), mounted at a height of 12 m, has an average output of 11 kWh per day. A wind farm of machines with this performance, one per 30 m × 30 m square, would have a power per unit area of 0.5 W/m².

C Planes II

What we need to do is to look at how you make air travel more energy efficient, how you develop the new fuels that will allow us to burn less energy and emit less.

Tony Blair

Hoping for the best is not a policy, it is a delusion.

Emily Armistead, Greenpeace

Figure C.1. Birds: two Arctic terns, a bar-tailed godwit, and a Boeing 747.

What are the fundamental limits of travel by flying? Does the physics of flight require an unavoidable use of a certain amount of energy, per ton, per kilometre flown? What's the maximum distance a 300-ton Boeing 747 can fly? What about a 1-kg bar-tailed godwit or a 100-gram Arctic tern?

Just as Chapter 3, in which we estimated consumption by cars, was followed by Chapter A, offering a model of where the energy goes in cars, this chapter fills out Chapter 5, discussing where the energy goes in planes. The only physics required is Newton's laws of motion, which I'll describe when they're needed.

This discussion will allow us to answer questions such as "would air travel consume much less energy if we travelled in slower propellor-driven planes?" There's a lot of equations ahead: I hope you enjoy them!

How to fly

Planes (and birds) move through air, so, just like cars and trains, they experience a drag force, and much of the energy guzzled by a plane goes into pushing the plane along against this force. Additionally, unlike cars and trains, planes have to expend energy *in order to stay up*.

Planes stay up by throwing air down. When the plane pushes down on air, the air pushes up on the plane (because Newton's third law tells it to). As long as this upward push, which is called lift, is big enough to balance the downward weight of the plane, the plane avoids plummeting downwards.

When the plane throws air down, it gives that air kinetic energy. So creating lift requires energy. The total power required by the plane is the sum of the power required to create lift and the power required to overcome ordinary drag. (The power required to create lift is usually called "induced drag," by the way. But I'll call it the lift power, P_{lift}.)

The two equations we'll need, in order to work out a theory of flight, are Newton's second law:

$$\text{force} = \text{rate of change of momentum,} \qquad \text{(C.1)}$$

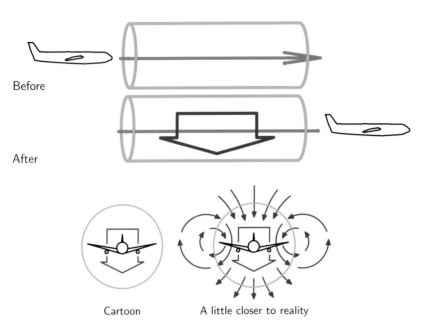

Before

After

Figure C.2. A plane encounters a stationary tube of air. Once the plane has passed by, the air has been thrown downwards by the plane. The force exerted by the plane on the air to accelerate it downwards is equal and opposite to the upwards force exerted on the plane by the air.

Cartoon   A little closer to reality

Figure C.3. Our cartoon assumes that the plane leaves a sausage of air moving down in its wake. A realistic picture involves a more complex swirling flow. For the real thing, see figure C.4.

and Newton's third law, which I just mentioned:

$$\text{force exerted on A by B} = -\text{force exerted on B by A.} \qquad (C.2)$$

If you don't like equations, I can tell you the punchline now: we're going to find that the power required to create lift turns out to be *equal* to the power required to overcome drag. So the requirement to "stay up" *doubles* the power required.

Let's make a cartoon of the lift force on a plane moving at speed v. In a time t the plane moves a distance vt and leaves behind it a sausage of downward-moving air (figure C.2). We'll call the cross-sectional area of this sausage A_s. This sausage's diameter is roughly equal to the wingspan w of the plane. (Within this large sausage is a smaller sausage of swirling turbulent air with cross-sectional area similar to the frontal area of the plane's body.) Actually, the details of the air flow are much more interesting than this sausage picture: each wing tip leaves behind it a vortex, with the air between the wingtips moving down fast, and the air beyond (outside) the wingtips moving up (figures C.3 & C.4). This upward-moving air is exploited by birds flying in formation: just behind the tip of a bird's wing is a sweet little updraft. Anyway, let's get back to our sausage.

The sausage's mass is

Figure C.4. Air flow behind a plane. Photo by NASA Langley Research Center.

$$m_{\text{sausage}} = \text{density} \times \text{volume} = \rho v t A_s. \qquad (C.3)$$

Let's say the whole sausage is moving down with speed u, and figure out what u needs to be in order for the plane to experience a lift force equal to its weight mg. The downward momentum of the sausage created in time t is

$$\text{mass} \times \text{velocity} = m_{\text{sausage}}u = \rho v t A_{\text{s}} u. \tag{C.4}$$

And by Newton's laws this must equal the momentum delivered by the plane's weight in time t, namely,

$$mgt. \tag{C.5}$$

Rearranging this equation,

$$\rho v t A_{\text{s}} u = mgt, \tag{C.6}$$

we can solve for the required downward sausage speed,

$$u = \frac{mg}{\rho v A_{\text{s}}}.$$

Interesting! The sausage speed is *inversely* related to the plane's speed v. A slow-moving plane has to throw down air harder than a fast-moving plane, because it encounters less air per unit time. That's why landing planes, travelling slowly, have to extend their flaps: so as to create a larger and steeper wing that deflects air more.

What's the energetic cost of pushing the sausage down at the required speed u? The power required is

$$P_{\text{lift}} = \frac{\text{kinetic energy of sausage}}{\text{time}} \tag{C.7}$$

$$= \frac{1}{t}\frac{1}{2}m_{\text{sausage}}u^2 \tag{C.8}$$

$$= \frac{1}{2t}\rho v t A_{\text{s}}\left(\frac{mg}{\rho v A_{\text{s}}}\right)^2 \tag{C.9}$$

$$= \frac{1}{2}\frac{(mg)^2}{\rho v A_{\text{s}}}. \tag{C.10}$$

The total power required to keep the plane going is the sum of the drag power and the lift power:

$$P_{\text{total}} = P_{\text{drag}} + P_{\text{lift}} \tag{C.11}$$

$$= \frac{1}{2}c_{\text{d}}\rho A_{\text{p}}v^3 + \frac{1}{2}\frac{(mg)^2}{\rho v A_{\text{s}}}, \tag{C.12}$$

where A_{p} is the frontal area of the plane and c_{d} is its drag coefficient (as in Chapter A).

The fuel-efficiency of the plane, expressed as the energy per distance travelled, would be

$$\left.\frac{\text{energy}}{\text{distance}}\right|_{\text{ideal}} = \frac{P_{\text{total}}}{v} = \frac{1}{2}c_{\text{d}}\rho A_{\text{p}}v^2 + \frac{1}{2}\frac{(mg)^2}{\rho v^2 A_{\text{s}}}, \tag{C.13}$$

if the plane turned its fuel's power into drag power and lift power perfectly efficiently. (Incidentally, another name for "energy per distance travelled" is "force," and we can recognize the two terms above as the drag force $\frac{1}{2}c_d\rho A_p v^2$ and the lift-related force $\frac{1}{2}\frac{(mg)^2}{\rho v^2 A_s}$. The sum is the force, or "thrust," that specifies exactly how hard the engines have to push.)

Real jet engines have an efficiency of about $\epsilon = 1/3$, so the energy-per-distance of a plane travelling at speed v is

$$\frac{\text{energy}}{\text{distance}} = \frac{1}{\epsilon}\left(\frac{1}{2}c_d\rho A_p v^2 + \frac{1}{2}\frac{(mg)^2}{\rho v^2 A_s}\right). \qquad (C.14)$$

This energy-per-distance is fairly complicated; but it simplifies greatly if we assume that the plane is *designed* to fly at the speed that *minimizes* the energy-per-distance. The energy-per-distance, you see, has got a sweet-spot as a function of v (figure C.5). The sum of the two quantities $\frac{1}{2}c_d\rho A_p v^2$ and $\frac{1}{2}\frac{(mg)^2}{\rho v^2 A_s}$ is smallest when the two quantities are equal. This phenomenon is delightfully common in physics and engineering: two things that don't obviously *have* to be equal *are* actually equal, or equal within a factor of 2.

So, this equality principle tells us that the optimum speed for the plane is such that

$$c_d\rho A_p v^2 = \frac{(mg)^2}{\rho v^2 A_s}, \qquad (C.15)$$

i.e.,

$$\rho v_{\text{opt}}^2 = \frac{mg}{\sqrt{c_d A_p A_s}}, \qquad (C.16)$$

This defines the optimum speed if our cartoon of flight is accurate; the cartoon breaks down if the engine efficiency ϵ depends significantly on speed, or if the speed of the plane exceeds the speed of sound (330 m/s); above the speed of sound, we would need a different model of drag and lift.

Let's check our model by seeing what it predicts is the optimum speed for a 747 and for an albatross. We must take care to use the correct air-density: if we want to estimate the optimum cruising speed for a 747 at 30 000 feet, we must remember that air density drops with increasing altitude z as $\exp(-mgz/kT)$, where m is the mass of nitrogen or oxygen molecules, and kT is the thermal energy (Boltzmann's constant times absolute temperature). The density is about 3 times smaller at that altitude.

The predicted optimal speeds (table C.6) are more accurate than we have a right to expect! The 747's optimal speed is predicted to be 540 mph, and the albatross's, 32 mph – both very close to the true cruising speeds of the two birds (560 mph and 30–55 mph respectively).

Let's explore a few more predictions of our cartoon. We can check whether the force (C.13) is compatible with the known thrust of the 747. Remembering that at the optimal speed, the two forces are equal, we just

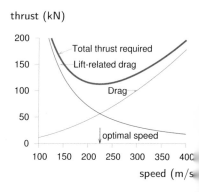

thrust (kN)

Figure C.5. The force required to keep a plane moving, as a function of its speed v, is the sum of an ordinary drag force $\frac{1}{2}c_d\rho A_p v^2$ – which increases with speed – and the lift-related force (also known as the induced drag) $\frac{1}{2}\frac{(mg)^2}{\rho v^2 A_s}$ – which decreases with speed. There is an ideal speed, v_{optimal}, at which the force required is minimized. The force is an energy per distance, so minimizing the force also minimizes the fuel per distance. To optimize the fuel efficiency, fly at v_{optimal}. This graph shows our cartoon's estimate of the thrust required, in kilonewtons, for a Boeing 747 of mass 319 t, wingspan 64.4 m, drag coefficient 0.03 and frontal area 180 m², travelling in air of density $\rho = 0.41$ kg/m³ (the density at a height of 10 km), as a function of its speed v in m/s. Our model has an optimal speed $v_{\text{optimal}} = 220$ m/s (540 mph). For a cartoon based on sausages, this is a good match to real life!

BIRD		747	Albatross
Designer		Boeing	natural selection
Mass (fully-laden)	m	363 000 kg	8 kg
Wingspan	w	64.4 m	3.3 m
Area*	A_p	180 m²	0.09 m²
Density	ρ	0.4 kg/m³	1.2 kg/m³
Drag coefficient	c_d	0.03	0.1
Optimum speed	v_{opt}	220 m/s	14 m/s
		= 540 mph	= 32 mph

Table C.6. Estimating the optimal speeds for a jumbo jet and an albatross.
★ Frontal area estimated for 747 by taking cabin width (6.1 m) times estimated height of body (10 m) and adding double to allow for the frontal area of engines, wings, and tail; for albatross, frontal area of 1 square foot estimated from a photograph.

need to pick one of them and double it:

$$\text{force} = \left.\frac{\text{energy}}{\text{distance}}\right|_{\text{ideal}} = \frac{1}{2}c_d\rho A_p v^2 + \frac{1}{2}\frac{(mg)^2}{\rho v^2 A_s} \tag{C.17}$$

$$= c_d\rho A_p v_{opt}^2 \tag{C.18}$$

$$= c_d\rho A_p \frac{mg}{\rho(c_d A_p A_s)^{1/2}} \tag{C.19}$$

$$= \left(\frac{c_d A_p}{A_s}\right)^{1/2} mg. \tag{C.20}$$

Let's define the filling factor f_A to be the area ratio:

$$f_A = \frac{A_p}{A_s}. \tag{C.21}$$

(Think of f_A as the fraction of the square occupied by the plane in figure C.7.) Then

$$\text{force} = (c_d f_A)^{1/2}(mg). \tag{C.22}$$

Figure C.7. Frontal view of a Boeing 747, used to estimate the frontal area A_p of the plane. The square has area A_s (the square of the wingspan).

Interesting! Independent of the density of the fluid through which the plane flies, the required thrust (for a plane travelling at the optimal speed) is just a dimensionless constant $(c_d f_A)^{1/2}$ times the weight of the plane. This constant, by the way, is known as the drag-to-lift ratio of the plane. (The lift-to-drag ratio has a few other names: the glide number, glide ratio, aerodynamic efficiency, or finesse; typical values are shown in table C.8.)

Taking the jumbo jet's figures, $c_d \simeq 0.03$ and $f_A \simeq 0.04$, we find the required thrust is

$$(c_d f_A)^{1/2} mg = 0.036\, mg = 130\, \text{kN}. \tag{C.23}$$

Airbus A320	17
Boeing 767-200	19
Boeing 747-100	18
Common Tern	12
Albatross	20

How does this agree with the 747's spec sheets? In fact each of the 4 engines has a maximum thrust of about 250 kN, but this maximum thrust is used only during take-off. During cruise, the thrust is much smaller:

Table C.8. Lift-to-drag ratios.

produce 300 kg of hydrogen per day. Hydrogen contains 39 kWh per kg, so this algae-to-hydrogen facility would deliver a power per unit area of 4.4 W/m². Taking into account the estimated electricity required to run the facility, the net power delivered would be reduced to 3.6 W/m². That strikes me as still quite a promising number – compare it with the Bavarian solar photovoltaic farm, for example (5 W/m²).

Food for humans or other animals

Grain crops such as wheat, oats, barley, and corn have an energy density of about 4 kWh per kg. In the UK, wheat yields of 7.7 tons per hectare per year are typical. If the wheat is eaten by an animal, the power per unit area of this process is 0.34 W/m². If 2800 m² of Britain (that's all agricultural land) were devoted to the growth of crops like these, the chemical energy generated would be about 24 kWh/d per person.

Incineration of agricultural by-products

We found a moment ago that the power per unit area of a biomass power station burning the best energy crops is 0.2 W/m². If instead we grow crops for food, and put the left-overs that we don't eat into a power station – or if we feed the food to chickens and put the left-overs that come out of the chickens' back ends into a power station – what power could be delivered per unit area of farmland? Let's make a rough guess, then take a look at some real data. For a wild guess, let's imagine that by-products are harvested from half of the area of Britain (2000 m² per person) and trucked to power stations, and that general agricultural by-products deliver 10% as much power per unit area as the best energy crops: 0.02 W/m². Multiplying this by 2000 m² we get 1 kWh per day per person.

Have I been unfair to agricultural garbage in making this wild guess? We can re-estimate the plausible production from agricultural left-overs by scaling up the prototype straw-burning power station at Elean in East Anglia. Elean's power output is 36 MW, and it uses 200 000 tons per year from land located within a 50-mile radius. If we assume this density can be replicated across the whole country, the Elean model offers 0.002 W/m². At 4000 m² per person, that's 8 W per person, or 0.2 kWh/day per person.

Let's calculate this another way. UK straw production is 10 million tons per year, or 0.46 kg per day per person. At 4.2 kWh per kg, this straw has a chemical energy of 2 kWh per day per person. If all the straw were burned in 30%-efficient power stations – a proposal that wouldn't go down well with farm animals, who have other uses for straw – the electricity generated would be 0.6 kWh/d per person.

Landfill methane gas

At present, much of the methane gas leaking out of rubbish tips comes from biological materials, especially waste food. So, as long as we keep throwing away things like food and newspapers, landfill gas is a sustainable energy source – plus, burning that methane might be a good idea from a climate-change perspective, since methane is a stronger greenhouse-gas than CO_2. A landfill site receiving 7.5 million tons of household waste per year can generate $50\,000\,m^3$ per hour of methane.

In 1994, landfill methane emissions were estimated to be $0.05\,m^3$ per person per day, which has a chemical energy of $0.5\,kWh/d$ per person, and would generate $0.2\,kWh(e)/d$ per person, if it were all converted to electricity with 40% efficiency. Landfill gas emissions are declining because of changes in legislation, and are now roughly 50% lower.

Burning household waste

SELCHP ("South East London Combined Heat and Power") [www.selchp.com] is a 35 MW power station that is paid to burn 420 kt per year of black-bag waste from the London area. They burn the waste as a whole, without sorting. After burning, ferrous metals are removed for recycling, hazardous wastes are filtered out and sent to a special landfill site, and the remaining ash is sent for reprocessing into recycled material for road building or construction use. The calorific value of the waste is $2.5\,kWh/kg$, and the thermal efficiency of the power station is about 21%, so each 1 kg of waste gets turned into 0.5 kWh of electricity. The carbon emissions are about $1000\,g\,CO_2$ per kWh. Of the 35 MW generated, about 4 MW is used by the plant itself to run its machinery and filtering processes.

Scaling this idea up, if every borough had one of these, and if everyone sent 1 kg per day of waste, then we'd get $0.5\,kWh(e)$ per day per person from waste incineration.

This is similar to the figure estimated above for methane capture at landfill sites. And remember, we can't have both. More waste incineration means less methane gas leaking out of landfill sites. See figure 27.2, p206, and figure 27.3, p207, for further data on waste incineration.

Figure D.4. SELCHP – your trash is their business.

Notes and further reading

page no.

283   *The power per unit area of using willow, miscanthus, or poplar, for electricity is $0.2\,W/m^2$.* Source: Select Committee on Science and Technology Minutes of Evidence – Memorandum from the Biotechnology & Biological Sciences Research Council [www.publications.parliament.uk/pa/ld200304/ldselect/ldsctech/126/4032413.htm]. "Typically a sustainable crop of 10

dry t/ha/y of woody biomass can be produced in Northern Europe. ...
Thus an area of 1 km² will produce 1000 dry t/y – enough for a power output 150 kWe at low conversion efficiencies or 300 kWe at high conversion efficiencies." This means 0.15–0.3 W(e)/m².
See also Layzell et al. (2006), [3ap71c].

283   *Oilseed rape.* Sources: Bayer Crop Science (2003), Evans (2007), `www.defra.gov.uk`.

–   *Sugar beet.* Source: `statistics.defra.gov.uk/esg/default.asp`

284   *Bioethanol from corn.* Source: Shapouri et al. (1995).

–   *Bioethanol from cellulose.* See also Mabee et al. (2006).

–   *Jatropha.* Sources: Francis et al. (2005), Asselbergs et al. (2006).

285   *In America, in ponds fed with concentrated CO_2, algae can grow at 30 grams per square metre per day, producing 0.01 litres of biodiesel per square metre per day.* Source: Putt (2007). This calculation has ignored the energy cost of running the algae ponds and processing the algae into biodiesel. Putt describes the energy balance of a proposed design for a 100-acre algae farm, powered by methane from an animal litter digester. The farm described would in fact produce less power than the methane power input. The 100-acre farm would use 2600 kW of methane, which corresponds to an input power density of 6.4 W/m². To recap, the power density of the output, in the form of biodiesel, would be just 4.2 W/m². All proposals to make biofuels should be approached with a critical eye!

286   *A research study from the National Renewable Energy Laboratory predicted that genetically-modified green algae, covering an area of 11 hectares, could produce 300 kg of hydrogen per day.* Source: Amos (2004).

–   *Elean power station.* Source: Government White Paper (2003). Elean Power Station (36 MW) – the UK's first straw-fired power plant. *Straw production:* `www.biomassenergycentre.org.uk`.

287   *Landfill gas.* Sources: Matthew Chester, City University, London, personal communication; Meadows (1996), Aitchison (1996); Alan Rosevear, UK Representative on Methane to Markets Landfill Gas Sub-Committee, May 2005 [4hamks].

E Heating II

A perfectly sealed and insulated building would hold heat for ever and thus would need no heating. The two dominant reasons why buildings lose heat are:

1. **Conduction** – heat flowing directly through walls, windows and doors;

2. **Ventilation** – hot air trickling out through cracks, gaps, or deliberate ventilation ducts.

In the standard model for heat loss, both these heat flows are proportional to the temperature difference between the air inside and outside. For a typical British house, conduction is the bigger of the two losses, as we'll see.

Conduction loss

The rate of conduction of heat through a wall, ceiling, floor, or window is the product of three things: the area of the wall, a measure of conductivity of the wall known in the trade as the "U-value" or thermal transmittance, and the temperature difference –

$$\text{power loss} = \text{area} \times U \times \text{temperature difference.}$$

The U-value is usually measured in $\text{W/m}^2/\text{K}$. (One kelvin (1 K) is the same as one degree Celsius (1 °C).) Bigger U-values mean bigger losses of power. The thicker a wall is, the smaller its U-value. Double-glazing is about as good as a solid brick wall. (See table E.2.)

The U-values of objects that are "in series," such as a wall and its inner lining, can be combined in the same way that electrical conductances combine:

$$u_{\text{series combination}} = 1 \left/ \left(\frac{1}{u_1} + \frac{1}{u_2} \right) \right. .$$

There's a worked example using this rule on page 296.

Ventilation loss

To work out the heat required to warm up incoming cold air, we need the heat capacity of air: $1.2\,\text{kJ/m}^3/\text{K}$.

In the building trade, it's conventional to describe the power-losses caused by ventilation of a space as the product of the number of changes N of the air per hour, the volume V of the space in cubic metres, the heat capacity C, and the temperature difference ΔT between the inside and

kitchen	2
bathroom	2
lounge	1
bedroom	0.5

Table E.1. Air changes per hour: typical values of N for draught-proofed rooms. The worst draughty rooms might have $N = 3$ air changes per hour. The recommended minimum rate of air exchange is between 0.5 and 1.0 air changes per hour, providing adequate fresh air for human health, for safe combustion of fuels and to prevent damage to the building fabric from excess moisture in the air (EST 2003).

	U-values (W/m²/K)		
	old buildings	modern standards	best methods
Walls		0.45–0.6	0.12
solid masonry wall	2.4		
outer wall: 9 inch solid brick	2.2		
11 in brick-block cavity wall, unfilled	1.0		
11 in brick-block cavity wall, insulated	0.6		
Floors		0.45	0.14
suspended timber floor	0.7		
solid concrete floor	0.8		
Roofs		0.25	0.12
flat roof with 25 mm insulation	0.9		
pitched roof with 100mm insulation	0.3		
Windows			1.5
single-glazed	5.0		
double-glazed	2.9		
double-glazed, 20 mm gap	1.7		
triple-glazed	0.7–0.9		

Table E.2. U-values of walls, floors, roofs, and windows.

outside of the building.

$$\begin{aligned} \text{power (watts)} &= C\frac{N}{1\,\text{h}}V(\text{m}^3)\Delta T(\text{K}) & \text{(E.1)} \\[4pt] &= (1.2\,\text{kJ/m}^3/\text{K})\frac{N}{3600\,\text{s}}V(\text{m}^3)\Delta T(\text{K}) & \text{(E.2)} \\[4pt] &= \frac{1}{3}NV\Delta T. & \text{(E.3)} \end{aligned}$$

Energy loss and temperature demand (degree-days)

Since energy is power × time, you can write the energy lost by *conduction* through an area in a short duration as

$$\text{energy loss} = \text{area} \times U \times (\Delta T \times \text{duration}),$$

and the energy lost by *ventilation* as

$$\frac{1}{3}NV \times (\Delta T \times \text{duration}).$$

Both these energy losses have the form

$$\text{Something} \times (\Delta T \times \text{duration}),$$

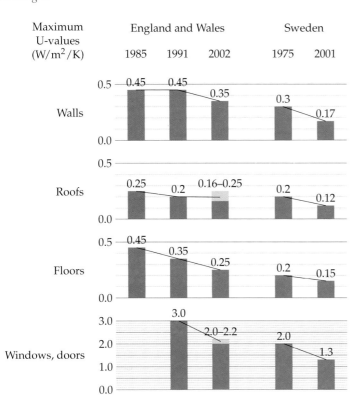

Figure E.3. U-values required by British and Swedish building regulations.

where the "Something" is measured in watts per °C. As day turns to night, and seasons pass, the temperature difference ΔT changes; we can think of a long period as being chopped into lots of small durations, during each of which the temperature difference is roughly constant. From duration to duration, the temperature difference changes, but the Somethings don't change. When predicting a space's total energy loss due to conduction and ventilation over a long period we thus need to multiply two things:

1. the sum of all the Somethings (adding area $\times U$ for all walls, roofs, floors, doors, and windows, and $\frac{1}{3}NV$ for the volume); and

2. the sum of all the Temperature difference $\times$ duration factors (for all the durations).

The first factor is a property of the building measured in watts per °C. I'll call this the *leakiness* of the building. (This leakiness is sometimes called the building's *heat-loss coefficient*.) The second factor is a property of the weather; it's often expressed as a number of "degree-days," since temperature difference is measured in degrees, and days are a convenient unit for thinking about durations. For example, if your house interior is at 18 °C, and the outside temperature is 8 °C for a week, then we say that that

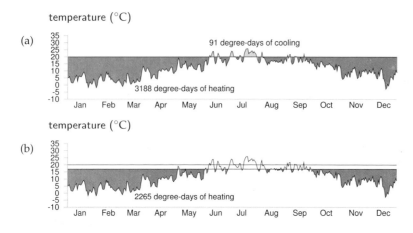

temperature (°C)

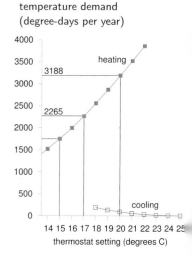

Figure E.4. The temperature demand in Cambridge, 2006, visualized as an area on a graph of daily average temperatures. (a) Thermostat set to 20 °C, including cooling in summer; (b) winter thermostat set to 17 °C.

week contributed $10 \times 7 = 70$ degree-days to the ($\Delta T \times$ duration) sum. I'll call the sum of all the ($\Delta T \times$ duration) factors the *temperature demand* of a period.

$$\text{energy lost} = \text{leakiness} \times \text{temperature demand}.$$

We can reduce our energy loss by reducing the leakiness of the building, or by reducing our temperature demand, or both. The next two sections look more closely at these two factors, using a house in Cambridge as a case-study.

There is a third factor we must also discuss. The lost energy is replenished by the building's heating system, and by other sources of energy such as the occupants, their gadgets, their cookers, and the sun. Focussing on the heating system, the energy *delivered* by the heating is not the same as the energy *consumed* by the heating. They are related by the *coefficient of performance* of the heating system.

$$\text{energy consumed} = \text{energy delivered}/\text{coefficient of performance}.$$

For a condensing boiler burning natural gas, for example, the coefficient of performance is 90%, because 10% of the energy is lost up the chimney.

To summarise, we can reduce the energy consumption of a building in three ways:

1. by reducing temperature demand;

2. by reducing leakiness; or

3. by increasing the coefficient of performance.

We now quantify the potential of these options. (A fourth option – increasing the building's incidental heat gains, especially from the sun – may also be useful, but I won't address it here.)

Figure E.5. Temperature demand in Cambridge, in degree-days per year, as a function of thermostat setting (°C). Reducing the winter thermostat from 20 °C to 17 °C reduces the temperature demand of heating by 30%, from 3188 to 2265 degree-days. Raising the summer thermostat from 20 °C to 23 °C reduces the temperature demand of cooling by 82%, from 91 to 16 degree-days.

Temperature demand

We can visualize the temperature demand nicely on a graph of external temperature versus time (figure E.4). For a building held at a temperature of 20 °C, the total temperature demand is the *area* between the horizontal line at 20 °C and the external temperature. In figure E.4a, we see that, for one year in Cambridge, holding the temperature at 20 °C year-round had a temperature demand of 3188 degree-days of heating and 91 degree-days of cooling. These pictures allow us easily to assess the effect of turning down the thermostat and living without air-conditioning. Turning the winter thermostat down to 17 °C, the temperature demand for heating drops from 3188 degree-days to 2265 degree-days (figure E.4b), which corresponds to a 30% reduction in heating demand. Turning the thermostat down to 15 °C reduces the temperature demand from 3188 to 1748 degree days, a 45% reduction.

These calculations give us a ballpark indication of the benefit of turning down thermostats, but will give an exact prediction only if we take into account two details: first, buildings naturally absorb energy from the sun, boosting the inside above the outside temperature, even without any heating; and second, the occupants and their gadget companions emit heat, so further cutting down the artificial heating requirements. The temperature demand of a location, as conventionally expressed in degree-days, is a bit of an unwieldy thing. I find it hard to remember numbers like "3500 degree-days." And academics may find the degree-day a distressing unit, since they already have another meaning for degree days (one involving dressing up in gowns and mortar boards). We can make this quantity more meaningful and perhaps easier to work with by dividing it by 365, the number of days in the year, obtaining the temperature demand in "degree-days per day," or, if you prefer, in plain "degrees." Figure E.6 shows this replotted temperature demand. Expressed this way, the temperature demand is simply the *average* temperature difference between inside and outside. The highlighted temperature demands are: 8.7 °C, for a thermostat setting of 20 °C; 6.2 °C, for a setting of 17 °C; and 4.8 °C, for a setting of 15 °C.

Leakiness – example: my house

My house is a three-bedroom semi-detached house built about 1940 (figure E.7). By 2006, its kitchen had been slightly extended, and most of the windows were double-glazed. The front door and back door were both still single-glazed.

My estimate of the leakiness in 2006 is built up as shown in table E.8. The total leakiness of the house was 322 W/°C (or 7.7 kWh/d/°C), with conductive leakiness accounting for 72% and ventilation leakiness for 28% of the total. The conductive leakiness is roughly equally divided into three parts: windows; walls; and floor and ceiling.

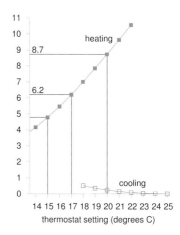

Figure E.6. The temperature demand in Cambridge, 2006, replotted in units of degree-days per day, also known as degrees. In these units, the temperature demand is just the average of the temperature difference between inside and outside.

Figure E.7. My house.

CONDUCTIVE LEAKINESS	area (m²)	U-value (W/m²/°C)	leakiness (W/°C)
Horizontal surfaces			
Pitched roof	48	0.6	28.8
Flat roof	1.6	3	4.8
Floor	50	0.8	40
Vertical surfaces			
Extension walls	24.1	0.6	14.5
Main walls	50	1	50
Thin wall (5in)	2	3	6
Single-glazed doors and windows	7.35	5	36.7
Double-glazed windows	17.8	2.9	51.6
Total conductive leakiness			232.4

VENTILATION LEAKINESS	volume (m³)	N (air-changes per hour)	leakiness (W/°C)
Bedrooms	80	0.5	13.3
Kitchen	36	2	24
Hall	27	3	27
Other rooms	77	1	25.7
Total ventilation leakiness			90

Table E.8. Breakdown of my house's conductive leakiness, and its ventilation leakiness, pre-2006. I've treated the central wall of the semi-detached house as a perfect insulating wall, but this may be wrong if the gap between the adjacent houses is actually well-ventilated.

I've highlighted the parameters that I altered after 2006, in modifications to be described shortly.

To compare the leakinesses of two buildings that have different floor areas, we can divide the leakiness by the floor area; this gives the *heat-loss parameter* of the building, which is measured in W/°C/m². The heat-loss parameter of this house (total floor area 88 m²) is

$$3.7 \, W/°C/m^2.$$

Let's use these figures to estimate the house's daily energy consumption on a cold winter's day, and year-round.

On a cold day, assuming an external temperature of $-1\,°C$ and an internal temperature of $19\,°C$, the temperature difference is $\Delta T = 20\,°C$. If this difference is maintained for 6 hours per day then the energy lost per day is

$$322 \, W/°C \times 120 \, degree\text{-}hours \simeq 39 \, kWh.$$

If the temperature is maintained at $19\,°C$ for 24 hours per day, the energy lost per day is

$$155 \, kWh/d.$$

To get a year-round heat-loss figure, we can take the temperature demand of Cambridge from figure E.5. With the thermostat at $19\,°C$, the temperature demand in 2006 was 2866 degree-days. The average rate of heat loss, if the house is always held at $19\,°C$, is therefore:

$$7.7\,\text{kWh/d/°C} \times 2866\,\text{degree-days/y}/(365\,\text{days/y}) = 61\,\text{kWh/d}.$$

Turning the thermostat down to $17\,°C$, the average rate of heat loss drops to $48\,\text{kWh/d}$. Turning it up to a tropical $21\,°C$, the average rate of heat loss is $75\,\text{kWh/d}$.

Effects of extra insulation

During 2007, I made the following modifications to the house:

1. Added cavity-wall insulation (which was missing in the main walls of the house) – figure 21.5.

2. Increased the insulation in the roof.

3. Added a new front door outside the old – figure 21.6.

4. Replaced the back door with a double-glazed one.

5. Double-glazed the one window that was still single-glazed.

What's the predicted change in heat loss?

The total leakiness before the changes was $322\,\text{W/°C}$.

Adding cavity-wall insulation (new U-value 0.6) to the main walls reduces the house's leakiness by $20\,\text{W/°C}$. The improved loft insulation (new U-value 0.3) should reduce the leakiness by $14\,\text{W/°C}$. The glazing modifications (new U-value 1.6–1.8) should reduce the conductive leakiness by $23\,\text{W/°C}$, and the ventilation leakiness by something like $24\,\text{W/°C}$. That's a total reduction in leakiness of 25%, from roughly 320 to $240\,\text{W/°C}$ (7.7 to $6\,\text{kWh/d/°C}$). Table E.9 shows the predicted savings from each of the modifications.

The heat-loss parameter of this house (total floor area $88\,\text{m}^2$) is thus hopefully reduced by about 25%, from 3.7 to $2.7\,\text{W/°C/m}^2$. (This is a long way from the $1.1\,\text{W/°C/m}^2$ required of a "sustainable" house in the new building codes.)

– Cavity-wall insulation (applicable to two-thirds of the wall area)	4.8 kWh/d
– Improved roof insulation	3.5 kWh/d
– Reduction in conduction from double-glazing two doors and one window	1.9 kWh/d
– Ventilation reductions in hall and kitchen from improvements to doors and windows	2.9 kWh/d

Table E.9. Break-down of the predicted reductions in heat loss from my house, on a cold winter day.

It's frustratingly hard to make a really big dent in the leakiness of an already-built house! As we saw a moment ago, a much easier way of achieving a big dent in heat loss is to turn the thermostat down. Turning down from 20 to 17 °C gave a reduction in heat loss of 30%.

Combining these two actions – the physical modifications and the turning-down of the thermostat – this model predicts that heat loss should be reduced by nearly 50%. Since some heat is generated in a house by sunshine, gadgets, and humans, the reduction in gas consumption should be more than 50%.

I made all these changes to my house and monitored my meters every week. I can confirm that my heating bill indeed went down by more than 50%. As figure 21.4 showed, my gas consumption has gone down from 40 kWh/d to 13 kWh/d – a reduction of 67%.

Leakiness reduction by internal wall-coverings

Can you reduce your walls' leakiness by covering the *inside* of the wall with insulation? The answer is yes, but there may be two complications. First, the thickness of internal covering is bigger than you might expect. To transform an existing nine-inch solid brick wall (U-value 2.2 W/m^2/K) into a decent 0.30 W/m^2/K wall, roughly 6 cm of insulated lining board is required. [65h3cb] Second, condensation may form on the hidden surface of such internal insulation layers, leading to damp problems.

If you're not looking for such a big reduction in wall leakiness, you can get by with a thinner internal covering. For example, you can buy 1.8-cm-thick insulated wallboards with a U-value of 1.7 W/m^2/K. With these over the existing wall, the U-value would be reduced from 2.2 W/m^2/K to:

$$1 \left/ \left(\frac{1}{2.2} + \frac{1}{1.7} \right) \right. \simeq 1\,\text{W/m}^2/\text{K}.$$

Definitely a worthwhile reduction.

Air-exchange

Once a building is really well insulated, the principal loss of heat will be through ventilation (air changes) rather than through conduction. The heat loss through ventilation can be reduced by transferring the heat from the outgoing air to the incoming air. Remarkably, a great deal of this heat can indeed be transferred without any additional energy being required. The trick is to use a nose, as discovered by natural selection. A nose warms incoming air by cooling down outgoing air. There's a temperature gradient along the nose; the walls of a nose are coldest near the nostrils. The longer your nose, the better it works as a counter-current heat exchanger. In nature's noses, the direction of the air-flow usually alternates. Another way to organize a nose is to have two air-passages, one for in-flow and one for out-flow, separate from the point of view of air, but tightly coupled with each other so that heat can easily flow between the two passages. This is how the noses work in buildings. It's conventional to call these noses heat-exchangers.

An energy-efficient house

In 1984, an energy consultant, Alan Foster, built an energy-efficient house near Cambridge; he kindly gave me his thorough measurements. The house is a timber-framed bungalow based on a Scandinavian "Heatkeeper Serrekunda" design (figure E.10), with a floor area of 140 m^2, composed of three bedrooms, a study, two bathrooms, a living room, a kitchen, and a lobby. The wooden outside walls were supplied in kit form by a Scottish company, and the main parts of the house took only a few days to build.

The walls are 30 cm thick and have a U-value of 0.28 W/m^2/°C. From the inside out, they consist of 13 mm of plasterboard, 27 mm airspace, a vapour barrier, 8 mm of plywood, 90 mm of rockwool, 12 mm of bitumen-impregnated fibreboard, 50 mm cavity, and 103 mm of brick. The ceiling construction is similar with 100–200 mm of rockwool insulation. The ceiling has a U-value of 0.27 W/m^2/°C, and the floor, 0.22 W/m^2/°C. The windows are double-glazed (U-value 2 W/m^2/°C), with the inner panes' outer surfaces specially coated to reduce radiation. The windows are arranged to give substantial solar gain, contributing about 30% of the house's space-heating.

The house is well sealed, every door and window lined with neoprene gaskets. The house is heated by warm air pumped through floor grilles; in winter, pumps remove used air from several rooms, exhausting it to the outside, and they take in air from the loft space. The incoming air and outgoing air pass through a heat exchanger (figure E.11), which saves 60% of the heat in the extracted air. The heat exchanger is a passive device, using no energy: it's like a big metal nose, warming the incoming air with the outgoing air. On a cold winter's day, the outside air temperature was −8 °C, the temperature in the loft's air intake was 0 °C, and the air coming out of the heat exchanger was at +8 °C.

For the first decade, the heat was supplied entirely by electric heaters, heating a 150-gallon heat store during the overnight economy period. More recently a gas supply was brought to the house, and the space heating is now obtained from a condensing boiler.

The heat loss through conduction and ventilation is 4.2 kWh/d/°C. The *heat loss parameter* (the leakiness per square metre of floor area) is 1.25 W/m^2/°C (cf. my house's 2.7 W/°C/m^2).

With the house occupied by two people, the average space-heating consumption, with the thermostat set at 19 or 20 °C during the day, was 8100 kWh per year, or 22 kWh/d; the total energy consumption for all purposes was about 15 000 kWh per year, or 40 kWh/d. Expressed as an aver-

Figure E.10. The Heatkeeper Serrekunda.

Figure E.11. The Heatkeeper's heat-exchanger.

age power per unit area, that's $6.6\,W/m^2$.

Figure E.12 compares the power consumption per unit area of this Heatkeeper house with my house (before and after my efficiency push) and with the European average. My house's post-efficiency-push consumption is close to that of the Heatkeeper, thanks to the adoption of lower thermostat settings.

Benchmarks for houses and offices

The German Passivhaus standard aims for power consumption for heating and cooling of $15\,kWh/m^2/y$, which is $1.7\,W/m^2$; and total power consumption of $120\,kWh/m^2/y$, which is $13.7\,W/m^2$.

The average energy consumption of the UK service sector, per unit floor area, is $30\,W/m^2$.

An energy-efficient office

The National Energy Foundation built themselves a low-cost low-energy building. It has solar panels for hot water, solar photovoltaic (PV) panels generating up to 6.5 kW of electricity, and is heated by a 14-kW groundsource heat pump and occasionally by a wood stove. The floor area is $400\,m^2$ and the number of occupants is about 30. It is a single-storey building. The walls contain 300 mm of rockwool insulation. The heat pump's coefficient of performance in winter was 2.5. The energy used is 65 kWh per year per square metre of floor area ($7.4\,W/m^2$). The PV system delivers almost 20% of this energy.

Contemporary offices

New office buildings are often hyped up as being amazingly environment-friendly. Let's look at some numbers.

The William Gates building at Cambridge University holds computer science researchers, administrators, and a small café. Its area is $11\,110\,m^2$, and its energy consumption is 2392 MWh/y. That's a power per unit area of $215\,kWh/m^2/y$, or $25\,W/m^2$. This building won a RIBA award in 2001 for its predicted energy consumption. "The architects have incorporated many environmentally friendly features into the building." [5dhups]

But are these buildings impressive? Next door, the Rutherford building, built in the 1970s without any fancy eco-claims – indeed without even double glazing – has a floor area of $4998\ m^2$ and consumes 1557 MWh per year; that's $0.85\,kWh/d/m^2$, or $36\,W/m^2$. So the award-winning building is just 30% better, in terms of power per unit area, than its simple 1970s cousin. Figure E.12 compares these buildings and another new building, the Law Faculty, with the Old Schools, which are ancient offices built pre-

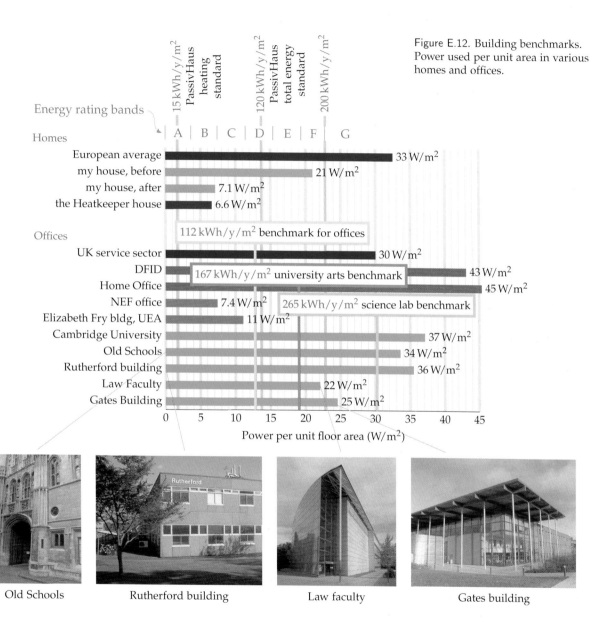

Figure E.12. Building benchmarks. Power used per unit area in various homes and offices.

Old Schools

Rutherford building

Law faculty

Gates building

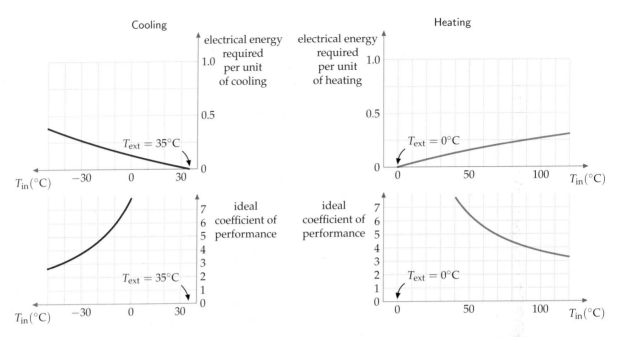

Figure E.13. Ideal heat pump efficiencies. Top left: ideal electrical energy required, according to the limits of thermodynamics, to pump heat *out* of a place at temperature T_{in} when the heat is being pumped to a place at temperature $T_{out} = 35\,°C$. Right: ideal electrical energy required to pump heat *into* a place at temperature T_{in} when the heat is being pumped from a place at temperature $T_{out} = 0\,°C$. Bottom row: the efficiency is conventionally expressed as a "coefficient of performance," which is the heat pumped per unit electrical energy. In practice, I understand that well-installed ground-source heat pumps and the best air-source heat pumps usually have a coefficient of performance of 3 or 4; however, government regulations in Japan have driven the coefficient of performance as high as 6.6.

1890. For all the fanfare, the difference between the new and the old is really quite disappointing!

Notice that the building power consumptions, per unit floor area, are in just the same units (W/m²) as the renewable powers per unit area that we discussed on pages 43, 47, and 177. Comparing these consumption and production numbers helps us realize how difficult it is to power modern buildings entirely from on-site renewables. The power per unit area of biofuels (figure 6.11, p43) is 0.5 W/m²; of wind farms, 2 W/m²; of solar photovoltaics, 20 W/m² (figure 6.18, p47); only solar hot-water panels come in at the right sort of power per unit area, 53 W/m² (figure 6.3, p39).

Improving the coefficient of performance

You might think that the coefficient of performance of a condensing boiler, 90%, sounds pretty hard to beat. But it can be significantly improved upon, by heat pumps. Whereas the condensing boiler takes chemical energy and turns 90% of it into useful heat, the heat pump takes some electrical energy and uses it to *move* heat from one place to another (for example, from outside a building to inside). Usually the amount of useful heat delivered is much bigger than the amount of electricity used. A coefficient of performance of 3 or 4 is normal.

Theory of heat pumps

Here are the formulae for the ideal efficiency of a heat pump, that is, the electrical energy required per unit of heat pumped. If we are pumping heat from an outside place at temperature T_1 into a place at higher temperature T_2, both temperatures being expressed relative to absolute zero (that is, T_2, in kelvin, is given in terms of the Celsius temperature T_{in}, by $273.15 + T_{in}$), the ideal efficiency is:

$$\text{efficiency} = \frac{T_2}{T_2 - T_1}.$$

If we are pumping heat out from a place at temperature T_2 to a warmer exterior at temperature T_1, the ideal efficiency is:

$$\text{efficiency} = \frac{T_2}{T_1 - T_2}.$$

These theoretical limits could only be achieved by systems that pump heat infinitely slowly. Notice that the ideal efficiency is bigger, the closer the inside temperature T_2 is to the outside temperature T_1.

While in theory ground-source heat pumps might have better performance than air-source, because the ground temperature is usually closer than the air temperature to the indoor temperature, in practice an air-source heat pump might be the best and simplest choice. In cities, there may be uncertainty about the future effectiveness of ground-source heat pumps, because the more people use them in winter, the colder the ground gets; this thermal fly-tipping problem may also show up in the summer in cities where too many buildings use ground-source (or should I say "ground-sink"?) heat pumps for air-conditioning.

Heat capacity:	$C = 820\,\text{J/kg/K}$
Conductivity:	$\kappa = 2.1\,\text{W/m/K}$
Density:	$\rho = 2750\,\text{kg/m}^3$
Heat capacity per unit volume:	
	$C_V = 2.3\,\text{MJ/m}^3/\text{K}$

Table E.14. Vital statistics for granite. (I use granite as an example of a typical rock.)

Heating and the ground

Here's an interesting calculation to do. Imagine having solar heating panels on your roof, and, whenever the water in the panels gets above 50°C, pumping the water through a large rock under your house. When a dreary grey cold month comes along, you could then use the heat in the rock to warm your house. Roughly how big a 50°C rock would you need to hold enough energy to heat a house for a whole month? Let's assume we're after 24 kWh per day for 30 days and that the house is at 16°C. The heat capacity of granite is $0.195 \times 4200\,\text{J/kg/K} = 820\,\text{J/kg/K}$. The mass of granite required is:

$$\begin{aligned}
\text{mass} &= \frac{\text{energy}}{\text{heat capacity} \times \text{temperature difference}} \\
&= \frac{24 \times 30 \times 3.6\,\text{MJ}}{(820\,\text{J/kg/°C})(50\,\text{°C} - 16\,\text{°C})} \\
&= 100\,000\,\text{kg},
\end{aligned}$$

100 tonnes, which corresponds to a cuboid of rock of size 6 m × 6 m × 1 m.

Ground storage without walls

OK, we've established the size of a useful ground store. But is it difficult to keep the heat in? Would you need to surround your rock cuboid with lots of insulation? It turns out that the ground itself is a pretty good insulator. A spike of heat put down a hole in the ground will spread as

$$\frac{1}{\sqrt{4\pi\kappa t}} \exp\left(-\frac{x^2}{4(\kappa/(C\rho))t}\right)$$

where κ is the conductivity of the ground, C is its heat capacity, and ρ is its density. This describes a bell-shaped curve with width

$$\sqrt{2\frac{\kappa}{C\rho}t};$$

for example, after six months ($t = 1.6 \times 10^7$ s), using the figures for granite ($C = 0.82\,\text{kJ/kg/K}$, $\rho = 2500\,\text{kg/m}^3$, $\kappa = 2.1\,\text{W/m/K}$), the width is 6 m.

Using the figures for water ($C = 4.2\,\text{kJ/kg/K}$, $\rho = 1000\,\text{kg/m}^3$, $\kappa = 0.6\,\text{W/m/K}$), the width is 2 m.

So if the storage region is bigger than 20 m × 20 m × 20 m then most of the heat stored will still be there in six months time (because 20 m is significantly bigger than 6 m and 2 m).

	(W/m/K)
water	0.6
quartz	8
granite	2.1
earth's crust	1.7
dry soil	0.14

Table E.15. Thermal conductivities. For more data see table E.18, p304.

Limits of ground-source heat pumps

The low thermal conductivity of the ground is a double-edged sword. Thanks to low conductivity, the ground holds heat well for a long time. But on the other hand, low conductivity means that it's not easy to shove heat in and out of the ground rapidly. We now explore how the conductivity of the ground limits the use of ground-source heat pumps.

Consider a neighbourhood with quite a high population density. Can *everyone* use ground-source heat pumps, without using active summer replenishment (as discussed on p152)? The concern is that if we all sucked heat from the ground at the same time, we might freeze the ground solid. I'm going to address this question by two calculations. First, I'll work out the natural flux of energy in and out of the ground in summer and winter.

temperature (°C)

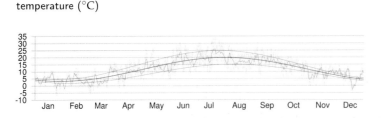

Figure E.16. The temperature in Cambridge, 2006, and a cartoon, which says the temperature is the sum of an annual sinusoidal variation between 3 °C and 20 °C, and a daily sinusoidal variation with range up to 10.3 °C. The average temperature is 11.5 °C.

If the flux we want to suck out of the ground in winter is much bigger than these natural fluxes then we know that our sucking is going to significantly alter ground temperatures, and may thus not be feasible. For this calculation, I'll assume the ground just below the surface is held, by the combined influence of sun, air, cloud, and night sky, at a temperature that varies slowly up and down during the year (figure E.16).

Response to external temperature variations

Working out how the temperature inside the ground responds, and what the flux in or out is, requires some advanced mathematics, which I've cordoned off in box E.19 (p306).

The payoff from this calculation is a rather beautiful diagram (figure E.17) that shows how the temperature varies in time at each depth. This diagram shows the answer for any material in terms of the *characteristic length-scale* z_0 (equation (E.7)), which depends on the conductivity κ and heat capacity C_V of the material, and on the frequency ω of the external temperature variations. (We can choose to look at either daily and yearly variations using the same theory.) At a depth of $2z_0$, the variations in temperature are one seventh of those at the surface, and lag them by about one third of a cycle (figure E.17). At a depth of $3z_0$, the variations in temperature are one twentieth of those at the surface, and lag them by half a cycle.

For the case of daily variations and solid granite, the characteristic length-scale is $z_0 = 0.16$ m. (So 32 cm of rock is the thickness you need to ride out external daily temperature oscillations.) For yearly variations and solid granite, the characteristic length-scale is $z_0 = 3$ m.

Let's focus on annual variations and discuss a few other materials. Characteristic length-scales for various materials are in the third column of table E.18. For damp sandy soils or concrete, the characteristic length-scale z_0 is similar to that of granite – about 2.6 m. In dry or peaty soils, the length-scale z_0 is shorter – about 1.3 m. That's perhaps good news because it means you don't have to dig so deep to find ground with a stable temperature. But it's also coupled with some bad news: the natural fluxes are smaller in dry soils.

The natural flux varies during the year and has a peak value (equation (E.9)) that is smaller, the smaller the conductivity.

For the case of solid granite, the peak flux is 8 W/m². For dry soils, the peak flux ranges from 0.7 W/m² to 2.3 W/m². For damp soils, the peak flux ranges from 3 W/m² to 8 W/m².

What does this mean? I suggest we take a flux in the middle of these numbers, 5 W/m², as a useful benchmark, giving guidance about what sort of power we could expect to extract, per unit area, with a ground-source heat pump. If we suck a flux significantly smaller than 5 W/m², the perturbation we introduce to the natural flows will be small. If on the

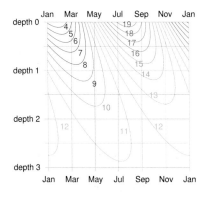

Figure E.17. Temperature (in °C) versus depth and time. The depths are given in units of the characteristic depth z_0, which for granite and annual variations is 3 m.

At "depth 2" (6 m), the temperature is always about 11 or 12 °C. At "depth 1" (3 m), it wobbles between 8 and 15 °C.

other hand we try to suck a flux bigger than $5\,W/m^2$, we should expect that we'll be shifting the temperature of the ground significantly away from its natural value, and such fluxes may be impossible to demand.

The population density of a typical English suburb corresponds to $160\,m^2$ per person (rows of semi-detached houses with about $400\,m^2$ per house, including pavements and streets). At this density of residential area, we can deduce that a ballpark limit for heat pump power delivery is

$$5\,W/m^2 \times 160\,m^2 = 800\,W = 19\,kWh/d \text{ per person.}$$

This is uncomfortably close to the sort of power we would like to deliver in winter-time: it's plausible that our peak winter-time demand for hot air and hot water, in an old house like mine, might be 40 kWh/d per person.

This calculation suggests that in a typical suburban area, *not everyone can use ground-source heat pumps*, unless they are careful to actively dump heat back into the ground during the summer.

Let's do a second calculation, working out how much power we could steadily suck from a ground loop at a depth of $h = 2\,m$. Let's assume that we'll allow ourselves to suck the temperature at the ground loop down to $\Delta T = 5\,°C$ below the average ground temperature at the surface, and let's assume that the surface temperature is constant. We can then deduce the heat flux from the surface. Assuming a conductivity of $1.2\,W/m/K$

	thermal conductivity κ (W/m/K)	heat capacity C_V (MJ/m^3/K)	length-scale z_0 (m)	flux $A\sqrt{C_V \kappa \omega}$ (W/m^2)
Air	0.02	0.0012		
Water	0.57	4.18	1.2	5.7
Solid granite	2.1	2.3	3.0	8.1
Concrete	1.28	1.94	2.6	5.8
Sandy soil				
dry	0.30	1.28	1.5	2.3
50% saturated	1.80	2.12	2.9	7.2
100% saturated	2.20	2.96	2.7	9.5
Clay soil				
dry	0.25	1.42	1.3	2.2
50% saturated	1.18	2.25	2.3	6.0
100% saturated	1.58	3.10	2.3	8.2
Peat soil				
dry	0.06	0.58	1.0	0.7
50% saturated	0.29	2.31	1.1	3.0
100% saturated	0.50	4.02	1.1	5.3

Table E.18. Thermal conductivity and heat capacity of various materials and soil types, and the deduced length-scale $z_0 = \sqrt{\frac{2\kappa}{C_V \omega}}$ and peak flux $A\sqrt{C_V \kappa \omega}$ associated with annual temperature variations with amplitude $A = 8.3\,°C$. The sandy and clay soils have porosity 0.4; the peat soil has porosity 0.8.

(typical of damp clay soil),

$$\text{Flux} = \kappa \times \frac{\Delta T}{h} = 3\,\text{W/m}^2.$$

If, as above, we assume a population density corresponding to $160\,\text{m}^2$ per person, then the maximum power per person deliverable by ground-source heat pumps, if everyone in a neighbourhood has them, is $480\,\text{W}$, which is $12\,\text{kWh/d}$ per person.

So again we come to the conclusion that in a typical suburban area composed of poorly insulated houses like mine, *not everyone can use ground-source heat pumps*, unless they are careful to actively dump heat back into the ground during the summer. And in cities with higher population density, ground-source heat pumps are unlikely to be viable.

I therefore suggest air-source heat pumps are the best heating choice for most people.

Thermal mass

Does increasing the thermal mass of a building help reduce its heating and cooling bills? It depends. The outdoor temperature can vary during the day by about $10\,^\circ\text{C}$. A building with large thermal mass – thick stone walls, for example – will naturally ride out those variations in temperature, and, without heating or cooling, will have a temperature close to the average outdoor temperature. Such buildings, in the UK, need neither heating nor cooling for many months of the year. In contrast, a poorly-insulated building with low thermal mass might be judged too hot during the day and too cool at night, leading to greater expenditure on cooling and heating.

However, large thermal mass is not always a boon. If a room is occupied in winter for just a couple of hours a day (think of a lecture room for example), the energy cost of warming the room up to a comfortable temperature will be greater, the greater the room's thermal mass. This extra invested heat will linger for longer in a thermally massive room, but if nobody is there to enjoy it, it's wasted heat. So in the case of infrequently-used rooms it makes sense to aim for a structure with low thermal mass, and to warm that small mass rapidly when required.

Notes and further reading

page no.

304   *Table E.18.* Sources: Bonan (2002),
      www.hukseflux.com/thermalScience/thermalConductivity.html

If we assume the ground is made of solid homogenous material with conductivity κ and heat capacity C_V, then the temperature at depth z below the
ground and time t responds to the imposed temperature at the surface in
accordance with the diffusion equation

$$\frac{\partial T(z,t)}{\partial t} = \frac{\kappa}{C_V}\frac{\partial^2 T(z,t)}{\partial z^2}.$$ (E.4)

For a sinusoidal imposed temperature with frequency ω and amplitude A at
depth $z = 0$,
$$T(0,t) = T_{\text{surface}}(t) = T_{\text{average}} + A\cos(\omega t),$$ (E.5)
the resulting temperature at depth z and time t is a decaying and oscillating
function
$$T(z,t) = T_{\text{average}} + A\,e^{-z/z_0}\cos(\omega t - z/z_0),$$ (E.6)
where z_0 is the characteristic length-scale of both the decay and the oscillation,

$$z_0 = \sqrt{\frac{2\kappa}{C_V\omega}}.$$ (E.7)

The flux of heat (the power per unit area) at depth z is

$$\kappa\frac{\partial T}{\partial z} = \kappa\frac{A}{z_0}\sqrt{2}e^{-z/z_0}\sin(\omega t - z/z_0 - \pi/4).$$ (E.8)

For example, at the surface, the peak flux is

$$\kappa\frac{A}{z_0}\sqrt{2} = A\sqrt{C_V\kappa\omega}.$$ (E.9)

Box E.19. Working out the natural
flux caused by sinusoidal temperature
variations.

F   Waves II

The physics of deep-water waves

Waves contain energy in two forms: potential energy, and kinetic energy. The potential energy is the energy required to move all the water from the troughs to the crests. The kinetic energy is associated with the water moving around.

People sometimes assume that when the crest of a wave moves across an ocean at 30 miles per hour, the water in that crest must also be moving at 30 miles per hour in the same direction. But this isn't so. It's just like a Mexican wave. When the wave rushes round the stadium, the humans who are making the wave aren't themselves moving round the stadium: they just bob up and down a little. The motion of a piece of water in the ocean is similar: if you focused on a bit of seaweed floating in the water as waves go by, you'd see that the seaweed moves up and down, and also a little to and fro in the direction of travel of the wave – the exact effect could be recreated in a Mexican wave if people moved like window-cleaners, polishing a big piece of glass in a circular motion. The wave has potential energy because of the elevation of the crests above the troughs. And it has kinetic energy because of the small circular bobbing motion of the water.

Our rough calculation of the power in ocean waves will require three ingredients: an estimate of the period T of the waves (the time between crests), an estimate of the height h of the waves, and a physics formula that tells us how to work out the speed v of the wave from its period.

The wavelength λ and period of the waves (the distance and time respectively between two adjacent crests) depend on the speed of the wind that creates the waves, as shown in figure F.1. The height of the waves doesn't depend on the windspeed; rather, it depends on how long the wind has been caressing the water surface.

You can estimate the period of ocean waves by recalling the time between waves arriving on an ocean beach. Is 10 seconds reasonable? For the height of ocean waves, let's assume an amplitude of 1 m, which means 2 m from trough to crest. In waves this high, a man in a dinghy can't see beyond the nearest crest when he's in a trough; I think this height is bigger than average, but we can revisit this estimate if we decide it's important. The speed of deep-water waves is related to the time T between crests by the physics formula (see Faber (1995), p170):

$$v = \frac{gT}{2\pi},$$

where g is the acceleration of gravity (9.8 m/s^2). For example, if $T = 10$ seconds, then $v = 16$ m/s. The wavelength of such a wave – the distance between crests – is $\lambda = vT = gT^2/2\pi = 160$ m.

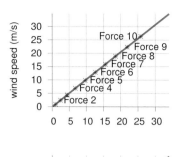

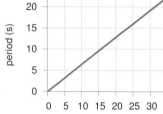

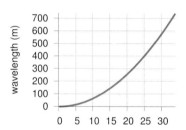

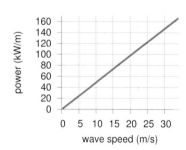

Figure F.1. Facts about deep-water waves. In all four figures the horizontal axis is the wave speed in m/s. From top to bottom the graphs show: wind speed (in m/s) required to make a wave with this wave speed; period (in seconds) of a wave; wavelength (in m) of a wave; and power density (in kW/m) of a wave with amplitude 1 m.

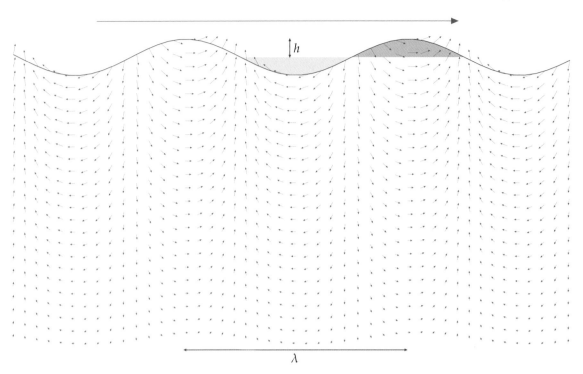

For a wave of wavelength λ and period T, if the height of each crest and depth of each trough is $h = 1\,\text{m}$, the potential energy passing per unit time, per unit length, is

$$P_{\text{potential}} \simeq m^* g \bar{h} / T, \quad\quad\quad\quad (\text{F.1})$$

where m^* is the mass per unit length, which is roughly $\frac{1}{2}\rho h(\lambda/2)$ (approximating the area of the shaded crest in figure F.2 by the area of a triangle), and $\bar{h}$ is the change in height of the centre-of-mass of the chunk of elevated water, which is roughly h. So

$$P_{\text{potential}} \simeq \frac{1}{2}\rho h \frac{\lambda}{2} g h / T. \quad\quad\quad\quad (\text{F.2})$$

(To find the potential energy properly, we should have done an integral here; it would have given the same answer.) Now λ/T is simply the speed at which the wave travels, v, so:

$$P_{\text{potential}} \simeq \frac{1}{4}\rho g h^2 v. \quad\quad\quad\quad (\text{F.3})$$

Waves have kinetic energy as well as potential energy, and, remarkably, these are exactly equal, although I don't show that calculation here; so the total power of the waves is double the power calculated from potential

Figure F.2. A wave has energy in two forms: potential energy associated with raising water out of the light-shaded troughs into the heavy-shaded crests; and kinetic energy of all the water within a few wavelengths of the surface – the speed of the water is indicated by the small arrows. The speed of the wave, travelling from left to right, is indicated by the much bigger arrow at the top.

energy.

$$P_{\text{total}} \simeq \frac{1}{2}\rho g h^2 v. \tag{F.4}$$

There's only one thing wrong with this answer: it's too big, because we've neglected a strange property of dispersive waves: the energy in the wave doesn't actually travel at the same speed as the crests; it travels at a speed called the group velocity, which for deep-water waves is *half* of the speed v. You can see that the energy travels slower than the crests by chucking a pebble in a pond and watching the expanding waves carefully. What this means is that equation (F.4) is wrong: we need to halve it. The correct power per unit length of wave-front is

$$P_{\text{total}} = \frac{1}{4}\rho g h^2 v. \tag{F.5}$$

Plugging in $v = 16\,\text{m/s}$ and $h = 1\,\text{m}$, we find

$$P_{\text{total}} = \frac{1}{4}\rho g h^2 v = 40\,\text{kW/m}. \tag{F.6}$$

This rough estimate agrees with real measurements in the Atlantic (Mollison, 1986). (See p75.)

The losses from viscosity are minimal: a wave of 9 seconds period would have to go three times round the world to lose 10% of its amplitude.

Real wave power systems

Deep-water devices

How effective are real systems at extracting power from waves? Stephen Salter's "duck" has been well characterized: a row of 16-m diameter ducks, feeding off Atlantic waves with an average power of 45 kW/m, would deliver 19 kW/m, including transmission to central Scotland (Mollison, 1986).

The Pelamis device, created by Ocean Power Delivery, has taken over the Salter duck's mantle as the leading floating deep-water wave device. Each snake-like device is 130 m long and is made of a chain of four segments, each 3.5 m in diameter. It has a maximum power output of 750 kW. The Pelamises are designed to be moored in a depth of about 50 m. In a wavefarm, 39 devices in three rows would face the principal wave direction, occupying an area of ocean, about 400 m long and 2.5 km wide (an area of 1 km^2). The effective cross-section of a single Pelamis is 7 m (i.e., for good waves, it extracts 100% of the energy that would cross 7 m). The company says that such a wave-farm would deliver about 10 kW/m.

Shallow-water devices

Typically 70% of energy in ocean waves is lost through bottom-friction as the depth decreases from 100 m to 15 m. So the average wave-power per unit length of coastline in shallow waters is reduced to about 12 kW/m. The Oyster, developed by Queen's University Belfast and Aquamarine Power Ltd [www.aquamarinepower.com], is a bottom-mounted flap, about 12 m high, that is intended to be deployed in waters about 12 m deep, in areas where the average incident wave power is greater than 15 kW/m. Its peak power is 600 kW. A single device would produce about 270 kW in wave heights greater than 3.5 m. It's predicted that an Oyster would have a bigger power per unit mass of hardware than a Pelamis.

Oysters could also be used to directly drive reverse-osmosis desalination facilities. "The peak freshwater output of an Oyster desalinator is between 2000 and 6000 m^3/day." That production has a value, going by the Jersey facility (which uses 8 kWh per m^3), equivalent to 600–2000 kW of electricity.

G   Tide II

Power density of tidal pools

To estimate the power of an artificial tide-pool, imagine that it's filled rapidly at high tide, and emptied rapidly at low tide. Power is generated in both directions, on the ebb and on the flood. (This is called two-way generation or double-effect generation.) The change in potential energy of the water, each six hours, is mgh, where h is the change in height of the centre of mass of the water, which is half the range. (The range is the difference in height between low and high tide; figure G.1.) The mass per unit area covered by tide-pool is $\rho \times (2h)$, where ρ is the density of water ($1000 \, \text{kg/m}^3$). So the power per unit area generated by a tide-pool is

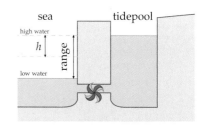

Figure G.1. A tide-pool in cross section. The pool was filled at high tide, and now it's low tide. We let the water out through the electricity generator to turn the water's potential energy into electricity.

$$\frac{2\rho hgh}{6 \, \text{hours}},$$

assuming perfectly efficient generators. Plugging in $h = 2 \, \text{m}$ (i.e., range 4 m), we find the power per unit area of tide-pool is $3.6 \, \text{W/m}^2$. Allowing for an efficiency of 90% for conversion of this power to electricity, we get

$$\text{power per unit area of tide-pool} \simeq 3 \, \text{W/m}^2.$$

So to generate 1 GW of power (on average), we need a tide-pool with an area of about $300 \, \text{km}^2$. A circular pool with diameter 20 km would do the trick. (For comparison, the area of the Severn estuary behind the proposed barrage is about $550 \, \text{km}^2$, and the area of the Wash is more than $400 \, \text{km}^2$.

If a tide-pool produces electricity in one direction only, the power per unit area is halved. The average power density of the tidal barrage at La Rance, where the mean tidal range is 10.9 m, has been $2.7 \, \text{W/m}^2$ for decades (p87).

The raw tidal resource

The tides around Britain are genuine tidal waves. (Tsunamis, which are called "tidal waves," have nothing to do with tides: they are caused by underwater landslides and earthquakes.) The location of the high tide (the crest of the tidal wave) moves much faster than the tidal flow – 100 miles per hour, say, while the water itself moves at just 1 mile per hour.

The energy we can extract from tides, using tidal pools or tide farms, can never be more than the energy of these tidal waves from the Atlantic. We can estimate the total power of these great Atlantic tidal waves in the same way that we estimate the power of ordinary wind-generated waves. The next section describes a standard model for the power arriving in

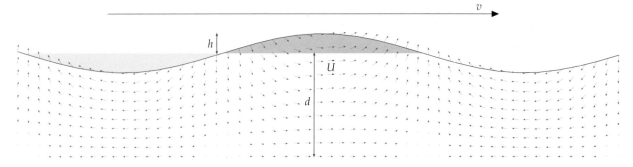

Figure G.2. A shallow-water wave. Just like a deep-water wave, the wave has energy in two forms: potential energy associated with raising water out of the light-shaded troughs into the heavy-shaded crests; and kinetic energy of all the water moving around as indicated by the small arrows. The speed of the wave, travelling from left to right, is indicated by the much bigger arrow at the top. For tidal waves, a typical depth might be 100 m, the crest velocity 30 m/s, the vertical amplitude at the surface 1 or 2 m, and the water velocity amplitude 0.3 or 0.6 m/s.

travelling waves in water of depth d that is shallow compared to the wavelength of the waves (figure G.2). The power per unit length of wavecrest of shallow-water tidal waves is

$$\rho g^{3/2}\sqrt{d}h^2/2. \qquad (G.1)$$

Table G.3 shows the power per unit length of wave crest for some plausible figures. If $d = 100$ m, and $h = 1$ or 2 m, the power per unit length of wave crest is 150 kW/m or 600 kW/m respectively. These figures are impressive compared with the raw power per unit length of ordinary Atlantic deep-water waves, 40 kW/m (Chapter F). Atlantic waves and the Atlantic tide have similar vertical amplitudes (about 1 m), but the raw power in tides is roughly 10 times bigger than that of ordinary wind-driven waves.

Taylor (1920) worked out a more detailed model of tidal power that includes important details such as the Coriolis effect (the effect produced by the earth's daily rotation), the existence of tidal waves travelling in the opposite direction, and the direct effect of the moon on the energy flow in the Irish Sea. Since then, experimental measurements and computer models have verified and extended Taylor's analysis. Flather (1976) built a detailed numerical model of the lunar tide, chopping the continental shelf around the British Isles into roughly 1000 square cells. Flather estimated that the total average power entering this region is 215 GW. According to his model, 180 GW enters the gap between France and Ireland. From Northern Ireland round to Shetland, the incoming power is 49 GW. Between Shetland and Norway there is a net loss of 5 GW. As shown in figure G.4, Cartwright et al. (1980) found experimentally that the average power transmission was 60 GW between Malin Head (Ireland) and Florø (Norway) and 190 GW between Valentia (Ireland) and the Brittany coast near Ouessant. The power entering the Irish Sea was found to be 45 GW, and entering the North Sea via the Dover Straits, 16.7 GW.

h (m)	$\rho g^{3/2}\sqrt{d}h^2/2$ (kW/m)
0.9	125
1.0	155
1.2	220
1.5	345
1.75	470
2.0	600
2.25	780

Table G.3. Power fluxes (power per unit length of wave crest) for depth $d = 100$ m.

The power of tidal waves

This section, which can safely be skipped, provides more details behind the formula for tidal power used in the previous section. I'm going to go into this model of tidal power in some detail because most of the official estimates of the UK tidal resource have been based on a model that I believe is incorrect.

Figure G.2 shows a model for a tidal wave travelling across relatively shallow water. This model is intended as a cartoon, for example, of tidal crests moving up the English channel or down the North Sea. It's important to distinguish the speed U at which the water itself moves (which might be about 1 mile per hour) from the speed v at which the high tide moves, which is typically 100 or 200 miles per hour.

The water has depth d. Crests and troughs of water are injected from the left hand side by the 12-hourly ocean tides. The crests and troughs move with velocity

$$v = \sqrt{gd}. \qquad (G.2)$$

We assume that the wavelength is much bigger than the depth, and we neglect details such as Coriolis forces and density variations in the water. Call the vertical amplitude of the tide h. For the standard assumption of nearly-vorticity-free flow, the horizontal velocity of the water is near-constant with depth. The horizontal velocity U is proportional to the surface displacement and can be found by conservation of mass:

$$U = vh/d. \qquad (G.3)$$

If the depth decreases, the wave velocity v reduces (equation (G.2)). For the present discussion we'll assume the depth is constant. Energy flows from left to right at some rate. How should this total tidal power be estimated? And what's the *maximum* power that could be extracted?

One suggestion is to choose a cross-section and estimate the average *flux of kinetic energy* across that plane, then assert that this quantity represents the power that could be extracted. This kinetic-energy-flux method was used by consultants Black and Veatch to estimate the UK resource. In our cartoon model, we can compute the total power by other means. We'll see that the kinetic-energy-flux answer is too small by a significant factor.

The peak kinetic-energy flux at any section is

$$K_{BV} = \frac{1}{2}\rho A U^3, \qquad (G.4)$$

where A is the cross-sectional area. (This is the formula for kinetic energy flux, which we encountered in Chapter B.)

The true total incident power is not equal to this kinetic-energy flux. The true total incident power in a shallow-water wave is a standard textbook calculation; one way to get it is to find the total energy present in one wavelength and divide by the period. The total energy per wavelength is the sum of the potential energy and the kinetic energy. The kinetic energy happens to be identical to the potential energy. (This is a standard feature of almost all things that wobble, be they masses on springs or children

Figure G.4. Average tidal powers measured by Cartwright et al. (1980).

on swings.) So to compute the total energy all we need to do is compute one of the two – the potential energy per wavelength, or the kinetic energy per wavelength – then double it. The potential energy of a wave (per wavelength and per unit width of wavefront) is found by integration to be

$$\frac{1}{4}\rho g h^2 \lambda. \tag{G.5}$$

So, doubling and dividing by the period, the true power of this model shallow-water tidal wave is

$$\text{power} = \frac{1}{2}(\rho g h^2 \lambda) \times w/T = \frac{1}{2}\rho g h^2 v \times w, \tag{G.6}$$

where w is the width of the wavefront. Substituting $v = \sqrt{gd}$,

$$\text{power} = \rho g h^2 \sqrt{gd} \times w/2 = \rho g^{3/2}\sqrt{d}h^2 \times w/2. \tag{G.7}$$

Let's compare this power with the kinetic-energy flux K_{BV}. Strikingly, the two expressions scale differently with the amplitude h. Using the amplitude conversion relation (G.3), the crest velocity (G.2), and $A = wd$, we can re-express the kinetic-energy flux as

$$K_{BV} = \frac{1}{2}\rho A U^3 = \frac{1}{2}\rho w d(vh/d)^3 = \rho\left(g^{3/2}/\sqrt{d}\right)h^3 \times w/2. \tag{G.8}$$

So the kinetic-energy-flux method suggests that the total power of a shallow-water wave scales as amplitude *cubed* (equation (G.8)); but the correct formula shows that the power scales as amplitude *squared* (equation (G.7)).
     The ratio is

$$\frac{K_{BV}}{\text{power}} = \frac{\rho w\left(g^{3/2}/\sqrt{d}\right)h^3}{\rho g^{3/2}h^2\sqrt{d}w} = \frac{h}{d}. \tag{G.9}$$

Because h is usually much smaller than d (h is about 1 m or 2 m, while d is 100 m or 10 m), estimates of tidal power resources that are based on the kinetic-energy-flux method may be *much too small*, at least in cases where this shallow-water cartoon of tidal waves is appropriate.
     Moreover, estimates based on the kinetic-energy-flux method incorrectly assert that the total available power at springs (the biggest tides) is eight times greater than at neaps (the smallest tides), assuming an amplitude ratio, springs to neaps, of two to one; but the correct answer is that the total available power of a travelling wave scales as its amplitude squared, so the springs-to-neaps ratio of total-incoming-power is four to one.

Effect of shelving of sea bed, and Coriolis force

If the depth d decreases gradually and the width remains constant such that there is minimal reflection or absorption of the incoming power, then

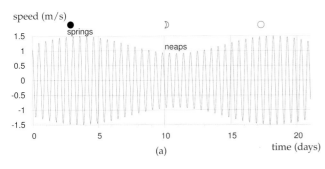

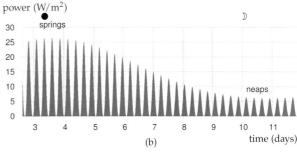

Figure G.5. (a) Tidal current over a 21-day period at a location where the maximum current at spring tide is 2.9 knots (1.5 m/s) and the maximum current at neap tide is 1.8 knots (0.9 m/s).
(b) The power per unit sea-floor area over a nine-day period extending from spring tides to neap tides. The power peaks four times per day, and has a maximum of about 27 W/m². The average power of the tide farm is 6.4 W/m².

the power of the wave will remain constant. This means $\sqrt{d}h^2$ is a constant, so we deduce that the height of the tide scales with depth as $h \sim 1/d^{1/4}$.

This is a crude model. One neglected detail is the Coriolis effect. The Coriolis force causes tidal crests and troughs to tend to drive on the right – for example, going up the English Channel, the high tides are higher and the low tides are lower on the French side of the channel. By neglecting this effect I may have introduced some error into the estimates.

Power density of tidal stream farms

Imagine sticking underwater windmills on the sea-bed. The flow of water will turn the windmills. Because the density of water is roughly 1000 times that of air, the power of water flow is 1000 times greater than the power of wind at the same speed.

What power could tidal stream farms extract? It depends crucially on whether or not we can add up the power contributions of tidefarms on *adjacent* pieces of sea-floor. For wind, this additivity assumption is believed to work fine: as long as the wind turbines are spaced a standard distance apart from each other, the total power delivered by 10 adjacent wind farms is the sum of the powers that each would deliver if it were alone.

Does the same go for tide farms? Or do underwater windmills interfere with each other's power extraction in a different way? I don't think the answer to this question is known in general. We can name two alternative assumptions, however, and identify cartoon situations in which each assumption seems valid. The "tide is like wind" assumption says that you can put tide-turbines all over the sea-bed, spaced about 5 diameters apart from each other, and they won't interfere with each other, no matter how much of the sea-bed you cover with such tide farms.

The "you can have only one row" assumption, in contrast, asserts that the maximum power extractable in a region is the power that would be delivered by a *single* row of turbines facing the flow. A situation where this assumption is correct is the special case of a hydroelectric dam: if the water from the dam passes through a single well-designed turbine, there's no point putting any more turbines behind that one. You can't get 100 times more power by putting 99 more turbines downstream from the first. The oomph gets extracted by the first one, and there isn't any more oomph left for the others. The "you can have only one row" assumption is the right assumption for estimating the extractable power in a place where water flows through a narrow channel from approximately stationary water at one height into another body of water at a lower height. (This case is analysed by Garrett and Cummins (2005, 2007).)

I'm now going to nail my colours to a mast. I think that in many places round the British Isles, the "tide is like wind" assumption is a good approximation. Perhaps some spots have some of the character of a narrow channel. In those spots, my estimates may be over-estimates.

Let's assume that the rules for laying out a sensible tide farm will be similar to those for wind farms, and that the efficiency of the tidemills will be like that of the best windmills, about 1/2. We can then steal the formula for the power of a wind farm (per unit land area) from p265. The power per unit sea-floor area is

$$\frac{\text{power per tidemill}}{\text{area per tidemill}} = \frac{\pi}{200}\frac{1}{2}\rho U^3$$

Using this formula, table G.6 shows this tide farm power for a few tidal currents.

Now, what are typical tidal currents? Tidal charts usually give the currents associated with the tides with the largest range (called spring tides) and the tides with the smallest range (called neap tides). Spring tides occur shortly after each full moon and each new moon. Neap tides occur shortly after the first and third quarters of the moon. The power of a tide farm would vary throughout the day in a completely predictable manner. Figure G.5 illustrates the variation of power density of a tide farm with a maximum current of 1.5 m/s. The average power density of this tide farm would be 6.4 W/m². There are many places around the British Isles where the power per unit area of tide farm would be 6 W/m² or more. This power density is similar to our estimates of the power densities of wind farms (2–3 W/m²) and of photovoltaic solar farms (5–10 W/m²).

We'll now use this "tide farms are like wind farms" theory to estimate the extractable power from tidal streams in promising regions around the British Isles. As a sanity check, we'll also work out the total tidal power crossing each of these regions, using the "power of tidal waves" theory, to check our tide farm's estimated power isn't bigger than the total power available. The main locations around the British Isles where tidal currents are large are shown in figure G.7.

I estimated the typical peak currents at six locations with large currents by looking at tidal charts in *Reed's Nautical Almanac*. (These estimates could easily be off by 30%.) Have I over-estimated or under-estimated the area of each region? I haven't surveyed the sea floor so I don't know if some regions might be unsuitable in some way – too deep, or too shallow, or too

U (m/s)	(knots)	tide farm power (W/m²)
0.5	1	1
1	2	8
2	4	60
3	6	200
4	8	500
5	10	1000

Table G.6. Tide farm power density (in watts per square metre of sea-floor) as a function of flow speed U. (1 knot = 1 nautical mile per hour = 0.514 m/s.) The power density is computed using $\frac{\pi}{200}\frac{1}{2}\rho U^3$ (equation (G.10)).

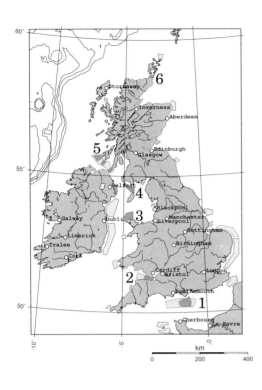

Figure G.7. Regions around the British Isles where peak tidal flows exceed 1 m/s. The six darkly-coloured regions are included in table G.8:

1. the English channel (south of the Isle of Wight);

2. the Bristol channel;

3. to the north of Anglesey;

4. to the north of the Isle of Man;

5. between Northern Ireland, the Mull of Kintyre, and Islay; and

6. the Pentland Firth (between Orkney and mainland Scotland), and within the Orkneys.

There are also enormous currents around the Channel Islands, but they are not governed by the UK. Runner-up regions include the North Sea, from the Thames (London) to the Wash (Kings Lynn). The contours show water depths greater than 100 m. Tidal data are from Reed's Nautical Almanac and DTI Atlas of UK Marine Renewable Energy Resources (2004).

tricky to build on.

Admitting all these uncertainties, I arrive at an estimated total power of 9 kWh/d per person from tidal stream-farms. This corresponds to 9% of the raw incoming power mentioned on p83, 100 kWh per day per person. (The extraction of 1.1 kWh/d/p in the Bristol channel, region 2, might conflict with power generation by the Severn barrage; it would depend on whether the tide farm significantly *adds* to the existing natural friction created by the channel, or *replaces* it.)

Region	U (knots) N	S	power density (W/m^2)	area (km^2)	average power (kWh/d/p)	d (m)	w (km)	raw power N	S (kWh/d/p)
1	1.7	3.1	7	400	1.1	30	30	2.3	7.8
2	1.8	3.2	8	350	1.1	30	17	1.5	4.7
3	1.3	2.3	2.9	1000	1.2	50	30	3.0	9.3
4	1.7	3.4	9	400	1.4	30	20	1.5	6.3
5	1.7	3.1	7	300	0.8	40	10	1.2	4.0
6	5.0	9.0	170	50	3.5	70	10	24	78
Total					9				
			(a)					(b)	

Table G.8. (a) Tidal power estimates assuming that stream farms are like wind farms. The power density is the average power per unit area of sea floor. The six regions are indicated in figure G.7. N = Neaps. S = Springs. (b) For comparison, this table shows the raw incoming power estimated using equation (G.1) (p312).

v (m/s)	v (knots)	Friction power density (W/m²)		tide farm power density (W/m²)
		$R_1 = 0.01$	$R_1 = 0.003$	
0.5	1	1.25	0.4	1
1	2	10	3	8
2	4	80	24	60
3	6	270	80	200
4	8	640	190	500
5	10	1250	375	1000

Table G.9. Friction power density $R_1\rho U^3$ (in watts per square metre of sea-floor) as a function of flow speed, assuming $R_1 = 0.01$ or 0.003. Flather (1976) uses $R_1 = 0.0025$–0.003; Taylor (1920) uses 0.002. (1 knot = 1 nautical mile per hour = 0.514 m/s.) The final column shows the tide farm power estimated in table G.6. For further reading see Kowalik (2004), Sleath (1984).

Estimating the tidal resource via bottom friction

Another way to estimate the power available from tide is to compute how much power is already dissipated by friction on the sea floor. A coating of turbines placed just above the sea floor could act as a substitute bottom, exerting roughly the same drag on the passing water as the sea floor used to exert, and extracting roughly the same amount of power as friction used to dissipate, without significantly altering the tidal flows.

So, what's the power dissipated by "bottom friction"? Unfortunately, there isn't a straightforward model of bottom friction. It depends on the roughness of the sea bed and the material that the bed is made from – and even given this information, the correct formula to use is not settled. One widely used model says that the magnitude of the stress (force per unit area) is $R_1\rho U^2$, where U is the average flow velocity and R_1 is a dimensionless quantity called the shear friction coefficient. We can estimate the power dissipated per unit area by multiplying the stress by the velocity. Table G.9 shows the power dissipated in friction, $R_1\rho U^3$, assuming $R_1 = 0.01$ or $R_1 = 0.003$. For values of the shear friction coefficient in this range, the friction power is very similar to the estimated power that a tide farm would deliver. This is good news, because it suggests that planting a forest of underwater windmills on the sea-bottom, spaced five diameters apart, won't radically alter the flow. The natural friction already has an effect that is in the same ballpark.

Tidal pools with pumping

"The pumping trick" artificially increases the amplitude of the tides in a tidal pool so as to amplify the power obtained. The energy cost of pumping *in* extra water at high tide is repaid with interest when the same water is let out at low tide; similarly, extra water can be pumped *out* at low tide, then let back in at high tide. The pumping trick is sometimes used at La Rance, boosting its net power generation by about 10% (Wilson and Balls, 1990). Let's work out the theoretical limit for this technology. I'll assume

tidal amplitude (half-range) h (m)	optimal boost height b (m)	power with pumping (W/m^2)	power without pumping (W/m^2)
1.0	6.5	3.5	0.8
2.0	13	14	3.3
3.0	20	31	7.4
4.0	26	56	13

Table G.10. Theoretical power density from tidal power using the pumping trick, assuming no constraint on the height of the basin's walls.

that generation has an efficiency of $\epsilon_g = 0.9$ and that pumping has an efficiency of $\epsilon_p = 0.85$. Let the tidal range be $2h$. I'll assume for simplicity that the prices of buying and selling electricity are the same at all times, so that the optimal height boost b to which the pool is pumped above high water is given by (marginal cost of extra pumping = marginal return of extra water):

$$b/\epsilon_p = \epsilon_g(b + 2h).$$

Defining the round-trip efficiency $\epsilon = \epsilon_g\epsilon_p$, we have

$$b = 2h\frac{\epsilon}{1 - \epsilon}.$$

For example, with a tidal range of $2h = 4\,\text{m}$, and a round-trip efficiency of $\epsilon = 76\%$, the optimal boost is $b = 13\,\text{m}$. This is the maximum height to which pumping can be justified if the price of electricity is constant.

Let's assume the complementary trick is used at low tide. (This requires the basin to have a vertical range of 30 m!) The delivered power per unit area is then

$$\left(\frac{1}{2}\rho g\epsilon_g(b + 2h)^2 - \frac{1}{2}\rho g\frac{1}{\epsilon_p}b^2\right)\Big/ T,$$

where T is the time from high tide to low tide. We can express this as the maximum possible power density without pumping, $\epsilon_g 2\rho gh^2/T$, scaled up by a boost factor

$$\left(\frac{1}{1 - \epsilon}\right),$$

which is roughly a factor of 4. Table G.10 shows the theoretical power density that pumping could deliver. Unfortunately, this pumping trick will rarely be exploited to the full because of the economics of basin construction: full exploitation of pumping requires the total height of the pool to be roughly 4 times the tidal range, and increases the delivered power four-fold. But the amount of material in a sea-wall of height H scales as H^2, so the cost of constructing a wall four times as high will be more than four times as big. Extra cash would probably be better spent on enlarging a tidal pool horizontally rather than vertically.

The pumping trick can nevertheless be used for free on any day when the range of natural tides is smaller than the maximum range: the water

tidal amplitude (half-range) h (m)	boost height b (m)	power with pumping (W/m^2)	power without pumping (W/m^2)
1.0	1.0	1.6	0.8
2.0	2.0	6.3	3.3
3.0	3.0	14	7.4
4.0	4.0	25	13

Table G.11. Power density offered by the pumping trick, assuming the boost height is constrained to be the same as the tidal amplitude. This assumption applies, for example, at neap tides, if the pumping pushes the tidal range up to the springs range.

level at high tide can be pumped up to the maximum. Table G.11 gives the power delivered if the boost height is set to h, that is, the range in the pool is just double the external range. A doubling of vertical range is easy at neap tides, since neap tides are typically about half as high as spring tides. Pumping the pool at neaps so that the full springs range is used thus allows neap tides to deliver roughly twice as much power as they would offer without pumping. So a system with pumping would show two-weekly variations in power of just a factor of 2 instead of 4.

Getting "always-on" tidal power by using two basins

Here's a neat idea: have two basins, one of which is the "full" basin and one the "empty" basin; every high tide, the full basin is topped up; every low tide, the empty basin is emptied. These toppings-up and emptyings could be done either passively through sluices, or actively by pumps (using the trick mentioned above). Whenever power is required, water is allowed to flow from the full basin to the empty basin, or (better in power terms) between one of the basins and the sea. The capital cost of a two-basin scheme may be bigger because of the need for extra walls; the big win is that power is available all the time, so the facility can follow demand.

We can use power generated from the empty basin to pump extra water into the full basin at high tide, and similarly use power from the full basin to pump down the empty basin at low tide. This self-pumping would boost the total power delivered by the facility without ever needing to buy energy from the grid. It's a delightful feature of a two-pool solution that the optimal time to *pump* water into the high pool is high tide, which is also the optimal time to *generate* power from the low pool. Similarly, low tide is the perfect time to pump down the low pool, and it's the perfect time to generate power from the high pool. In a simple simulation, I've found that a two-lagoon system in a location with a natural tidal range of 4 m can, with an appropriate pumping schedule, deliver a *steady* power of 4.5 W/m^2 (MacKay, 2007a). One lagoon's water level is always kept above mean sea-level; the other lagoon's level is always kept below mean sea-level. This power density of 4.5 W/m^2 is 50% bigger than the maximum possible average power density of an ordinary tide-pool in the same lo-

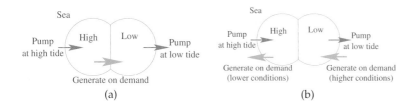

(a)   (b)

cation (3 W/m²). The steady power of the lagoon system would be more valuable than the intermittent and less-flexible power from the ordinary tide-pool.

A two-basin system could also function as a pumped-storage facility.

Notes

page no.

311 *Efficiency of 90%...* Turbines are about 90% efficient for heads of 3.7 m or more. Baker et al. (2006).

320 *Getting "always-on" tidal power by using two basins.* There is a two-basin tidal power plant at Haishan, Maoyan Island, China. A single generator located between the two basins, as shown in figure G.12(a), delivers power continuously, and generates 39 kW on average. [2bqapk].

Further reading: Shaw and Watson (2003b); Blunden and Bahaj (2007); Charlier (2003a,b).

For further reading on bottom friction and variation of flow with depth, see Sleath (1984).

For more on the estimation of the UK tidal resource, see MacKay (2007b).

For more on tidal lagoons, see MacKay (2007a).

Figure G.12. Different ways to use the tidal pumping trick. Two lagoons are located at sea-level. (a) One simple way of using two lagoons is to label one the high pool and the other the low pool; when the surrounding sea level is near to high tide, let water into the high pool, or actively pump it in (using electricity from other sources); and similarly, when the sea level is near to low tide, empty the low pool, either passively or by active pumping; then whenever power is sufficiently valuable, generate power on demand by letting water from the high pool to the low pool. (b) Another arrangement that might deliver more power per unit area has no flow of water between the two lagoons. While one lagoon is being pumped full or pumped empty, the other lagoon can deliver steady, demand-following power to the grid. Pumping may be powered by bursty sources such as wind, by spare power from the grid (say, nuclear power stations), or by the other half of the facility, using one lagoon's power to pump the other lagoon up or down.

H Stuff II

Imported energy

Dieter Helm and his colleagues estimated the footprint of each pound's worth of imports from country X using the average carbon intensity of country X's economy (that is, the ratio of their carbon emissions to their gross domestic product). They concluded that the embodied carbon in imports to Britain (which should arguably be added to Britain's official carbon footprint of 11 tons CO_2e per year per person) is roughly 16 tons CO_2e per year per person. A subsequent, more detailed study commissioned by DEFRA estimated that the embodied carbon in imports is smaller, but still very significant: about 6.2 tons CO_2e per year per person. In energy terms, 6 tons CO_2e per year is something like 60 kWh/d.

Here, let's see if we can reproduce these conclusions in a different way, using the weights of the imports.

Figure H.2 shows Britain's imports in the year 2006 in three ways: on the left, the total *value* of the imports is broken down by the country of origin. In the middle, the same total financial value is broken down by the type of stuff imported, using the categories of HM Revenue and Customs. On the right, all maritime imports to Britain are shown by *weight* and broken down by the categories used by the Department for Transport, which doesn't care whether something is leather or tobacco – it keeps track of how heavy stuff is, whether it is dry or liquid, and whether the stuff arrived in a container or a lorry.

The energy cost of the imported fuels (top right) *is* included in the standard accounts of British energy consumption; the energy costs of all the other imports are not. For most materials, the embodied energy per unit weight is greater than or equal to 10 kWh per kg – the same as the energy per unit weight of fossil fuels. This is true of all metals and alloys, all polymers and composites, most paper products, and many ceramics, for example. The exceptions are raw materials like ores; porous ceramics such as concrete, brick, and porcelain, whose energy cost is 10 times lower; wood and some rubbers; and glasses, whose energy cost is a whisker lower than 10 kWh per kg. [r22oz]

We can thus roughly estimate the energy footprint of our imports simply from the weight of their manufactured materials, if we exclude things like ores and wood. Given the crudity of the data with which we are working, we will surely slip up and inadvertently include some things made of wood and glass, but hopefully such slips will be balanced by our underestimation of the energy content of most of the metals and plastics and more complex goods, many of which have an embodied energy of not 10 but 30 kWh per kg, or even more.

For this calculation I'll take from the right-hand column in figure H.2

Figure H.1. Continuous casting of steel strands at Korea Iron and Steel Company.

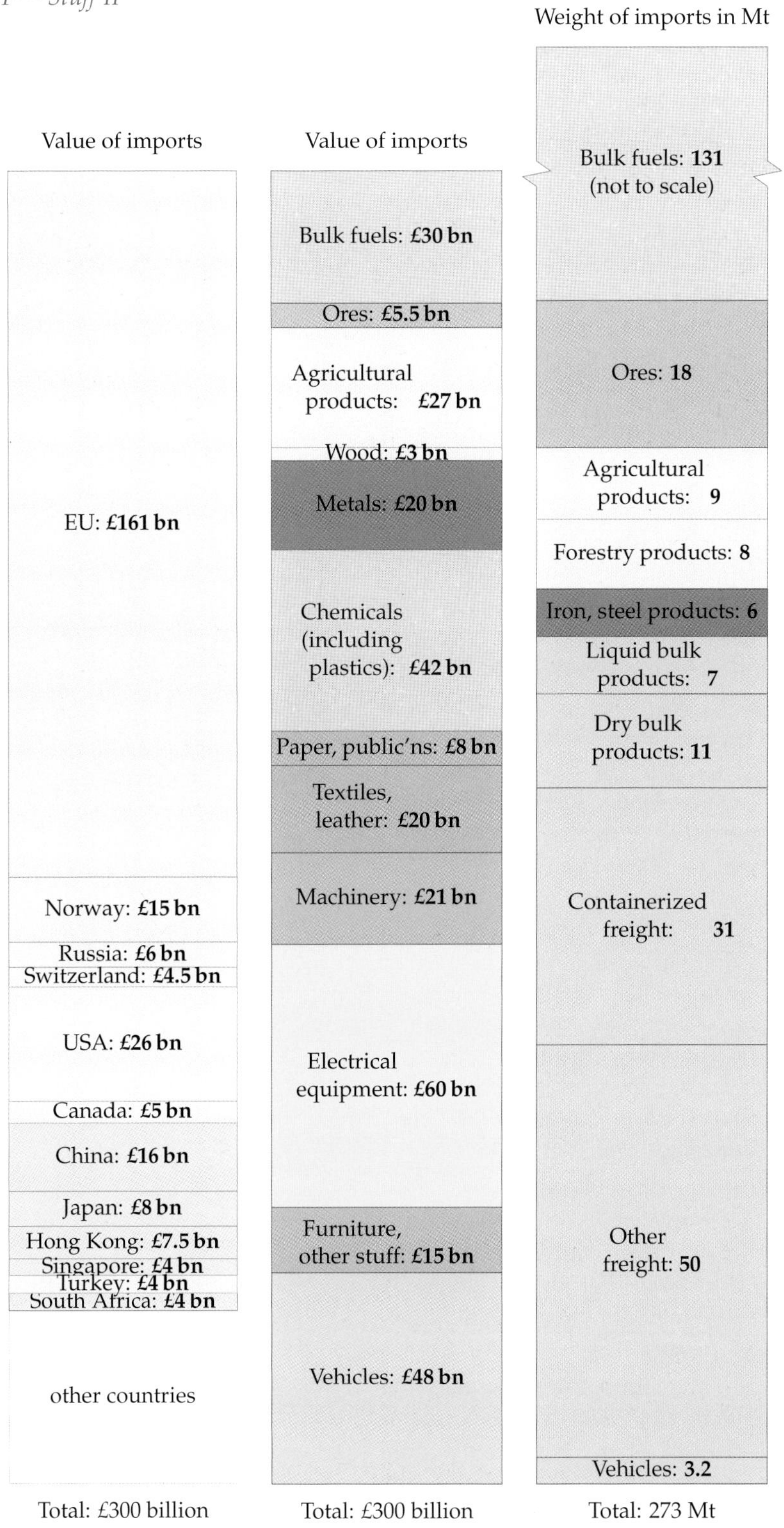

Figure H.2. Imports of stuff to the UK, 2006.

Weight of imports in Mt

Bulk fuels: **131**
(not to scale)

Ores: **18**

Agricultural products: **9**

Forestry products: **8**

Iron, steel products: **6**

Liquid bulk products: **7**

Dry bulk products: **11**

Containerized freight: **31**

Other freight: **50**

Vehicles: **3.2**

Total: 273 Mt

Value of imports

Bulk fuels: **£30 bn**

Ores: **£5.5 bn**

Agricultural products: **£27 bn**

Wood: **£3 bn**

Metals: **£20 bn**

Chemicals (including plastics): **£42 bn**

Paper, public'ns: **£8 bn**

Textiles, leather: **£20 bn**

Machinery: **£21 bn**

Electrical equipment: **£60 bn**

Furniture, other stuff: **£15 bn**

Vehicles: **£48 bn**

Total: £300 billion

Value of imports

EU: **£161 bn**

Norway: **£15 bn**

Russia: **£6 bn**

Switzerland: **£4.5 bn**

USA: **£26 bn**

Canada: **£5 bn**

China: **£16 bn**

Japan: **£8 bn**

Hong Kong: **£7.5 bn**

Singapore: **£4 bn**

Turkey: **£4 bn**

South Africa: **£4 bn**

other countries

Total: £300 billion

the iron and steel products, the dry bulk products, the containerized freight and the "other freight," which total 98 million tons per year. I'm leaving the vehicles to one side for a moment. I subtract from this an estimated 25 million tons of food which is presumably lurking in the "other freight" category (34 million tons of food were imported in 2006), leaving 73 million tons.

Converting 73 million tons to energy using the exchange rate suggested above, and sharing between 60 million people, we estimate that those imports have an embodied energy of 33 kWh/d per person.

For the cars, we can hand-wave a little less, because we know a little more: the number of imported vehicles in 2006 was 2.4 million. If we take the embodied energy per car to be 76 000 kWh (a number we picked up on p90) then these imported cars have an embodied energy of 8 kWh/d per person.

I left the "liquid bulk products" out of these estimates because I am not sure what sort of products they are. If they are actually liquid chemicals then their contribution might be significant.

We've arrived at a total estimate of 41 kWh/d per person for the embodied energy of imports – definitely in the same ballpark as the estimate of Dieter Helm and his colleagues.

I suspect that 41 kWh/d per person may be an underestimate because the energy intensity we assumed (10 kWh/d per person) is too low for most forms of manufactured goods such as machinery or electrical equipment. However, without knowing the weights of all the import categories, this is the best estimate I can make for now.

Figure H.3. Niobium open cast mine, Brazil.

Lifecycle analysis for buildings

Tables H.4 and H.5 show estimates of the *Process Energy Requirement* of building materials and building constructions. This includes the energy used in transporting the raw materials to the factory but not energy used to transport the final product to the building site.

Table H.6 uses these numbers to estimate the process energy for making a three-bedroom house. The *gross energy requirement* widens the boundary, including the embodied energy of urban infrastructure, for example, the embodied energy of the machinery that makes the raw materials. A rough rule of thumb to get the gross energy requirement of a building is to double the process energy requirement [3kmcks].

If we share 42 000 kWh over 100 years, and double it to estimate the gross energy cost, the total embodied energy of a house comes to about 2.3 kWh/d. This is the energy cost of the *shell* of the house only – the bricks, tiles, roof beams.

Material	Embodied energy	
	(MJ/kg)	(kWh/kg)
kiln-dried sawn softwood	3.4	0.94
kiln-dried sawn hardwood	2.0	0.56
air dried sawn hardwood	0.5	0.14
hardboard	24.2	6.7
particleboard	8.0	2.2
MDF	11.3	3.1
plywood	10.4	2.9
glue-laminated timber	11	3.0
laminated veneer lumber	11	3.0
straw	0.24	0.07
stabilised earth	0.7	0.19
imported dimension granite	13.9	3.9
local dimension granite	5.9	1.6
gypsum plaster	2.9	0.8
plasterboard	4.4	1.2
fibre cement	4.8	1.3
cement	5.6	1.6
in situ concrete	1.9	0.53
precast steam-cured concrete	2.0	0.56
precast tilt-up concrete	1.9	0.53
clay bricks	2.5	0.69
concrete blocks	1.5	0.42
autoclaved aerated concrete	3.6	1.0
plastics – general	90	25
PVC	80	22
synthetic rubber	110	30
acrylic paint	61.5	17
glass	12.7	3.5
fibreglass (glasswool)	28	7.8
aluminium	170	47
copper	100	28
galvanised steel	38	10.6
stainless steel	51.5	14.3

Table H.4. Embodied energy of building materials (assuming virgin rather than recycled product is used). (Dimension stone is natural stone or rock that has been selected and trimmed to specific sizes or shapes.) Sources: [3kmcks], Lawson (1996).

	Embodied energy (kWh/m^2)
Walls	
timber frame, timber weatherboard, plasterboard lining	52
timber frame, clay brick veneer, plasterboard lining	156
timber frame, aluminium weatherboard, plasterboard lining	112
steel frame, clay brick veneer, plasterboard lining	168
double clay brick, plasterboard lined	252
cement stabilised rammed earth	104
Floors	
elevated timber floor	81
110 mm concrete slab on ground	179
200 mm precast concrete T beam/infill	179
Roofs	
timber frame, concrete tile, plasterboard ceiling	70
timber frame, terracotta tile, plasterboard ceiling	75
timber frame, steel sheet, plasterboard ceiling	92

Table H.5. Embodied energy in various walls, floors, and roofs. Sources: [3kmcks], Lawson (1996).

	Area (m^2)	×	energy density (kWh/m^2)		energy (kWh)
Floors	100	×	81	=	8100
Roof	75	×	75	=	5600
External walls	75	×	252	=	19 000
Internal walls	75	×	125	=	9400
Total					42 000

Table H.6. Process energy for making a three-bedroom house.

Notes and further reading

page no.

322 *A subsequent more-detailed study commissioned by DEFRA estimated that the embodied carbon in imports is about 6.2 tons CO$_2$e per person.* Wiedmann et al. (2008).

Further resources: `www.greenbooklive.com` has life cycle assessments of building products.
Some helpful cautions about life-cycle analysis: `www.gdrc.org/uem/lca/life-cycle.html`.
More links: `www.epa.gov/ord/NRMRL/lcaccess/resources.htm`.

Figure H.7. Millau Viaduct in France the highest bridge in the world. Steel and concrete, 2.5 km long and 353 m high.

Part IV

Useful data

I   Quick reference

SI Units

> **The watt.** This SI unit is named after James Watt. As for all SI units whose names are derived from the proper name of a person, the first letter of its symbol is uppercase (W). But when an SI unit is spelled out, it should always be written in lowercase (watt), with the exception of the "degree Celsius."
>
> from wikipedia

SI stands for Système Internationale. SI units are the ones that all engineers should use, to avoid losing spacecraft.

SI units		
energy	one joule	$1\,\mathrm{J}$
power	one watt	$1\,\mathrm{W}$
force	one newton	$1\,\mathrm{N}$
length	one metre	$1\,\mathrm{m}$
time	one second	$1\,\mathrm{s}$
temperature	one kelvin	$1\,\mathrm{K}$

prefix	kilo	mega	giga	tera	peta	exa
symbol	k	M	G	T	P	E
factor	10^3	10^6	10^9	10^{12}	10^{15}	10^{18}

prefix	centi	milli	micro	nano	pico	femto
symbol	c	m	μ	n	p	f
factor	10^{-2}	10^{-3}	10^{-6}	10^{-9}	10^{-12}	10^{-15}

Table I.1. SI units and prefixes

My preferred units for energy, power, and transport efficiencies

	My preferred units, expressed in SI		
energy	one kilowatt-hour	$1\,\mathrm{kWh}$	$3\,600\,000\,\mathrm{J}$
power	one kilowatt-hour per day	$1\,\mathrm{kWh/d}$	$(1000/24)\,\mathrm{W} \simeq 40\,\mathrm{W}$
force	one kilowatt-hour per 100 km	$1\,\mathrm{kWh/100\,km}$	$36\,\mathrm{N}$
time	one hour	$1\,\mathrm{h}$	$3600\,\mathrm{s}$
	one day	$1\,\mathrm{d}$	$24 \times 3600\,\mathrm{s} \simeq 10^5\,\mathrm{s}$
	one year	$1\,\mathrm{y}$	$365.25 \times 24 \times 3600\,\mathrm{s} \simeq \pi \times 10^7\,\mathrm{s}$
force per mass	kilowatt-hour per ton-kilometre	$1\,\mathrm{kWh/t\text{-}km}$	$3.6\,\mathrm{m/s^2}\ (\simeq 0.37g)$

Additional units and symbols

Thing measured	unit name	symbol	value
humans	person	p	
mass	ton	t	$1\,t = 1000\,kg$
	gigaton	Gt	$1\,Gt = 10^9 \times 1000\,kg = 1\,Pg$
transport	person-kilometre	p-km	
transport	ton-kilometre	t-km	
volume	litre	l	$1\,l = 0.001\,m^3$
area	square kilometre	sq km, km^2	$1\,sq\,km = 10^6\,m^2$
	hectare	ha	$1\,ha = 10^4\,m^2$
	Wales		$1\,Wales = 21\,000\,km^2$
	London (Greater London)		$1\,London = 1580\,km^2$
energy	Dinorwig		$1\,Dinorwig = 9\,GWh$

Billions, millions, and other people's prefixes

Throughout this book "a billion" (1 bn) means a standard American billion, that is, 10^9, or a thousand million. A trillion is 10^{12}. The standard prefix meaning "billion" (10^9) is "giga."

In continental Europe, the abbreviations Mio and Mrd denote a million and billion respectively. Mrd is short for milliard, which means 10^9.

The abbreviation m is often used to mean million, but this abbreviation is incompatible with the SI – think of mg (milligram) for example. So I don't use m to mean million. Where some people use m, I replace it by M. For example, I use Mtoe for million tons of oil equivalent, and $Mt\,CO_2$ for million tons of CO_2.

Annoying units

There's a whole bunch of commonly used units that are annoying for various reasons. I've figured out what some of them mean. I list them here, to help you translate the media stories you read.

Homes
The "home" is commonly used when describing the power of renewable facilities. For example, "The £300 million Whitelee wind farm's 140 turbines will generate 322 MW – enough to power 200 000 homes." The "home" is defined by the British Wind Energy Association to be a power of 4700 kWh per year [www.bwea.com/ukwed/operational.asp]. That's 0.54 kW, or 13 kWh per day. (A few other organizations use 4000 kWh/y per household.)

The "home" annoys me because I worry that people confuse it with *the total power consumption of the occupants of a home* – but the latter is actually about 24 times bigger. The "home" covers the average domestic *electricity* consumption of a household, only. Not the household's home heating. Nor their workplace. Nor their transport. Nor all the energy-consuming things that society does for them.

Incidentally, when they talk of the CO_2 emissions of a "home," the official exchange rate appears to be 4 tons CO_2 per home per year.

Power stations

Energy saving ideas are sometimes described in terms of power stations. For example according to a BBC report on putting new everlasting LED lightbulbs in traffic lights, "The power savings would be huge – keeping the UK's traffic lights running requires the equivalent of two medium-sized power stations." `news.bbc.co.uk/1/low/sci/tech/specials/sheffield_99/449368.stm`

What is a medium-sized power station? 10 MW? 50 MW? 100 MW? 500 MW? I don't have a clue. A google search indicates that some people think it's 30 MW, some 250 MW, some 500 MW (the most common choice), and some 800 MW. What a useless unit!

Surely it would be clearer for the article about traffic lights to express what it's saying as a percentage? "Keeping the UK's traffic lights running requires 11 MW of electricity, which is 0.03% of the UK's electricity." This would reveal how "huge" the power savings are.

Figure I.2 shows the powers of the UK's 19 coal power stations.

Cars taken off the road

Some advertisements describe reductions in CO_2 pollution in terms of the "equivalent number of cars taken off the road." For example, Richard Branson says that if Virgin Trains' Voyager fleet switched to 20% biodiesel – incidentally, don't you feel it's outrageous to call a train a "green biodiesel-powered train" when it runs on 80% fossil fuels and just 20% biodiesel? – sorry, I got distracted. Richard Branson says that *if* Virgin Trains' Voyager fleet switched to 20% biodiesel – I emphasize the "*if*" because people like Beardie are always getting media publicity for announcing that they are *thinking of* doing good things, but some of these fanfared initiatives are later quietly cancelled, such as the idea of towing aircraft around airports to make them greener – sorry, I got distracted again. Richard Branson says that *if* Virgin Trains' Voyager fleet switched to 20% biodiesel, then there would be a reduction of 34 500 tons of CO_2 per year, which is equivalent to "23 000 cars taken off the road." This statement reveals the exchange rate:

"one car taken off the road" $\longleftrightarrow$ −1.5 tons per year of CO_2.

Calories

The calorie is annoying because the diet community call a kilocalorie a Calorie. 1 such food Calorie = 1000 calories.

2500 kcal = 3 kWh = 10 000 kJ = 10 MJ.

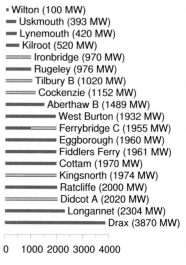

- Wilton (100 MW)
- Uskmouth (393 MW)
- Lynemouth (420 MW)
- Kilroot (520 MW)
- Ironbridge (970 MW)
- Rugeley (976 MW)
- Tilbury B (1020 MW)
- Cockenzie (1152 MW)
- Aberthaw B (1489 MW)
- West Burton (1932 MW)
- Ferrybridge C (1955 MW)
- Eggborough (1960 MW)
- Fiddlers Ferry (1961 MW)
- Cottam (1970 MW)
- Kingsnorth (1974 MW)
- Ratcliffe (2000 MW)
- Didcot A (2020 MW)
- Longannet (2304 MW)
- Drax (3870 MW)

0   1000 2000 3000 4000

Power (MW)

Figure I.2. Powers of Britain's coal power stations. I've highlighted in blue 8 GW of generating capacity that will close by 2015. 2500 MW, shared across Britain, is the same as 1 kWh per day per person.

Barrels

An annoying unit loved by the oil community, along with the ton of oil. Why can't they stick to one unit? A barrel of oil is 6.1 GJ or 1700 kWh.

Barrels are doubly annoying because there are multiple definitions of barrels, all having different volumes.

Here's everything you need to know about barrels of oil. One barrel is 42 U.S. gallons, or 159 litres. One barrel of oil is 0.1364 tons of oil. One barrel of crude oil has an energy of 5.75 GJ. One barrel of oil weighs 136 kg. One ton of crude oil is 7.33 barrels and 42.1 GJ. The carbon-pollution rate of crude oil is 400 kg of CO_2 per barrel. www.chemlink.com.au/conversions.htm This means that when the price of oil is $100 per barrel, oil energy costs 6¢ per kWh. If there were a carbon tax of $250 per ton of CO_2 on fossil fuels, that tax would increase the price of a barrel of oil by $100.

Gallons

The gallon would be a fine human-friendly unit, except the Yanks messed it up by defining the gallon differently from everyone else, as they did the pint and the quart. The US volumes are all roughly five-sixths of the correct volumes.

1 US gal = 3.785 l = 0.83 imperial gal. 1 imperial gal = 4.545 l.

Tons

Tons are annoying because there are short tons, long tons and metric tons. They are close enough that I don't bother distinguishing between them. 1 short ton (2000 lb) = 907 kg; 1 long ton (2240 lb) = 1016 kg; 1 metric ton (or tonne) = 1000 kg.

BTU and quads

British thermal units are annoying because they are neither part of the *Système Internationale*, nor are they of a useful size. Like the useless joule, they are too small, so you have to roll out silly prefixes like "quadrillion" (10^{15}) to make practical use of them.

1 kJ is 0.947 BTU. 1 kWh is 3409 BTU.

A "quad" is 1 quadrillion BTU = 293 TWh.

Funny units

Cups of tea

Is this a way to make solar panels sound good? "Once all the 7000 photovoltaic panels are in place, it is expected that the solar panels will create 180 000 units of renewable electricity each year – enough energy to make **nine million cups of tea**." This announcement thus equates 1 kWh to 50 cups of tea.

As a unit of volume, 1 US cup (half a US pint) is officially 0.24 l; but a cup of tea or coffee is usually about 0.18 l. To raise 50 cups of water, at 0.18 l per cup, from 15 °C to 100 °C requires 1 kWh.

So "nine million cups of tea per year" is another way of saying "20 kW."

Double-decker buses, Albert Halls and Wembley stadiums
"If everyone in the UK that could, installed cavity wall insulation, we could cut carbon dioxide emissions by a huge 7 million tons. That's enough carbon dioxide to fill nearly 40 million double-decker buses or fill the new Wembley stadium 900 times!"

From which we learn the helpful fact that one Wembley is 44 000 double decker buses. Actually, Wembley's bowl has a volume of 1 140 000 m³.

"If every household installed just one energy saving light bulb, there would be enough carbon dioxide saved to fill the Royal Albert Hall 1,980 times!" (An Albert Hall is 100 000 m³.)

Expressing amounts of CO_2 by volume rather than mass is a great way to make them sound big. Should "1 kg of CO_2 per day" sound too small, just say "200 000 litres of CO_2 per year"!

mass of CO_2 $\leftrightarrow$ volume	
2 kg CO_2 $\leftrightarrow$	1 m³
1 kg CO_2 $\leftrightarrow$	500 litres
44 g CO_2 $\leftrightarrow$	22 litres
2 g CO_2 $\leftrightarrow$	1 litre

Table I.3. Volume-to-mass conversion.

More volumes

A container is 2.4 m wide by 2.6 m high by (6.1 or 12.2) metres long (for the TEU and FEU respectively).

One TEU is the size of a small 20-foot container – an interior volume of about 33 m³. Most containers you see today are 40-foot containers with a size of 2 TEU. A 40-foot container weighs 4 tons and can carry 26 tons of stuff; its volume is 67.5 m³.

A swimming pool has a volume of about 3000 m³.

One double decker bus has a volume of 100 m³.

One hot air balloon is 2500 m³.

The great pyramid at Giza has a volume of 2 500 000 cubic metres.

Figure I.4. A twenty-foot container (1 TEU).

Areas

The area of the earth's surface is 500×10^6 km²; the land area is 150×10^6 km².

My typical British 3-bedroom house has a floor area of 88 m². In the USA, the average size of a single-family house is 2330 square feet (216 m²).

hectare	= 10^4 m²
acre	= 4050 m²
square mile	= 2.6 km²
square foot	= 0.093 m²
square yard	= 0.84 m²

Table I.5. Areas.

Powers

If we add the suffix "e" to a power, this means that we're explicitly talking about electrical power. So, for example, a power station's output might be 1 GW(e), while it uses chemical power at a rate of 2.5 GW. Similarly the

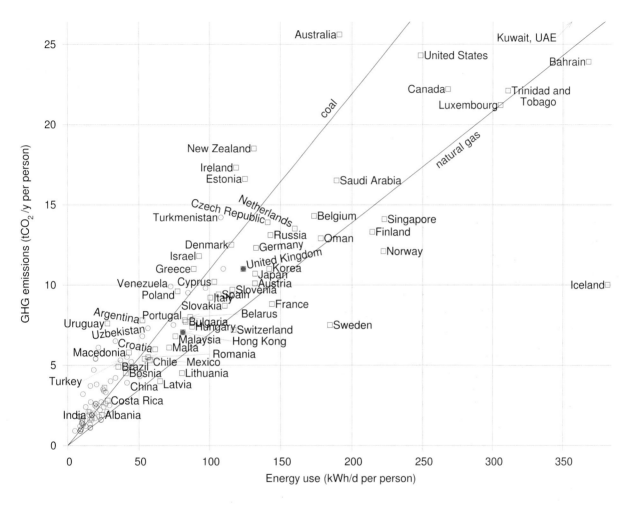

Figure I.12. Greenhouse-gas emissions per capita, versus power consumption per capita. The lines show the emission-intensities of coal and natural gas. Squares show countries having "high human development;" circles, "medium" or "low." See also figures 30.1 (p231) and 18.4 (p105). Source: UNDP Human Development Report, 2007.

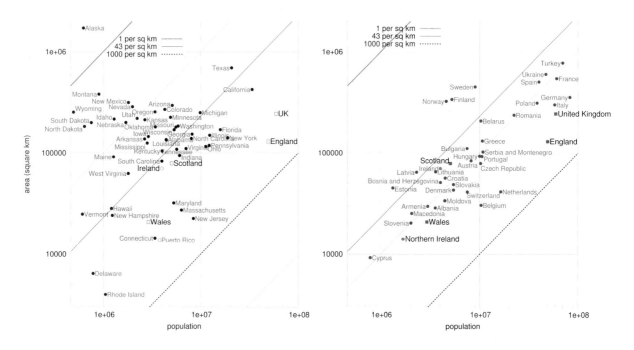

Figure J.4. Populations and areas of the States of America and regions around Europe.

Region	Population	Area (km²)	People per km²	Area per person (m²)
Afghanistan	29 900 000	647 000	46	21 600
Africa	**778 000 000**	**30 000 000**	26	38 600
Alaska	655 000	1 480 000	0.44	2 260 000
Albania	3 560 000	28 700	123	8 060
Algeria	32 500 000	2 380 000	14	73 200
Angola	11 100 000	1 240 000	9	111 000
Antarctica	4 000	**13 200 000**		
Argentina	39 500 000	2 760 000	14	69 900
Asia	**3 670 000 000**	**44 500 000**	82	12 100
Australia	**20 000 000**	**7 680 000**	2.6	382 000
Austria	8 180 000	83 800	98	10 200
Bangladesh	**144 000 000**	144 000	1 000	997
Belarus	10 300 000	207 000	50	20 100
Belgium	10 000 000	31 000	340	2 945
Bolivia	8 850 000	1 090 000	8	124 000
Bosnia & Herzegovina	4 020 000	51 100	79	12 700
Botswana	1 640 000	600 000	2.7	366 000
Brazil	**186 000 000**	**8 510 000**	22	45 700
Bulgaria	7 450 000	110 000	67	14 800
CAR	3 790 000	622 000	6	163 000
Canada	32 800 000	**9 980 000**	3.3	304 000
Chad	9 820 000	1 280 000	8	130 000
Chile	16 100 000	756 000	21	46 900
China	**1 300 000 000**	**9 590 000**	136	7 340
Colombia	42 900 000	1 130 000	38	26 500
Croatia	4 490 000	56 500	80	12 500
Czech Republic	10 200 000	78 800	129	7 700
DRC	**60 000 000**	2 340 000	26	39 000
Denmark	5 430 000	43 000	126	7 930
Egypt	**77 500 000**	1 000 000	77	12 900
England	49 600 000	130 000	380	2 630
Estonia	1 330 000	45 200	29	33 900
Ethiopia	**73 000 000**	1 120 000	65	15 400
Europe	**732 000 000**	**9 930 000**	74	13 500
European Union	**496 000 000**	4 330 000	115	8 720
Finland	5 220 000	338 000	15	64 700
France	**60 600 000**	547 000	110	9 010
Gaza Strip	1 370 000	360	3 820	261
Germany	**82 400 000**	357 000	230	4 330
Greece	10 600 000	131 000	81	12 300
Greenland	56 300	2 160 000	0.026	38 400 000
Hong Kong	6 890 000	1 090	6 310	158
Hungary	10 000 000	93 000	107	9 290
Iceland	296 000	103 000	2.9	347 000
India	**1 080 000 000**	3 280 000	328	3 040
Indonesia	**241 000 000**	1 910 000	126	7 930
Iran	**68 000 000**	1 640 000	41	24 200
Ireland	4 010 000	70 200	57	17 500
Italy	**58 100 000**	301 000	192	5 180
Japan	**127 000 000**	377 000	337	2 960
Kazakhstan	15 100 000	2 710 000	6	178 000
Kenya	33 800 000	582 000	58	17 200
Latin America	**342 000 000**	**17 800 000**	19	52 100
Latvia	2 290 000	64 500	35	28 200
Libya	5 760 000	1 750 000	3.3	305 000
Lithuania	3 590 000	65 200	55	18 100
Madagascar	18 000 000	587 000	31	32 500
Mali	12 200 000	1 240 000	10	100 000
Malta	398 000	316	1 260	792
Mauritania	3 080 000	1 030 000	3	333 000
Mexico	**106 000 000**	1 970 000	54	18 500
Moldova	4 450 000	33 800	131	7 590
Mongolia	2 790 000	1 560 000	1.8	560 000
Mozambique	19 400 000	801 000	24	41 300
Myanmar	42 900 000	678 000	63	15 800
Namibia	2 030 000	825 000	2.5	406 000
Netherlands	16 400 000	41 500	395	2 530
New Zealand	4 030 000	268 000	15	66 500
Niger	11 600 000	1 260 000	9	108 000
Nigeria	**128 000 000**	923 000	139	7 170
North America	**483 000 000**	**24 200 000**	20	50 200
Norway	4 593 000	324 000	14	71 000
Oceania	31 000 000	**7 680 000**	4	247 000
Pakistan	**162 000 000**	803 000	202	4 940
Peru	27 900 000	1 280 000	22	46 000
Philippines	**87 800 000**	300 000	292	3 410
Poland	39 000 000	313 000	124	8 000
Portugal	10 500 000	92 300	114	8 740
Republic of Macedonia	2 040 000	25 300	81	12 300
Romania	22 300 000	237 000	94	10 600
Russia	**143 000 000**	**17 000 000**	8	119 000
Saudi Arabia	26 400 000	1 960 000	13	74 200
Scotland	5 050 000	78 700	64	15 500
Serbia & Montenegro	10 800 000	102 000	105	9 450
Singapore	4 420 000	693	6 380	156
Slovakia	5 430 000	48 800	111	8 990
Slovenia	2 010 000	20 200	99	10 000
Somalia	8 590 000	637 000	13	74 200
South Africa	44 300 000	1 210 000	36	27 500
South Korea	48 400 000	98 400	491	2 030
Spain	40 300 000	504 000	80	12 500
Sudan	40 100 000	2 500 000	16	62 300
Suriname	438 000	163 000	2.7	372 000
Sweden	9 000 000	449 000	20	49 900
Switzerland	7 480 000	41 200	181	5 510
Taiwan	22 800 000	35 900	636	1 570
Tanzania	36 700 000	945 000	39	25 700
Thailand	**65 400 000**	514 000	127	7 850
Turkey	**69 600 000**	780 000	89	11 200
Ukraine	47 400 000	603 000	78	12 700
United Kingdom	*59 500 000*	*244 000*	*243*	*4 110*
USA (ex. Alaska)	**295 000 000**	**8 150 000**	36	27 600
Venezuela	25 300 000	912 000	28	35 900
Vietnam	**83 500 000**	329 000	253	3 940
Wales	2 910 000	20 700	140	7 110
Western Sahara	273 000	266 000	1	974 000
World	**6 440 000 000**	**148 000 000**	43	23 100
Yemen	20 700 000	527 000	39	25 400
Zambia	11 200 000	752 000	15	66 800

Table J.5. Regions and their population densities. Populations above 50 million and areas greater than 5 million km² are highlighted. These data are displayed graphically in figure J.1 (p338). Data are from 2005.

K UK energy history

Primary fuel	kWh/d/p	kWh(e)/d/p
Oil	43	
Natural gas	47	
Coal	20	
Nuclear	9 →	3.4
Hydro		0.2
Other renewables		0.8

Table K.1. Breakdown of primary energy sources in the UK (2004–2006

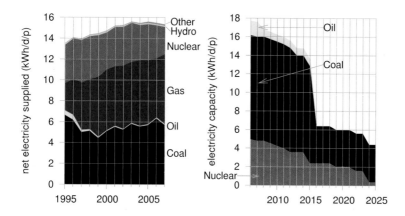

Figure K.2. Left: UK net electricity supplied, by source, in kWh per day per person. (Another 0.9 kWh/d/p is generated and used by the generators themselves.)
Right: the energy gap created by UK power station closures, as projected by energy company EdF. This graph shows the predicted *capacity* of nuclear, coal, and oil power stations, in kilowatt-hours per day per person. The capacity is the maximum deliverable power of a source.

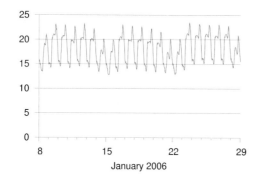

January 2006

Figure K.3. Electricity demand in Great Britain (in kWh/d per person) during two winter weeks of 2006. The peaks in January are at 6pm each day (If you'd like to obtain the national demand in GW, the top of the scale, 24 kWh/d per person, is the same as 60 GW per UK.)

	2006	2007
"Primary units" (the first 2 kWh/d)	10.73 p/kWh	17.43 p/kWh
"Secondary units" (the rest)	8.13 p/kWh	9.70 p/kWh

Table K.4. Domestic electricity charges (2006, 2007) for Powergen customers in Cambridge, including tax.

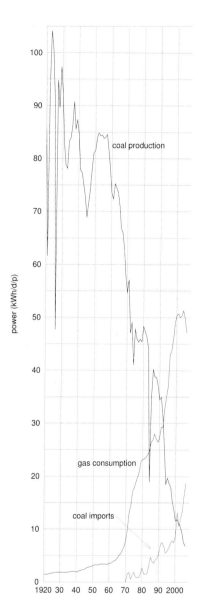

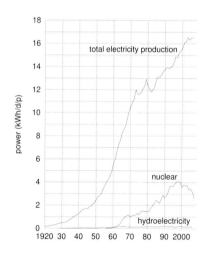

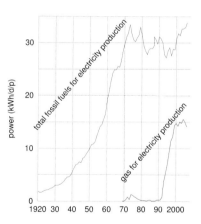

Figure K.5. History of UK production of electricity, hydroelectricity, and nuclear electricity.
Powers are expressed "per person" by dividing each power by 60 million.

Figure K.6. History of UK use of fossil fuels for electricity production.
Powers are expressed "per person" by dividing each power by 60 million.

Figure K.7. UK production and imports of coal, and UK consumption of gas.
Powers are expressed "per person" by dividing each power by 60 million.

List of web links

This section lists the full links corresponding to each of the tiny URLs mentioned in the text. Each item starts with the page number on which the tiny URL was mentioned. See also `http://tinyurl.com/yh8xse` (or `www.inference.phy.cam.ac.uk/sustainable/book/tex/cft.url.html`) for a clickable page with all URLs in this book.

If you find a URL doesn't work any more, you may be able to find the page on the Wayback Machine internet archive [f754].

p	tinyURL	Full web link.
18	ydoobr	`www.bbc.co.uk/radio4/news/anyquestions_transcripts_20060127.shtml`
19	2jhve6	`www.ft.com/cms/s/0/48e334ce-f355-11db-9845-000b5df10621.html`
19	25e59w	`news.bbc.co.uk/1/low/uk_politics/7135299.stm`
19	5o7mxk	`www.guardian.co.uk/environment/2007/dec/10/politics`
19	5c4olc	`www.foe.co.uk/resource/press_releases/green_solutions_undermined_10012008.html`
19	2fztd3	`www.jalopnik.com/cars/alternative-energy/now-thats-some-high-quality-h20-car-runs-on-water-177788.php`
19	26e8z	`news.bbc.co.uk/1/hi/sci/tech/3381425.stm`
19	ykhayj	`politics.guardian.co.uk/terrorism/story/0,,1752937,00.html`
20	16y5g	`www.grida.no/climate/ipcc_tar/wg1/fig3-1.htm`
20	5qfkaw	`www.nap.edu/catalog.php?record_id=12181`
21	2z2xg7	`assets.panda.org/downloads/2_vs_3_degree_impacts_1oct06_1.pdf`
21	yyxq2m	`www.bp.com/genericsection.do?categoryId=93&contentId=2014442`
21	dzcqq	`www.defra.gov.uk/environment/climatechange/internat/pdf/avoid-dangercc.pdf`
21	y98ys5	`news.bbc.co.uk/1/hi/business/4933190.stm`
30	5647rh	`www.dft.gov.uk/pgr/statistics/datatablespublications/tsgb/`
31	27jdc5	`www.dft.gov.uk/pgr/statistics/datatablespublications/energyenvironment/tsgb-chapter3energyandthenvi1863`
31	28abpm	`corporate.honda.com/environmentology/`
31	nmn41	`www.simetric.co.uk/si_liquids.htm`
31	2hcgdh	`cta.ornl.gov/data/appendix_b.shtml`
34	vxhhj	`www.cl.cam.ac.uk/research/dtg/weather/`
34	tdvml	`www.phy.hw.ac.uk/resrev/aws/awsarc.htm`
36	3fbufz	`www.ipcc.ch/ipccreports/sres/aviation/004.htm`
36	3asmgy	`news.independent.co.uk/uk/transport/article324294.ece`
36	9ehws	`www.boeing.com/commercial/747family/technical.html`
36	3exmgv	`www.ryanair.com/site/EN/about.php?page=About&sec=environment`
36	yrnmum	`www.grida.no/climate/ipcc/aviation/124.htm`
37	36w5gz	`www.rolls-royce.com/community/downloads/environment04/products/air.html`
44	?rqloc	`www.metoffice.gov.uk/climate/uk/location/scotland/index.html`
44	2szckw	`www.metoffice.gov.uk/climate/uk/stationdata/cambridgedata.txt`
45	5hrxls	`eosweb.larc.nasa.gov/cgi-bin/sse/sse.cgi?+s01`
45	6z9epq	`www.solarcentury.com/knowledge_base/images/solar_pv_orientation_diagram`
47	2t17t6	`www.reuk.co.uk/40-Percent-Efficiency-PV-Solar-Panels.htm`
47	6hobq2	`www.azonano.com/news.asp?newsID=4546`
47	21sx6t	`www.udel.edu/PR/UDaily/2008/jul/solar072307.html`
47	62ccou	`www.nrel.gov/news/press/2008/625.html`
48	5hzs5y	`www.ens-newswire.com/ens/dec2007/2007-12-26-093.asp`
48	39z5m5	`news.bbc.co.uk/1/hi/world/europe/6505221.stm`

48 2uk8q8 www.powerlight.com/about/press2006_page.php?id=59
48 2ahecp www.aps.org/meetings/multimedia/upload/The_Status_and_Outlook_for_the_Photovoltaics_Industry_David_E_Carlson.pdf
48 6kqq77 www.defra.gov.uk/erdp/pdfs/ecs/miscanthus-guide.pdf
58 ynjzej www.aceee.org/conf/06modeling/azevado.pdf
64 wbd8o www.ref.org.uk/energydata.php
66 25e59w news.bbc.co.uk/1/low/uk_politics/7135299.stm
66 2t2vjq www.guardian.co.uk/environment/2007/dec/11/windpower.renewableenergy
66 57984r www.businessgreen.com/business-green/news/2205496/critics-question-government
66 6oc3ja www.independent.co.uk/environment/green-living/donnachadh-mccarthy-my-carbonfree-year-767115.html
66 5soql2 www.housebuildersupdate.co.uk/2006/12/eco-bollocks-award-windsave-ws1000.html
66 6g2jm5 www.carbontrust.co.uk/technology/technologyaccelerator/small-wind
79 5h69fm www.thepoultrysite.com/articles/894/economic-approach-to-broiler-production
80 5pwojp www.fertilizer.org/ifa/statistics/STATSIND/pkann.asp
80 5bj8k3 www.walkerscarbonfootprint.co.uk/walkers_carbon_footprint.html
80 3s576h www.permatopia.com/transportation.html
87 6xrm5q www.edf.fr/html/en/decouvertes/voyage/usine/retour-usine.html
94 yx7zm4 www.cancentral.com/funFacts.cfm
94 r22oz www-materials.eng.cam.ac.uk/mpsite/interactive_charts/energy-cost/NS6Chart.html
94 yhrest www.transportation.anl.gov/pdfs/TA/106.pdf
94 y5as53 www.aluminum.org/Content/NavigationMenu/The_Industry/Government_Policy/Energy/Energy.htm
94 y2ktgg www.ssab.com/templates/Ordinary___573.aspx
95 6lbrab www.lindenau-shipyard.de/pages/newsb.html
95 5ctx4k www.wilhelmsen.com/SiteCollectionDocuments/WW_Miljorapport_engelsk.pdf
95 yqbz13 www.normanbaker.org.uk/downloads/Supermarkets Report Final Version.doc
102 yttg7p budget2007.treasury.gov.uk/page_09.htm
102 fcqfw www.mod.uk/DefenceInternet/AboutDefence/Organisation/KeyFactsAboutDefence/DefenceSpending.htm
102 2e4fcs press.homeoffice.gov.uk/press-releases/security-prebudget-report
102 33x5kc www.mod.uk/NR/rdonlyres/95BBA015-22B9-43EF-B2DC-DFF14482A590/0/gep_200708.pdf
102 35ab2c www.dasa.mod.uk/natstats/ukds/2007/c1/table103.html
102 yg5fsj siteresources.worldbank.org/DATASTATISTICS/Resources/GDP.pdf
102 yfgjna www.sipri.org/contents/milap/milex/mex_major_spenders.pdf/download
102 slbae www.wisconsinproject.org/countries/israel/plut.html
102 yh45h8 www.usec.com/v2001_02/HTML/Aboutusec_swu.asp
102 t2948 www.world-nuclear.org/info/inf28.htm
102 2ywzee www.globalsecurity.org/wmd/intro/u-centrifuge.htm
112 uzek2 www.dti.gov.uk/energy/inform/dukes/
112 3av4s9 hdr.undp.org/en/statistics/
112 6frj55 news.independent.co.uk/environment/article2086678.ece
129 5qhvcb www.tramwayinfo.com/Tramframe.htm?www.tramwayinfo.com/tramways/Articles/Compair2.htm
134 4qgg8q www.newsweek.com/id/112733/output/print
135 5o5x5m www.cambridgeenergy.com/archive/2007-02-08/cef08feb2007kemp.pdf
135 5o5x5m www.cambridgeenergy.com/archive/2007-02-08/cef08feb2007kemp.pdf
135 5fbeg9 www.cfit.gov.uk/docs/2001/racomp/racomp/pdf/racomp.pdf
135 679rpc www.tfl.gov.uk/assets/downloads/environmental-report-2007.pdf
136 5cp27j www.eaton.com/EatonCom/ProductsServices/Hybrid/SystemsOverview/HydraulicHLA/index.htm
137 4wm2w4 www.citroenet.org.uk/passenger-cars/psa/berlingo/berlingo-electrique.html
137 658ode www.greencarcongress.com/2008/02/mitsubishi-moto.html
139 czjjo corporate.honda.com/environment/fuel_cells.aspx?id=fuel_cells_fcx
139 5a3ryx automobiles.honda.com/fcx-clarity/specifications.aspx
154 yok2nw www.eca.gov.uk/etl/find/_P_Heatpumps/detail.htm?ProductID=9868&FromTechnology=S_WaterSourcePackaged

Bibliography

AITCHISON, E. (1996). Methane generation from UK landfill sites and its use as an energy resource. *Energy Conversion and Management*, 37(6/8):1111–1116. doi: doi:10.1016/0196-8904(95)00306-1. `www.ingentaconnect.com/content/els/01968904/1996/00000037/00000006/art00306`.

AMOS, W. A. (2004). Updated cost analysis of photobiological hydrogen production from Chlamydomonas reinhardtii green algae – milestone completion report. `www.nrel.gov/docs/fy04osti/35593.pdf`.

ANDERSON, K., BOWS, A., MANDER, S., SHACKLEY, S., AGNOLUCCI, P., and EKINS, P. (2006). Decarbonising modern societies: Integrated scenarios process and workshops. Technical Report 48, Tyndall Centre. `www.tyndall.ac.uk/research/theme2/final_reports/t3_24.pdf`.

ARCHER, M. D. and BARBER, J. (2004). Photosynthesis and photoconversion. In M. D. Archer and J. Barber, editors, *Molecular to Global Photosynthesis*. World Scientific. ISBN 978-1-86094-256-3. `www.worldscibooks.com/lifesci/p218.html`.

ASHWORTH, W. and PEGG, M. (1986). *The history of the British coal industry. Vol. 5, 1946–1982: the nationalized industry*. Clarendon, Oxford. ISBN 0198282958.

ASPLUND, G. (2004). Sustainable energy systems with HVDC transmission. In *Power Engineering Society General Meeting*, volume 2, pages 2299–2303. IEEE. doi: 10.1109/PES.2004.1373296. `www.trec-uk.org.uk/reports/HVDC_Gunnar_Asplund_ABB.pdf`.

ASSELBERGS, B., BOKHORST, J., HARMS, R., VAN HEMERT, J., VAN DER NOORT, L., TEN VELDEN, C., VERVUURT, R., WIJNEN, L., and VAN ZON, L. (2006). Size does matter – the possibilities of cultivating *jatropha curcas* for biofuel production in Cambodia. `environmental.scum.org/biofuel/jatropha/`.

BAER, P. and MASTRANDREA, M. (2006). High stakes: Designing emissions pathways to reduce the risk of dangerous climate change. `www.ippr.org/publicationsandreports/`.

BAHRMAN, M. P. and JOHNSON, B. K. (2007). The ABCs of HVDC transmission technology. *IEEE Power and Energy Magazine*, 5(2).

BAINES, J. A., NEWMAN, V. G., HANNA, I. W., DOUGLAS, T. H., CARLYLE, W. J., JONES, I. L., EATON, D. M., and ZERONIAN, G. (1983). Dinorwig pumped storage scheme. *Institution of Civil Engineers Proc. pt. 1*, 74:635–718.

BAINES, J. A., NEWMAN, V. G., HANNA, I. W., DOUGLAS, T. H., CARLYLE, W. J., JONES, I. L., EATON, D. M., and ZERONIAN, G. (1986). Dinorwig pumped storage scheme. *Institution of Civil Engineers Proc. pt. 1*, 80:493–536.

BAKER, C., WALBANCKE, J., and LEACH, P. (2006). Tidal lagoon power generation scheme in Swansea Bay. `www.dti.gov.uk/files/file30617.pdf`. A report on behalf of the Dept. of Trade and Industry and the Welsh Development Agency.

BAYER CROP SCIENCE. (2003). Potential of GM winter oilseed rape to reduce the environmental impact of farming whilst improving farmer incomes. `tinyurl.com/5j99df`.

BICKLEY, D. T. and RYRIE, S. C. (1982). A two-basin tidal power scheme for the Severn estuary. In *Conf. on new approaches to tidal power*.

BINDER, M., FALTENBACHER, M., KENTZLER, M., and SCHUCKERT, M. (2006). Clean urban transport for Europe. deliverable D8 final report. `www.fuel-cell-bus-club.com/`.

BLACK and VEATCH. (2005). The UK tidal stream resource and tidal stream technology. report prepared for the Carbon Trust Marine Energy Challenge. `www.carbontrust.co.uk/technology/technologyaccelerator/tidal_stream.htm`.

BLUNDEN, L. S. and BAHAJ, A. S. (2007). Tidal energy resource assessment for tidal stream generators. *Proc. IMechE*, 221 Part A: 137–146.

BONAN, G. B. (2002). *Ecological Climatology: Concepts and Applications*. Cambridge Univ. Press. ISBN 9780521804769.

BOYER, J. S. (1982). Plant productivity and environment. *Science*, 218 (4571):443–448. doi: 10.1126/science.218.4571.443.

BRASLOW, A. L. (1999). *A history of suction-type laminar-flow control with emphasis on flight research*. Number 13 in Monographs in Aerospace History. NASA. `www.nasa.gov/centers/dryden/pdf/88792main_Laminar.pdf`.

BROECKER, W. S. and KUNZIG, R. (2008). *Fixing Climate: What Past Climate Changes Reveal About the Current Threat–and How to Counter It*. Hill and Wang. ISBN 0809045028.

BURNHAM, A., WANG, M., and WU, Y. (2007). Development and applications of GREET 2.7 — the transportation vehicle-cycle model. `www.transportation.anl.gov/software/GREET/publications.html`.

CARBON TRUST. (2007). Micro-CHP accelerator – interim report. Technical Report CTC726. `www.carbontrust.co.uk/publications/publicationdetail.htm?productid=CTC726`.

CARLSSON, L. (2002). "Classical" HVDC: still continuing to evolve. *Modern Power Systems*.

CARTWRIGHT, D. E., EDDEN, A. C., SPENCER, R., and VASSIE, J. M. (1980). The tides of the northeast Atlantic Ocean. *Philos. Trans. R. Soc. Lond. Ser. A*, 298(1436):87–139.

CATLING, D. T. (1966). Principles and practice of train performance applied to London Transport's Victoria line. Paper 8, Convention on Guided Land Transport (London, 27-28 October 1966).

CHARLIER, R. H. (2003a). Sustainable co-generation from the tides: A review. *Renewable and Sustainable Energy Reviews*, 7:187213.

CHARLIER, R. H. (2003b). A "sleeper" awakes: tidal current power. *Renewable and Sustainable Energy Reviews*, 7:515529.

CHARNEY, J. G., ARAKAWA, A., BAKER, D. J., BOLIN, B., DICKINSON, R. E., GOODY, R. M., LEITH, C. E., STOMMEL, H. M., and WUNSCH, C. I. (1979). Carbon dioxide and climate: A scientific assessment. `www.nap.edu/catalog.php?record_id=12181`.

CHISHOLM, S. W., FALKOWSKI, P. G., and CULLEN, J. J. (2001). Discrediting ocean fertilisation. *Science*, 294(5541):309–310.

CHITRAKAR, R., KANOH, H., MIYAI, Y., and OOI, K. (2001). Recovery of lithium from seawater using manganese oxide adsorbent ($H_{1.6}Mn_{1.6}O_4$) derived from $Li_{1.6}Mn_{1.6}O_4$. *Ind. Eng. Chem. Res.*, 40(9):2054–2058. pubs.acs.org/cgi-bin/abstract.cgi/iecred/2001/40/i09/abs/ie000911h.html.

CHURCH, R. A., HALL, A., and KANEFSKY, J. (1986). *The history of the British coal industry. Vol. 3, 1830–1913: Victorian pre-eminence.* Clarendon, Oxford. ISBN 0198282842.

COHEN, B. L. (1983). Breeder reactors: A renewable energy source. *American Journal of Physics*, 51(1):75–76. sustainablenuclear.org/PADs/pad11983cohen.pdf.

COLEY, D. (2001). Emission factors for walking and cycling. www.centres.ex.ac.uk/cee/publications/reports/91.html.

COMMITTEE ON RADIOACTIVE WASTE MANAGEMENT. (2006). Managing our radioactive waste safely. www.corwm.org.uk/Pages/Current%20Publications/700%20-%20CoRWM%20July%202006%20Recommendations%20to%20Government.pdf.

CUTE. (2006). Clean urban transport for Europe. detailed summary of achievements. www.fuel-cell-bus-club.com/.

DAVID, J. and HERZOG, H. (2000). The cost of carbon capture. sequestration.mit.edu/pdf/David_and_Herzog.pdf. presented at the Fifth International Conf. on Greenhouse Gas Control Technologies, Cairns, Australia, August 13 - August 16 (2000).

DAVIDSON, E. A. and JANSSENS, I. A. (2006). Temperature sensitivity of soil carbon decomposition and feedbacks to climate change. *Nature*, 440:165–173. doi: doi:10.1038/nature04514. www.nature.com/nature/journal/v440/n7081/full/nature04514.html.

DEFFEYES, K. S. and MACGREGOR, I. D. (1980). World uranium resources. *Scientific American*, pages 66–76.

DENHOLM, P., KULCINSKI, G. L., and HOLLOWAY, T. (2005). Emissions and energy efficiency assessment of baseload wind energy systems. *Environ Sci Technol*, 39(6):1903–1911. ISSN 0013-936X. www.ncbi.nlm.nih.gov/entrez/query.fcgi?cmd=Retrieve&db=pubmed&dopt=Abstract&list_uids=15819254.

DENISON, R. A. (1997). Life-cycle assessment for paper products. In E. Ellwood, J. Antle, G. Eyring, and P. Schulze, editors, *Wood in Our Future: The Role of Life-Cycle Analysis: Proc. a Symposium.* National Academy Press. ISBN 0309057450. books.nap.edu/openbook.php?record_id=5734.

DENNIS, C. (2006). Solar energy: Radiation nation. *Nature*, 443:23–24. doi: 10.1038/443023a.

DEPT. FOR TRANSPORT. (2007). Transport statistics Great Britain. www.dft.gov.uk/pgr/statistics/datatablespublications/tsgb/.

DEPT. OF DEFENSE. (2008). More fight – less fuel. Report of the Defense Science Board Task Force on DoD Energy Strategy.

DEPT. OF TRADE AND INDUSTRY. (2004). DTI Atlas of UK marine renewable energy resources. www.offshore-sea.org.uk/.

DEPT. OF TRADE AND INDUSTRY. (2002a). Energy consumption in the United Kingdom. www.berr.gov.uk/files/file11250.pdf.

DEPT. OF TRADE AND INDUSTRY. (2002b). Future offshore. www.berr.gov.uk/files/file22791.pdf.

DEPT. OF TRADE AND INDUSTRY. (2007). Impact of banding the renewables obligation – costs of electricity production. www.berr.gov.uk/files/file39038.pdf.

DESSLER, A. E. and PARSON, E. A. (2006). *The Science and Politics of Global Climate Change – A Guide to the Debate.* Cambridge Univ. Press, Cambridge. ISBN 9780521539418.

DI PRAMPERO, P. E., CORTILI, G., MOGNONI, P., and SAIBENE, F. (1979). Equation of motion of a cyclist. *J. Appl. Physiology*, 47:201–206. jap.physiology.org/cgi/content/abstract/47/1/201.

DIAMOND, J. (2004). *Collapse: How Societies Choose to Fail or Succeed.* Penguin.

E4TECH. (2007). A review of the UK innovation system for low carbon road transport technologies. www.dft.gov.uk/pgr/scienceresearch/technology/lctis/e4techlcpdf.

ECKHARTT, D. (1995). Nuclear fuels for low-beta fusion reactors: Lithium resources revisited. *Journal of Fusion Energy*, 14(4):329–341. ISSN 0164-0313 (Print) 1572-9591 (Online). doi: 10.1007/BF02214511. www.springerlink.com/content/35470543rj8t2gk1/.

EDDINGTON, R. (2006). Transport's role in sustaining the UK's productivity and competitiveness.

EDEN, R. and BENDING, R. (1985). Gas/electricity competition in the UK. Technical Report 85/6, Cambridge Energy Research Group, Cambridge.

ELLIOTT, D. L., WENDELL, L. L., and GOWER, G. L. (1991). An assessment of windy land area and wind energy potential in the contiguous United States. www.osti.gov/energycitations/servlets/purl/5252760-ccuOpk/.

ENERGY FOR SUSTAINABLE DEVELOPMENT LTD. (2003). English partnerships sustainable energy review. www.englishpartnerships.co.uk.

ERDINCLER, A. U. and VESILIND, P. A. (1993). Energy recovery from mixed waste paper. *Waste Management & Research*, 11(6):507–513. doi: 10.1177/0734242X9301100605.

ETHERIDGE, D., STEELE, L., LANGENFELDS, R., FRANCEY, R., BARNOLA, J.-M., and MORGAN, V. (1998). Historical CO_2 records from the Law Dome DE08, DE08-2, and DSS ice cores. In *Trends: A Compendium of Data on Global Change.* Carbon Dioxide Information Analysis Center, Oak Ridge National Laboratory, US Dept. of Energy, Oak Ridge, Tenn., USA. cdiac.ornl.gov/trends/co2/lawdome.html.

EUROPEAN COMMISSION. (2007). Concentrating solar power - from research to implementation. www.solarpaces.org/Library/library.htm.

EVANS, D. G. (2007). Liquid transport biofuels – technology status report. www.nnfcc.co.uk/.

EVANS, R. K. (2008). An abundance of lithium. www.worldlithium.com.

FABER, T. E. (1995). *Fluid dynamics for physicists.* Cambridge Univ. Press, Cambridge.

FAIMAN, D., RAVIV, D., and ROSENSTREICH, R. (2007). Using solar energy to arrest the increasing rate of fossil-fuel consumption: The southwestern states of the USA as case studies. *Energy Policy*, 35: 567576.

FIES, B., PETERSON, T., and POWICKI, C. (2007). Solar photovoltaics – expanding electric generation options. mydocs.epri.com/docs/SEIG/1016279_Photovoltaic_White_Paper_1207.pdf.

FISHER, K., WALLÉN, E., LAENEN, P. P., and COLLINS, M. (2006). Battery waste management life cycle assessment. www.defra.gov.uk/environment/waste/topics/batteries/pdf/erm-lcareport0610.pdf.

FLATHER, R. A. (1976). A tidal model of the north-west European continental shelf. *Memoires Société Royale des Sciences de Liège*, 10 (6):141–164.

FLINN, M. W. and STOKER, D. (1984). *The history of the British coal industry. Vol. 2, 1700–1830: the Industrial Revolution.* Clarendon, Oxford. ISBN 0198282834.

FRANCIS, G., EDINGER, R., and BECKER, K. (2005). A concept for simultaneous wasteland reclamation, fuel production, and socio-economic development in degraded areas in India: Need, potential and perspectives of Jatropha plantations. *Natural Resources Forum*, 29(1):12–24. doi: 10.1111/j.1477-8947.2005.00109.x.

FRANKLIN, J. (2007). Principles of cycle planning. www.cyclenetwork.org.uk/papers/071119principles.pdf.

FREESTON, D. H. (1996). Direct uses of geothermal energy 1995. geoheat.oit.edu/bulletin/bull17-1/art1.pdf.

GABRIELLI, G. and VON KÁRMÁN, T. (1950). What price speed? *Mechanical Engineering*, 72(10).

GARRETT, C. and CUMMINS, P. (2005). The power potential of tidal currents in channels. *Proc. Royal Society A*, 461(2060):2563–2572. dx.doi.org/10.1098/rspa.2005.1494.

GARRETT, C. and CUMMINS, P. (2007). The efficiency of a turbine in a tidal channel. *J Fluid Mech*, 588:243–251. journals.cambridge.org/production/action/cjoGetFulltext?fulltextid=1346064.

GELLINGS, C. W. and PARMENTER, K. E. (2004). Energy efficiency in fertilizer production and use. In C. W. Gellings and K. Blok, editors, *Efficient Use and Conservation of Energy*, Encyclopedia of Life Support Systems. Eolss Publishers, Oxford, UK. www.eolss.net.

GERMAN AEROSPACE CENTER (DLR) INSTITUTE OF TECHNICAL THERMODYNAMICS SECTION SYSTEMS ANALYSIS AND TECHNOLOGY ASSESSMENT. (2006). Concentrating solar power for the Mediterranean region. www.dlr.de/tt/med-csp. Study commissioned by Federal Ministry for the Environment, Nature Conservation and Nuclear Safety, Germany.

GOODSTEIN, D. (2004). *Out of Gas*. W. W. Norton and Company, New York. ISBN 0393058573.

GREEN, J. E. (2006). Civil aviation and the environment – the next frontier for the aerodynamicist. *Aeronautical Journal*, 110(1110): 469–486.

GRUBB, M. and NEWBERY, D. (2008). Pricing carbon for electricity generation: national and international dimensions. In M. Grubb, T. Jamasb, and M. G. Pollitt, editors, *Delivering a Low Carbon Electricity System: Technologies, Economics and Policy*. Cambridge Univ. Press, Cambridge.

GUMMER, J., GOLDSMITH, Z., PECK, J., EGGAR, T., HURD, N., MIRAJ, A., NORRIS, S., NORTHCOTE, B., OLIVER, T., STRONG, D., TWITCHEN, K., and WILKIE, K. (2007). Blueprint for a green economy. www.qualityoflifechallenge.com.

HALKEMA, J. A. (2006). Wind energy: Facts and fiction. www.countryguardian.net/halkema-windenergyfactfiction.pdf.

HAMMOND, G. and JONES, C. (2006). Inventory of carbon & energy (ICE). www.bath.ac.uk/mech-eng/sert/embodied/. version 1.5a Beta.

HAMMONS, T. J. (1993). Tidal power. *Proc. IEEE*, 8(3):419–433.

HANSEN, J., SATO, M., KHARECHA, P., RUSSELL, G., LEA, D., and SIDDALL, M. (2007). Climate change and trace gases. *Phil. Trans. Royal. Soc. A*, 365:1925–1954. doi: 10.1098/rsta.2007.2052. pubs.giss.nasa.gov/abstracts/2007/Hansen_etal_2.html.

HASTINGS, R. and WALL, M. (2006). *Sustainable Solar Housing: Strategies And Solutions*. Earthscan. ISBN 1844073254.

HATCHER, J. (1993). *The History of the British Coal Industry: Towards the Age of Coal: Before 1700 Vol 1*. Clarendon Press.

HEATON, E., VOIGT, T., and LONG, S. (2004). A quantitative review comparing the yields of two candidate C4 perennial biomass crops in relation to nitrogen, temperature, and water. *Biomass and Bioenergy*, 27:21–30.

HELM, D., SMALE, R., and PHILLIPS, J. (2007). Too good to be true? The UK's climate change record. www.dieterhelm.co.uk/publications/Carbon_record_2007.pdf.

HELWEG-LARSEN, T. and BULL, J. (2007). Zero carbon Britain – an alternative energy strategy. zerocarbonbritain.com/.

HERRING, J. (2004). Uranium and thorium resource assessment. In C. J. Cleveland, editor, *Encyclopedia of Energy*. Boston Univ., Boston, USA. ISBN 0-12-176480-X.

HERZOG, H. (2003). Assessing the feasibility of capturing CO_2 from the air. web.mit.edu/coal/working_folder/pdfs/Air_Capture_Feasibility.pdf.

HERZOG, H. (2001). What future for carbon capture and sequestration? *Environmental Science and Technology*, 35:148A–153A. sequestration.mit.edu/.

HIRD, V., EMERSON, C., NOBLE, E., LONGFIELD, J., WILLIAMS, V., GOETZ, D., HOSKINS, R., PAXTON, A., and DUPEE, G. (1999). Still on the road to ruin? An assessment of the debate over the unnecessary transport of food, five years on from the food miles report.

HODGSON, P. (1999). *Nuclear Power, Energy and the Environment*. Imperial College Press.

HOPFIELD, J. J. and GOLLUB, J. (1978). Introduction to solar energy. www.inference.phy.cam.ac.uk/sustainable/solar/HopfieldGollub78/scan.html.

HORIE, H., TANJO, Y., MIYAMOTO, T., and KOGA, Y. (1997). Development of a lithium-ion battery pack system for EV. *JSAE Review*, 18 (3):295–300.

HPTCJ. (2007). Heat pumps: Long awaited way out of the global warming. www.hptcj.or.jp/about_e/contribution/index.html.

INDERMUHLE, A., STOCKER, T., JOOS, F., FISCHER, H., SMITH, H., WAHLEN, M., DECK, B., MASTROIANNI, D., TSCHUMI, J., BLUNIER, T., MEYER, R., and STAUFFER, B. (1999). Holocene carbon-cycle dynamics based on CO_2 trapped in ice at Taylor Dome, Antarctica *Nature*, 398:121–126.

INTERNATIONAL ENERGY AGENCY. (2001). Things that go blip in the night – standby power and how to limit it. `www.iea.org/textbase/nppdf/free/2000/blipinthenight01.pdf`.

JACKSON, P. and KERSHAW, S. (1996). Reducing long term methane emissions resulting from coal mining. *Energy Conversion and Management*, 37(6-8):801–806. doi: 10.1016/0196-8904(95)00259-6.

JEVONS, W. S. (1866). *The Coal Question; An Inquiry concerning the Progress of the Nation, and the Probable Exhaustion of our Coal-mines*. Macmillan and Co., London, second edition. `oll.libertyfund.org/`.

JONES, I. S. F. (2008). The production of additional marine protein by nitrogen nourishment. `www.oceannourishment.com/files/Jc08.pdf`.

JONES, P. M. S. (1984). Statistics and nuclear energy. *The Statistician*, 33(1):91–102. `www.jstor.org/pss/2987717`.

JUDD, B., HARRISON, D. P., and JONES, I. S. F. (2008). Engineering ocean nourishment. In *World Congress on Engineering WCE 2008*, pages 1315–1319. IAENG. ISBN 978-988-98671-9-5.

JUNIPER, T. (2007). *How Many Lightbulbs does it take To Change a Planet?* Quercus, London.

KAMMEN, D. M. and HASSENZAHL, D. M. (1999). *Should We Risk It? Exploring Environmental, Health, and Technological Problem Solving*. Princeton Univ. Press.

KANEKO, T., SHIMADA, M., KUJIRAOKA, S., and KOJIMA, T. (2004). Easy maintenance and environmentally-friendly train traction system. *Hitachi Review*, 53(1):15–19. `www.hitachi.com/ICSFiles/afieldfile/2004/05/25/r2004_01_103.pdf`.

KEELING, C. and WHORF, T. (2005). Atmospheric CO_2 records from sites in the SIO air sampling network. In *Trends: A Compendium of Data on Global Change*. Carbon Dioxide Information Analysis Center, Oak Ridge National Laboratory, US Dept. of Energy, Oak Ridge, Tenn., USA.

KEITH, D. W., HA-DUONG, M., and STOLAROFF, J. K. (2005). Climate strategy with CO_2 capture from the air. *Climatic Change*. doi: 10.1007/s10584-005-9026-x. `www.ucalgary.ca/~keith/papers/51.Keith.2005.ClimateStratWithAirCapture.e.pdf`.

KING, J. (2007). The King review of low-carbon cars. Part I: the potential for CO_2 reduction. `hm-treasury.gov.uk/king`.

KING, J. (2008). The King review of low-carbon cars. Part II: recommendations for action. `hm-treasury.gov.uk/king`.

KOOMEY, J. G. (2007). Estimating total power consumption by servers in the US and the world. `blogs.business2.com/greenwombat/files/serverpowerusecomplete-v3.pdf`.

KOWALIK, Z. (2004). Tide distribution and tapping into tidal energy. *Oceanologia*, 46(3):291–331.

KUEHR, R. (2003). *Computers and the Environment: Understanding and Managing their Impacts (Eco-Efficiency in Industry and Science)*. Springer. ISBN 1402016808.

LACKNER, K. S., GRIMES, P., and ZIOCK, H.-J. (2001). Capturing carbon dioxide from air. `www.netl.doe.gov/publications/proceedings/01/carbon_seq/7b1.pdf`. Presented at First National Conf. on Carbon Sequestration, Washington DC.

LAWSON, B. (1996). Building materials, energy and the environment: Towards ecologically sustainable development.

LAYZELL, D. B., STEPHEN, J., and WOOD, S. M. (2006). Exploring the potential for biomass power in Ontario. `www.biocap.ca/files/Ont_bioenergy_OPA_Feb23_final.pdf`.

LE QUÉRÉ, C., RÖDENBECK, C., BUITENHUIS, E., CONWAY, T. J., LANGENFELDS, R., GOMEZ, A., LABUSCHAGNE, C., RAMONET, M., NAKAZAWA, T., METZL, N., GILLETT, N., and HEIMANN, M. (2007). Saturation of the southern ocean CO_2 sink due to recent climate change. *Science*, 316:1735–1738. doi: 10.1126/science.1136188. `lgmacweb.env.uea.ac.uk/e415/publications.html`.

LEMOFOUET-GATSI, S. (2006). *Investigation and optimisation of hybrid electricity storage systems based on compressed air and supercapacitors*. PhD thesis, EPFL. `library.epfl.ch/theses/?nr=3628`.

LEMOFOUET-GATSI, S. and RUFER, A. (2005). Hybrid energy systems based on compressed air and supercapacitors with maximum efficiency point tracking. `leiwww.epfl.ch/publications/lemofouet_rufer_epe_05.pdf`.

LOMBORG, B. (2001). *The skeptical environmentalist: measuring the real state of the world*. Cambridge Univ. Press, Cambridge. ISBN 0-521-80447-7.

MABEE, W. E., SADDLER, J. N., NIELSEN, C., HENRIK, L., and STEEN JENSEN, E. (2006). Renewable-based fuels for transport. `www.risoe.dk/rispubl/Energy-report5/ris-r-1557_49-52.pdf`. Riso Energy Report 5.

MACDONALD, J. M. (2008). The economic organization of US broiler production. `www.ers.usda.gov/Publications/EIB38/EIB38.pdf`. Economic Information Bulletin No. 38. Economic Research Service, US Dept. of Agriculture.

MACDONALD, P., STEDMAN, A., and SYMONS, G. (1992). The UK geothermal hot dry rock R&D programme. In *Seventeenth Workshop on Geothermal Reservoir Engineering*.

MACKAY, D. J. C. (2007a). Enhancing electrical supply by pumped storage in tidal lagoons. `www.inference.phy.cam.ac.uk/mackay/abstracts/Lagoons.html`.

MACKAY, D. J. C. (2007b). Under-estimation of the UK tidal resource. `www.inference.phy.cam.ac.uk/mackay/abstracts/TideEstimate.html`.

MACLEAY, I., HARRIS, K., and MICHAELS, C. (2007). Digest of United Kingdom energy statistics 2007. `www.berr.gov.uk`.

MALANIMA, P. (2006). Energy crisis and growth 1650–1850: the European deviation in a comparative perspective. *Journal of Global History*, 1:101–121. doi: 10.1017/S1740022806000064.

MARLAND, G., BODEN, T., and ANDRES, R. J. (2007). Global, regional, and national CO_2 emissions. In *Trends: A Compendium of Data on Global Change*. Carbon Dioxide Information Analysis Center, Oak Ridge National Laboratory, US Dept. of Energy, Oak Ridge, Tenn., USA. `cdiac.ornl.gov/trends/emis/tre_glob.htm`.

MASSACHUSETTS INSTITUTE OF TECHNOLOGY. (2006). The future of geothermal energy. `geothermal.inel.gov/publications/future_of_geothermal_energy.pdf`.

MCBRIDE, J. P., MOORE, R. E., WITHERSPOON, J. P., and BLANCO, R. E. (1978). Radiological impact of airborne effluents of coal and nu-

TAYLOR, G. I. (1920). Tidal friction in the Irish Sea. *R. Soc. Lond. Ser. A*, 220:1–33. doi: 10.1098/rsta.1920.0001.

TAYLOR, G. K. (2002a). Are you missing the boat? the ekranoplan in the 21st century – its possibilities and limitations. www.hypercraft-associates.com/areyoumissingtheboat2002.pdf. Presented at the 18th Fast Ferry Conf., Nice, France.

TAYLOR, S. J. (2002b). The Severn barrage – definition study for a new appraisal of the project. www.dti.gov.uk/files/file15363.pdf. ETSU REPORT NO. T/09/00212/00/REP.

TENNEKES, H. (1997). *The Simple Science of Flight*. MIT Press.

THAKUR, P. C., LITTLE, H. G., and KARIS, W. G. (1996). Global coalbed methane recovery and use. *Energy Conversion and Management*, 37 (6/8):789–794.

THE EARTHWORKS GROUP. (1989). *50 Simple things you can do to save the earth*. The Earthworks Press, Berkeley, California. ISBN 0-929634-06-3.

TRELOAR, G. J., LOVE, P. E. D., and CRAWFORD, R. H. (2004). Hybrid life-cycle inventory for road construction and use. *J. Constr. Engrg. and Mgmt.*, 130(1):43–49.

TRIEB, F. and KNIES, G. (2004). A renewable energy and development partnership EU-ME-NA for large scale solar thermal power and desalination in the Middle East and in North Africa. www.gezen.nl/wordpress/wp-content/uploads/2006/09/sanaa-paper-and-annex_15-04-2004.pdf.

TSURUTA, T. (2005). Removal and recovery of lithium using various microorganisms. *Journal of Bioscience and Bioengineering*, 100(5):562–566. www.jstage.jst.go.jp/article/jbb/100/5/100_562/_article.

TURKENBURG, W. C. (2000). Renewable energy technologies. In *World Energy Assessment – Energy and the challenge of sustainability*, chapter 7. UNDP, New York, USA. www.undp.org/energy/activities/wea/draft-start.html.

UCUNCU, A. (1993). Energy recovery from mixed paper waste. Technical report, NC, USA. www.p2pays.org/ref/11/10059.pdf.

VAN DEN BERG, G. (2004). Effects of the wind profile at night on wind turbine sound. *Journal of Sound and Vibration*, 277:955–970. www.nowap.co.uk/docs/windnoise.pdf.

VAN VOORTHUYSEN, E. D. M. (2008). Two scenarios for a solar world economy. *Int. J. Global Environmental Issues*, 8(3):233247.

VENTOUR, L. (2008). The food we waste. news.bbc.co.uk/1/shared/bsp/hi/pdfs/foodwewaste_fullreport08_05_08.pdf.

WARWICK HRI. (2007). Direct energy use in agriculture: opportunities for reducing fossil fuel inputs. www2.warwick.ac.uk/fac/sci/whri/research/climatechange/energy/direct_energy_use_in_agriculture.pdf.

WATER UK. (2006). Towards sustainability 2005–2006. www.water.org.uk/home/policy/reports/sustainability/indicators-2005-06/towards-sustainability-2005-2006.pdf.

WATSON, J., HERTIN, J., RANDALL, T., and GOUGH, C. (2002). Renewable energy and combined heat and power resources in the UK. Technical report. www.tyndall.ac.uk/publications/working_papers/wp22.pdf. Working Paper 22.

WAVEGEN. (2002). Islay Limpet project monitoring – final report. www.wavegen.co.uk/pdf/art.1707.pdf.

WEBER, C. L. and MATTHEWS, H. S. (2008). Food-miles and the relative climate impacts of food choices in the United States. *Environ. Sci. Technol.*, 42(10):3508–3513. doi: 10.1021/es702969f.

WEIGHTMAN, M. (2007). Report of the investigation into the leak of dissolver product liquor at the Thermal Oxide Reprocessing Plant (THORP), Sellafield, notified to HSE on 20 April 2005. www.hse.gov.uk/nuclear/thorpreport.pdf.

WIEDMANN, T., WOOD, R., LENZEN, M., MINX, J., GUAN, D., and BARRETT, J. (2008). Development of an embedded carbon emissions indicator producing a time series of input-output tables and embedded carbon dioxide emissions for the UK by using a MRIO data optimisation system. randd.defra.gov.uk/Document.aspx?Document=EV02033_7331_FRP.pdf.

WILLIAMS, D. and BAVERSTOCK, K. (2006). Chernobyl and the future: Too soon for a final diagnosis. *Nature*, 440:993–994. doi: 10.1038/440993a.

WILLIAMS, E. (2004). Energy intensity of computer manufacturing: hybrid assessment combining process and economic input-output methods. *Environ Sci Technol*, 38(22):6166–6174. ISSN 0013-936X. \url{www.ncbi.nlm.nih.gov/entrez/query.fcgi?cmd=Retrieve&db=pubmed&dopt=Abstract&list_uids=15573621}.

WILLIAMS, R. H. (2000). Advanced energy supply technologies. In *World Energy Assessment – Energy and the challenge of sustainability*, chapter 8. UNDP, New York, USA. www.undp.org/energy/activities/wea/draft-start.html.

WILSON, E. M. and BALLS, M. (1990). Tidal power generation. In P. Novak, editor, *Developments in Hydraulic Engineering*, chapter 2. Taylor & Francis. ISBN 185166095X.

WOOD, B. (1985). Economic district heating from existing turbines. *Institution of Civil Engineers Proc. pt. 1*, 77:27–48.

YAROS, B. (1997). Life-cycle thinking for wood and paper products. In E. Ellwood, J. Antle, G. Eyring, and P. Schulze, editors, *Wood in Our Future: The Role of Life-Cycle Analysis: Proc. a Symposium*.

ZALESKI, C. P. (2005). The future of nuclear power in France, the EU and the world for the next quarter-century. www.npec-web.org/Essays/Essay050120Zalenski-FutureofNuclearPower.pdf. tinyurl.com/32louu.

ZHU, X.-G., LONG, S. P., and ORT, D. R. (2008). What is the maximum efficiency with which photosynthesis can convert solar energy into biomass? *Current Opinion in Biotechnology*, 19:153159.

Index

Errata

Page 2, page 18 Apologies to Jonathon Porritt for misspelling his first name.

Page 43 Figure 6.11 *corn to ethanol* "0.02 W/m^2" should be "0.048 W/m^2".

(See erratum for page 284, below, for details.)

Page 47 add closing parenthesis: "band-gap is lost." should read "band-gap is lost.)"

Page 55 Map showing Kinlochewe and Bedford: Kinlochewe should be located about 60 km further north.

Page 56 (note 56, line 8) "has a per" should read "has a power per".

Page 62 line 14 from the bottom, "0.14 million tons" should read "140 million tons".

Page 85 In the map of Northern Ireland the place-name "Downpatrick" is missing its first letter.

Page 120 **trolleybuses**… "270 kWh per vehicle-km" should read "270 kWh per 100 vehicle-km"

Page 133 "Rijnsdam" should read "Rijndam".

Page 167 After discussing the cost of cleaning up nuclear sites, add: "Moreover, most of this nuclear clean-up cost is associated with weapons-making facilities, not civilian power stations."

Page 169–170 "after 1000 years, the radioactivity of the high-level waste is about the same as that of uranium ore." should read "if we *reprocess* the waste, separating off the uranium and plutonium for reuse in new nuclear fuel, then after 1000 years, the radioactivity of the high-level waste is about the same as that of uranium ore."

Page 192 Table 26.7, Columns 2 and 3. All volumes (40, 40, 100...) and depths (20, 10, 20...) should be doubled (to 80, 80, 200... and 40, 20, 40... respectively).

Page 204 Figure 27.1. The red box marked Transport **20 kWh/d** and the adjacent blue box marked Electricity **18 kWh/d** were both accidentally drawn 10% too tall.

Page 205 Paragraph 2, last line: "2 kWh/d/p of solar hot water" should read "1 kWh/d/p of solar hot water".

Page 217 "the cost of decommissioning the UK's nuclear power stations" – add – "and nuclear-weapon factories".

Page 246 "To pulverized the rocks" should read "To pulverize the rocks".

Page 281 Paragraph 1, line 2: "depends only" should read "depends only on".

Page 284 *Bioethanol from corn in the USA*: "0.02 W/m^2" should read "0.2 W/m^2".

To make this section more informative I would discuss processing costs too, as follows:

1 acre produces 122 bushels of corn per year, which makes 122 × 2.6 US gallons of ethanol, which at 84 000 BTU per gallon would mean a power per unit area of 0.2 W/m^2; however, the energy *inputs* required to process the corn into ethanol amount to 83 000 BTU per gallon; so 99% of the energy produced is used up by the processing, and the *net* power per unit area is about 0.002 W/m^2. The only way to get significant net power from the corn-to-ethanol process is to ensure that all co-products are exploited; including the energy in the co-products, the net power per unit area is about 0.05 W/m^2.

Page 286 Paragraph 2, line 4: "If 2800 m^2 of Britain (that's all agricultural land)…" should read "If 2800 m^2 per person of Britain (that's all agricultural land)…"

Page 298, 299 The top line of page 298 gives 6.6 W/m^2 as the total power per unit area of the Heat-keeper house. This is incorrect. 6.6 W/m^2 is the heating power only. The total power per unit area is 12.2 W/m^2. This error is repeated in figure E.12 (p299).

(The equivalent breakdown of power consumption in my house, "after" the austerity measures were introduced, is 6.2 W/m^2 of gas and 7.1 W/m^2 total.)

Page 299 Another niggle with figure E.12 is that the PassivHaus standards use a different convention for defining power: they define power in terms of "primary energy consumption," which requires knowledge of the primary sources of electricity and fuel, and of conversion efficiencies. This means that the PassivHaus standards are actually more stringent than the figure shows; exactly how much more stringent depends on the fuel mix.

Page 316 Add the equation number (G.10) to the equation on this page.

Page 324 line 22:

"(10 kWh/d per person)"

should read "(10 kWh per kg)".

Page 353 SCHLAICH, J.: Correct bibliography entries are:

SCHLAICH J., BERGERMANN R., SCHIEL W., and WEINREBE G. (2005). Design of Commercial Solar Updraft Tower Systems – Utilization of Solar Induced Convective Flows for Power Generation. *Journal of Solar Energy Engineering* **127** (1): 117-124. doi:10.1115/1.1823493.

SCHLAICH J. and SCHIEL W. (2001). Solar Chimneys. In R.A. Meyers (ed), Encyclopedia of Physical Science and Technology, 3rd Edition, Academic Press, London. ISBN 0-12-227410-5.

Thanks to: François-Marie Lefevere, Tim Stone, Martin Zeidler, Vince O'Farrell, Dankrad Feist, Seb Wills, Jeanne Warren, Iztok Tiselj, and Tim Paine.

Sustainable Energy – without the hot air
David JC MacKay

About the author

David MacKay is a Professor in the Department of Physics at the University of Cambridge. He studied Natural Sciences at Cambridge and then obtained his PhD in Computation and Neural Systems at the California Institute of Technology. He returned to Cambridge as a Royal Society research fellow at Darwin College. He is internationally known for his research in machine learning, information theory, and communication systems, including the invention of Dasher, a software interface that enables efficient communication in any language with any muscle. He has taught Physics in Cambridge since 1995. Since 2005, he has devoted much of his time to public teaching about energy. He is a member of the World Economic Forum Global Agenda Council on Climate Change.

In October 2009 he was seconded from the University of Cambridge to become Chief Scientific Advisor to the UK's Department of Energy and Climate Change; his role is to ensure that the Department's policies and operations, and its contributions to wider Government issues, are underpinned by the best science and engineering advice available.

The author, July 2008.
Photo by David Stern.

Power translation chart

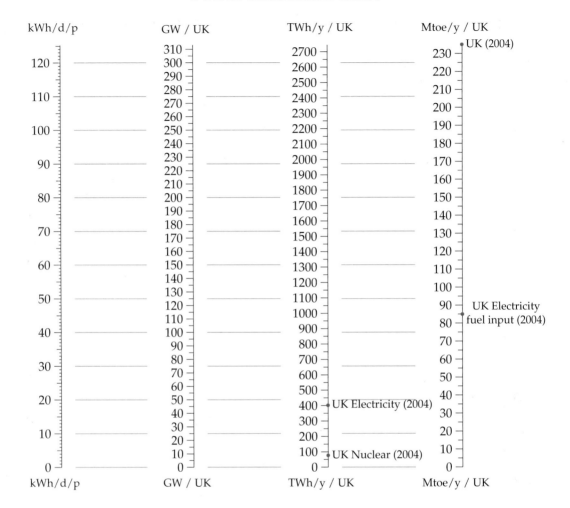

kWh/d/p	GW / UK	TWh/y / UK	Mtoe/y / UK

UK (2004)

UK Electricity
fuel input (2004)

UK Electricity (2004)

UK Nuclear (2004)

kWh/d/p	GW / UK	TWh/y / UK	Mtoe/y / UK

1 kWh/d   the same as $^1/_{24}$ kW

GW   often used for 'capacity' (peak output)

TWh/y   often used for average output

1 Mtoe   'one million tons of oil equivalent'

"UK"   = 60 million people

USA energy consumption: 250 kWh/d per person

Europe energy consumption: 125 kWh/d per person

The most commonly used units in public documents discussing power options are:

terawatt-hours per year (TWh/y).

gigawatts (GW).

million tons of oil equivalent per year (Mtoe/y).

1000 TWh/y per United Kingdom is roughly equal to 45 kWh/d per person.

2.5 GW per UK is 1 kWh/d per person.

2 Mtoe/y per UK is roughly 1 kWh/d per person.

Carbon translation chart

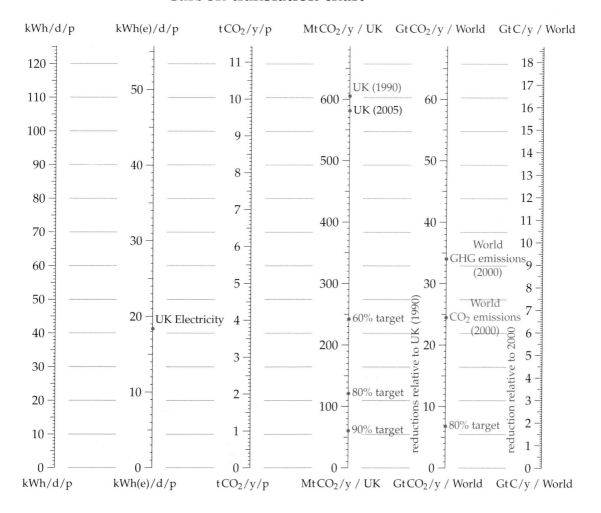

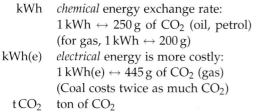

kWh	*chemical* energy exchange rate:
	$1\,kWh \leftrightarrow 250\,g$ of CO_2 (oil, petrol)
	(for gas, $1\,kWh \leftrightarrow 200\,g$)
kWh(e)	*electrical* energy is more costly:
	$1\,kWh(e) \leftrightarrow 445\,g$ of CO_2 (gas)
	(Coal costs twice as much CO_2)
$t\,CO_2$	ton of CO_2
$Mt\,C$	million tons of carbon

"UK" = 60 million people
"World" = 6 billion people

Web site for this book

Register your book: receive updates, notifications about author appearances, and discounts on new editions.
www.uit.co.uk/register

Resources for your book: links to further reading, corrections, reviews, commentary by the author on topical issues, and more.
www.uit.co.uk/resources

News: forthcoming titles, events, reviews, interviews, podcasts, etc. *www.uit.co.uk/news*

Join our mailing lists: get e-mail newsletters on topics of interest. *www.uit.co.uk/subscribe*

Order books: order online. If you are a bookstore, find out about our distributors or contact us to discuss your particular requirements. *www.uit.co.uk/order*

Press/media/trade: for cover images, or to arrange author interviews, author visits, etc. *www.uit.co.uk/order*

Send us a book proposal: if you want to write – even if you have just the kernel of an idea at present – we'd love to hear from you. We pride ourselves on supporting our authors and making the process of book-writing as satisfying and as easy as possible. *www.uit.co.uk/for-authors*

UIT Cambridge Ltd.
PO Box 145
Cambridge
CB4 1GQ
England

E-mail: *inquiries@uit.co.uk*
Phone: **+44 1223 302 041**